Advanced Power System Analysis and Dynamics

L.P. Singh
Professor
Department of Electrical Engineering
Indian Institute of Technology Kanpur
India

A HALSTED PRESS BOOK

JOHN WILEY & SONS

NEW YORK CHICHESTER BRISBANE TORONTO SINGAPORE

Published in the Western Hemisphere by
Halsted Press, a Division of
John Wiley & Sons, Inc., New York

Library of Congress Cataloging in Publication Data

Singh, L.P. (Lakneshwar Prakash)
Advanced power system, analysis and dynamics.

1. Electric power systems. 2. Electric power
sytems—Mathematical models. 1. Title.
TK1001.S53 1982 621.31 82-15497

ISBN 0-470-27349-6 (U.S.)

Printed in India at Prabhat Press, Meerut.

To

My parents

Preface

During March 1966, Department of Electrical Engineering, I.I.T. Kanpur organized an All India Course on "Computer Methods in Power System Analysis". The course was first of its kind in the country and it was well participated. Several distinguished scholars including Prof. A.H. El-Abiad from Purdue University delivered lectures in this course. A course based upon the topics covered in this All India course was started in the academic year 1967–68 by the Deptt. of Elect. Engg., I.I.T. Kanpur. The author also taught this course for a number of years to the elect. engg. students at I.I.T. Kanpur. The present book is the result of teaching by the author courses like EE 538 (Computer Methods in Power System Analysis), EE 630 (Power System Dynamics) and EE 631 (Computer Simulations of Power Systems) to the undergraduate as well as graduate students of I.I.T. Kanpur for several years. In addition, the draft of the book was also tested in several Q.I.P. courses offered by the author.

The important topics covered in different chapters of this book are as follows:

Chapter 2 deals with the formulation of network equations, graph theory, development of network matrices directly by inspection, tests, and by graph theoretic approach and also by building algorithm. Chapter 3 deals with the representations and modelling of synchronous machines and transmission networks. This chapter also deals with the development of symmetrical components and Clarke's components transformation solely based upon symmetries inherent in the power system network and also the analysis of higher order i.e. 6–phase systems.

Chapter 4 deals with the short circuit studies. Algorithms have been developed in this chapter to simulate different types of faults such as single line to the ground, line to line, double line to ground and 3-phase fault with or without fault impedance i.e. bolted fault on the digital computer. Chapter 5 deals with the different numerical techniques to solve linear and non-linear algebraic equations and differential equations along with solved examples. Chapter 6 deals with the formulation of load flow problem by different methods such as Gauss and Gauss Seidel iterative techniques, Newton Raphson's method and Newton's method due to W.F. Tinney. In addition, modelling of transformer is also included in this chapter. Chapter 7 deals with the economic operation of thermal and hydro-thermal stations. Transmission loss formula has also been developed in this chapter. Chapter 8 deals with the sparsity techniques and also optimal ordering to reduce both the computational efforts as well as the computer memory along with flow diagrams. Chapter 9 deals with the dynamic analysis of synchronous machines and also of induction machines. This chapter starts with the most simplified model of synchronous machines and end up with the detailed model of synchronous machines including that of exciter and continuously acting automatic voltage regulator, primemover and governor. In addition, the modelling of induction machines is also included in this chapter. Chapter 10 deals with the stability studies including steady state, dynamic state and transient state stability studies.

In the preparation of the manuscript of this book, the author gratefully acknowledge the assistance received from the Director and Administration of I.I.T. Kanpur and the financial assistance received from the Q.I.P. Cell at I.I.T. Kanpur. The author is also thankful to his friends and colleagues of the Department of Electrical Engineering and others of this Institute and abroad, especially the former Directors of this Institute, viz., Professor P.K. Kelkar and Professor M.S Muthana and other friends such as Professors M.A. Pai, H.K. Kesavan, R.P. Aggarwal, K.R. Padiyar, V.P. Sinha, V. Rajaraman, W.F. Tinney, Harman Dommel, Ahmed, H. El-Abiad, and also his graduate students for the encouragement and assistance he received from them from time to time. The pains taken by his office staff particularly Shri C.M. Abraham to type the original manuscript and also drawings is gratefully acknowledged. Finally, the author expresses his indebtedness to his family members particularly his parents, brothers and sisters-in-law, and his wife Tara and children Alok, Ashok, Ajit and Aneeta for the patience and co-operation they have shown and the assistance they have provided without which the book would not have come to the present form.

November, 1982 L.P. SINGH
I.I.T. Kanpur

Contents

Chapter 1

Introduction

1.1 Introductory Remarks

In the early days, there used to be small power stations for each locality such as urban power systems. But with the growth in the demand of electricity for different purposes such as industrial, agricultural, commercial and domestic together with the guarantee regarding continuity of supply to the different consumers, has forced the power system engineers to develop grid systems which are the interconnections of different generating stations located at different places. Thus the power system of today, is a complex network consisting of several sub-networks such as generation sub-networks, transmission sub-networks and distribution sub-networks. Planning the operation of such systems under existing conditions as well as its improvements and future expansion, requires load flow i.e. planning studies, short circuit studies i.e. fault analysis and stability studies.

The load flow studies are very important for planning the future expansion of the power system as well as determining its best operating conditions. The informations usually obtained from the load flow studies are the magnitude and phase angle of voltages at each bus and active and reactive power flow in each line i.e. in each element of the power system networks. The short circuit studies i.e. fault calculations and stability studies (both transient state and dynamic stability studies) are important in designing the adequate protective schemes of the electrical power system.

The mathematical formulation of the power system problems for these studies viz. load flow, short circuit and stability studies are a set of nonlinear algebraic and or differential equations, which require extensive caculations, and in most of the cases these cannot be solved by hand computation. And therefore, extensive calculations required for these studies, led to the design of the special purpose analog computer, called a.c. network analyser some times in the year 1929. The operation of the power system under existing conditions as well as the proposed future expansion could be simulated by this device. The digital computer for the power system studies gained importance during the beginning of 1950 and the first planning studies on the digital computer was completed by the year 1956. This change from the network analyser to the digital computer has resulted in a greater flexibility, economy, accuracy and quicker operation.

1.2 Digital Simulation of Power System

In order to apply the digital computer for the solution of power system problems as mentioned in the earlier section, a number of distinct steps are required. These steps are given below in sequence:

1. Proper definition of the problem and also the objective to be achieved.

2. Mathematical modelling of the problem. After the problem has been properly defined, the next step is to construct the suitable mathematical model of the given power system. These mathematical models for a given system, depend upon the types of studies to be performed and hence these may be different for the different studies.

3. Numerical technique. After a suitable mathematical model of a given power system has been constructed, solution techniques are needed to solve these mathematical equations. In most of the cases, the mathematical model (i.e. mathematical equations) of the power system problems, are a set of linear or nonlinear algebraic equations and or differential equations and since the digital computer can only perform four basic operations such as addition, subtraction, multiplication and division (actually only two operations viz. addition and subtraction), numerical techniques are needed to transform these mathematical equations to the set of these four basic operations as mentioned in order to solve the equations i.e. mathematical model by the digital computer.

4. Programming. This is the last but the most important stage in the sequence. However, certain amount of planning is necessary before any problem can be solved on the digital computer. This planning is normally aided by the flow chart which is actually the sequence in which the problem is to be solved. After the flow chart is made for solving the problem, it is necessary to translate instructions for solving the problem and also the datas into a language which can be decoded by the machine and thus the problem is solved. This is known as programming. Most of the programs are written in higher level languages such as Fortran IV, Algol, Cobol and Snobol etc. As there are many possibilities of errors in the development of program, it is therefore necessary to check the program before it is used.

1.3 Power System Components

Power system network is a complex network consisting of the following sub-networks:

1. Generation sub-network,
2. Transmission sub-network, and
3. Distribution sub-network.

Generation sub-network usually consists of 3-phase synchronous generators which are designed and constructed to generate 3-phase balanced voltages. Thus the generation sub-network is balanced i.e. symmetric. Transmission and distribution sub-networks consist usually of 3-phase lines, which are either arranged symmetrically on supporting poles or transposed at regular intervals, with the result transmission and distribution sub-networks are also balanced. Hence the power system network, on the whole, is balanced i.e. symmetric. When faults occur on the power system, even though excitations may become unbalanced but network remains normally balanced i.e. symmetric. These symmetries, which are inherent in the power system satisfy axioms of mathematical theory of groups. Solely based upon these symmetries and using mathematical theory of groups, linear power invariant transformations with complex elements similar to the symmetrical components and with, real elements similar to the Clarke's components, can be derived. This precisely has been done in the Chapter 3 of this book. This fundamental yet very significant result has not been reported so far in any text book. In addition, higher order i.e. 6-phase system which may become a reality in near future because of restrictions due to right of way, has also been discussed in this book.

1.4 Concluding Remarks

This book has been written after successfully class testing it in the under-graduate and graduate classes at I.I.T. Kanpur for a number of years and also in several Q.I.P. courses The material covered in the book can easily be understood by readers and at the same time, the book deals with nearly all aspects of modern power system analysis and dynamics such as formulation of network equations, graph theory, development of network matrices using graph theory and also building algorithms, dynamic analysis and mathematical modelling of power system components such as synchronous machines, transmission networks, transformers, induction machines and composite loads, power system studies such as load flow, short circuit and stability studies. Modern but associated topics such as numerical solution of algebraic and differential equations, sparsity techniques and economic operations are also included in this book. The book will be very useful to undergraduate as well as graduate students in Electrical Engineering as well as to the engineers working in the R and D section of power company.

Chapter 2

Network Formulation

2.1 Introductory Remarks

For performing any power system studies on the digital computer, the first step is to construct a suitable mathematical model of the power system network. The mathematical model of the power system network for the steady state analysis is the network equations which describes, both, the characteristics of the individual elements of the power system network as well their interconnections. Usually data available from any power company or electricity board are in the form of primitive network matrix. The primitive network matrix, while giving the complete information regarding the characteristics of the individual elements of the power system network, does not give any information about their interconnections. And hence, the primitive network (i.e. corresponding primitive network matrix) must be transformed either by singular or by nonsingular transformations to the network (matrix) equations as the network equations which describes both the characteristics as well as the interconnections of the different elements, are the suitable mathematical model of the power system network. The primitive network is a set of uncoupled elements of the network.

These network equations can be formed either in the bus frame of reference, or loop frame of reference (sometimes also in the branch frame of reference) using either impedance or admittance parameters. We have discussed in this chapter, the different methods of formulating these network equations and the corresponding network matrices. We have shown that in certain situations, some of these network equations and thus the corresponding notwork matrices can be developed directly by inspections. The network matrices can also be developed by performing open circuit or short circuit tests or by graph theoretic approach. Since graph theory plays an important role both in the development of network matrices as well as in the analysis of power systems, we have briefly described graph theory in this chapter.

2.2 Network Equations

The mathematical model of the power system network is the network equations which can be established in Bus (i.e. nodel), loop (i.e. mesh) or branch frame of reference. In the *bus frame of reference*, the performance of the interconnected network is described by $n-1$ linear independent equations where n is the number of nodes and

$n-1$ is the number of buses. In counting the number of buses, the reference node (which is normally the ground) is always neglected. And hence, for n-node systems, the number of buses are equal to $n-1$. Therefore, the number of linear independent equations are equal to the number of buses. Thus the performance equation in the bus frame of reference in the admittance form is given by

$$\bar{I}_{BUS} = [Y_{BUS}]\,\bar{E}_{BUS} \tag{2.1}$$

where $\bar{I}_{BUS}$ is the vector of injected bus currents (i.e. external current sources), it is positive when flowing towards the bus and it is negative if flowing away from the bus.

$\bar{E}_{BUS}$ is the vector of bus voltages (i.e. nodal voltages) measured from the reference node (which is normally ground) and Y_{BUS} is the Bus admittance matrix of the power system network. The equation (2.1) can be written as follows:

$$\begin{bmatrix} I_1 \\ I_2 \\ \vdots \\ I_{n-1} \end{bmatrix} = \begin{bmatrix} Y_{11} & Y_{12} & \cdots & Y_{1,\,n-1} \\ Y_{21} & Y_{22} & \cdots & Y_{2,\,n-1} \\ \vdots & \vdots & & \\ Y_{(n-1),\,1} & Y_{(n-1),\,2} & & Y_{(n-1),\,(n-1)} \end{bmatrix} \begin{bmatrix} E_1 \\ E_2 \\ \vdots \\ E_{n-1} \end{bmatrix} \tag{2.2}$$

Here the nth bus is the reference bus and hence there are $n-1$ linear independent equations to be solved. Now Y_{BUS} is a non-singular square matrix of the order $(n-1)(n-1)$ and hence it has a unique inverse i.e.

$$Y_{BUS}^{-1} = Z_{BUS}$$

Thus the performance equations in the bus frame of reference in the impedance form will be,

$$\bar{E}_{BUS} = [Z_{BUS}]\,\bar{I}_{BUS} \tag{2.3}$$

Z_{BUS} is again of the order $(n-1)(n-1)$ for a system having n nodes (i.e. $n-1$ buses).

Here, in the formulation of Y_{BUS} or Z_{BUS} matrices, ground is included and is taken as reference node (in this case we have taken nth node as the ground i.e. reference node). And therefore, the elements of the Y_{BUS} or Z_{BUS} matrix will include the shunt connections such as transformer magnetising circuit, static loads, and line charging etc. The bus volltges i.e. E_{BUS} are also measured with respect to ground which is the reference node.

However, if ground is not included in the development of Y_{BUS} or Z_{BUS} matrices a bus known as slack or swing bus is taken as reference bus and and all the variables are measured w.r.t. this reference bus. Since the shunt connections between buses and the ground are not included in the elements of Z_{BUS} or Y_{BUS} matrices, their effect is taken into account by treating them as external current sources which will be discussed later.

Assuming the slack bus voltage as E_S, the performance equations in this case will be

$$\bar{I}_{BUS} = [Y_{BUS}](\bar{E}_{BUS} - E_S) \tag{2.4}$$

or

$$\bar{E}_{BUS} - E_S = [Z_{BUS}]\,\bar{I}_{BUS} \tag{2.5}$$

For n nodes systems (including ground as one of the node), there will be $n-2$ linear independent equations to be solved.

The performance equations in the loop frame of reference, will be given by

$$\bar{E}_{loop} = [Z_{loop}]\,\bar{I}_{loop} \tag{2.6}$$

Here we obtain "l" number of linear independent equations. "l" is the number of loops i.e. meshes and

$\bar{E}_{loop}$ is the vector of the basic loop voltage i.e. resultant voltage sources in the loop.

$\bar{I}_{loop}$ is the vector of unknown loop i.e. mesh currents

Z_{loop} is the loop impedance matrix of the order $l \times l$

However Z_{loop} is a nonsingular square matrix which has an unique inverse, i.e.

$$Z_{loop}^{-1} = Y_{loop}$$

and hence network equations in the loop frame of reference in the admittance form is

$$\bar{I}_{loop} = [Y_{loop}]\,\bar{E}_{loop} \tag{2.7}$$

The performance equations in the branch frame of reference is expressed as shown below:

$$\bar{E}_{BR} = [Z_{BR}]\,I_{BR} \tag{2.8}$$

or

$$\bar{I}_{BR} = [Y_{BR}]\,\bar{E}_{BR} \tag{2.9}$$

Here Z_{BR} (where $Z_{BR}^{-1} = Y_{BR}$) is the branch impedance matrix of the branches of the tree of the connected power system network and is of the order $b \times b$ where b is the number of branches. $\bar{E}_{BR}$ is the vector of branch voltages and $\bar{I}_{BR}$ is the vector of currents through the branches. There will be $n - 1\ (= b)$ linear independent equations to be solved.

Now we discuss the methods to develop the impedance and admittance matrices in the different reference frames.

2.2.1 Development of Bus Admittance and Impedance Matrices

Bus admittance matrix is developed by applying Kirchhoff's current law at every bus. In this way, systematic nodal equations are written for every node except for the reference node which is normally the ground node. Assuming ground as the reference node, let us write the nodal equations using Kirchhoff's current law for the following power system network (Fig. 2.1). Here I_1 and I_3 are the external current sources at

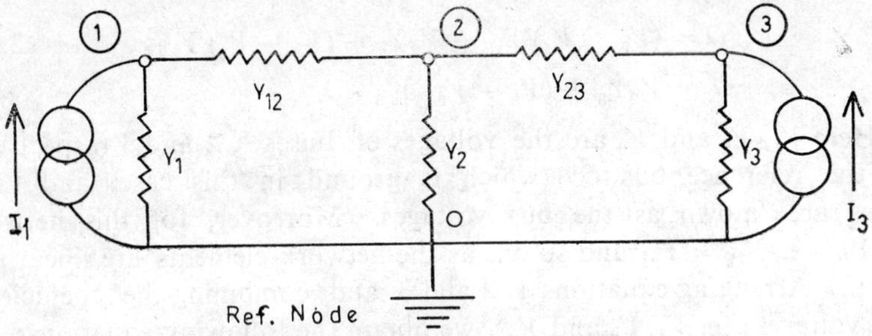

Fig. 2.1 (a) Power system network

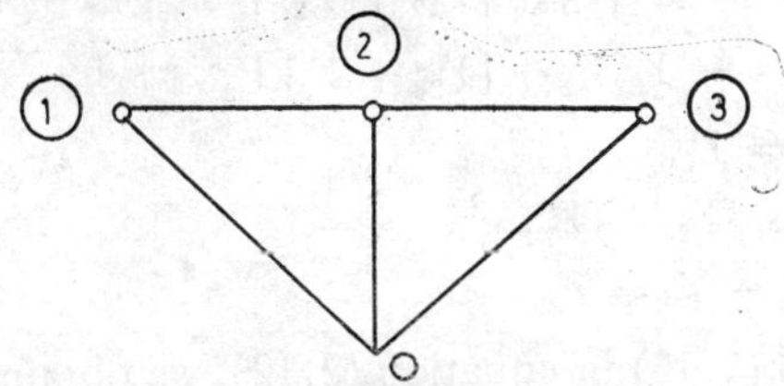

Fig. 2.1 (b) Corresponding graph

the bus 1 and 3. In the nodal formulation, all the voltage sources with the series impedances which is usually the case in the power system network, are replaced by the equivalent current sources with shunt impedance by the following method:

The two sources (refer to Fig. 2.2) are equivalent if

(i) $E_g = I_S Z_S$

and (ii) $Z_g = Z_S$ (2.10)

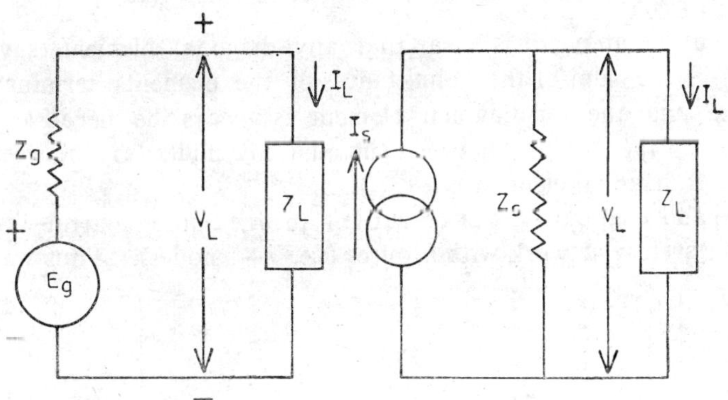

(a) Ideal voltage source (b) Ideal current source

Fig. 2.2

Now, going back to Fig. (2.1) and applying Kirchhoff's current law at the buses 1, 2 and 3, we will obtain the following nodal equations,

$$
\begin{aligned}
I_1 &= V_1 Y_1 + (V_1 - V_2)\, Y_{12} \\
O &= (V_2 - V_1)\, Y_{12} + V_2 Y_2 + (V_2 - V_3)\, Y_{23} \\
I_3 &= V_3 Y_3 + (V_3 - V_2)\, Y_{23}
\end{aligned} \tag{2.11}
$$

Here V_1, V_2 and V_3 are the voltages of buses 1, 2 and 3 respectively w.r.t. the reference bus 'O' (which is ground in this case) and these voltages are known as the bus voltages. Moreover, for this network $Y_{ij} = Y_{ji}$ i.e. $Y_{12} = Y_{21}$ and so on, as the network elements are linear and bilateral. Arranging equations 1, 2 and 3 and combining the coefficients of bus voltages (i.e. V_1, V_2 and V_3), we obtain the following equations:

$$
\begin{aligned}
I_1 &= (Y_1 + Y_{12})\, V_1 - Y_{12} V_2 \\
O &= -\, Y_{21} V_1 + [Y_{12} + Y_2 + Y_{23}]\, V_2 - Y_{23} V_3 \\
I_3 &= -\, Y_{32} V_2 + [Y_{23} + Y_3]\, V_3
\end{aligned} \tag{2.12}
$$

Let $Y_1 + Y_{12} = Y_{11}$

$Y_{12} + Y_2 + Y_{23} = Y_{22}$

and $\quad Y_{23} + Y_3 = Y_{33}$ (2.13)

Substituting relation (2.13) in equation (2.12), we obtain

$$
\begin{aligned}
I_1 &= Y_{11} V_1 - Y_{12} V_2 \\
O &= -\, Y_{21} V_1 + Y_{22} V_2 - Y_{23} V_3 \\
I_3 &= -\, Y_{32} V_2 + Y_{33} V_3
\end{aligned} \tag{2.14}
$$

These equations can be written in the following matrix form:

$$
\begin{bmatrix} I_1 \\ O \\ I_3 \end{bmatrix} = \begin{bmatrix} Y_{11} & -Y_{12} & O \\ -Y_{21} & Y_{22} & -Y_{23} \\ O & -Y_{32} & Y_{33} \end{bmatrix} \begin{bmatrix} V_1 \\ V_2 \\ V_3 \end{bmatrix} \tag{2.15}
$$

From the above it is clear that any diagonal element (say $Y_{11} = Y_1 + Y_{12}$) is the sum of the admittances of the elements terminating at that node and the off diagonal element is always the negative of the admittance of the elements between the adjacent nodes (say between node 1 and 2, off diagonal element is $-Y_{12}$).

The above equations can be written in the more general form for any power system network with n buses (i.e. $n+1$ nodes). Thus we have,

$$
\begin{bmatrix} I_1 \\ I_2 \\ I_3 \\ I_n \end{bmatrix} = \begin{bmatrix} Y_{11} & Y_{12} & Y_{13} & \cdots & Y_{1n} \\ Y_{21} & Y_{22} & Y_{23} & \cdots & Y_{2n} \\ & & & & \\ Y_{n1} & Y_{n2} & Y_{n3} & & Y_{nn} \end{bmatrix} \begin{bmatrix} V_1 \\ V_2 \\ \vdots \\ V_n \end{bmatrix} \tag{2.16}
$$

and in the symbolic form, we have

$$
\bar{I}_{\text{BUS}} = [Y_{\text{BUS}}]\, \bar{V}_{\text{BUS}} \tag{2.17}
$$

Now for $n+1$ nodes system, we obtain n linear independent nodal equations i.e. number of linear independent equations are one less than the number of nodes. Thus the Y_{BUS} matrix is:

$$Y_{BUS} = \begin{bmatrix} Y_{11} & Y_{12} & Y_{13} & \cdots & Y_{1n} \\ Y_{21} & Y_{22} & Y_{23} & \cdots & Y_{2n} \\ Y_{n1} & Y_{n2} & Y_{n3} & \cdots & Y_{nn} \end{bmatrix} \quad (2.18)$$

As we have seen, its diagonal element Y_{ii} is the sum of admittances of the elements terminating at the node i. These elements are known as *short circuit driving point admittances* and they correspond to self admittances.

Similarly off diagonal elements Y_{ij} is the negative of the admittances of elements connected between nodes i and j. These are known as the *short circuit transfer admittances* and they correspond to the mutual admittance. Thus Y_{BUS} matrix can be found *by inspection* provided mutual coupling between the elements of the given power system network (known as primitive network) is neglected.

The elements of Y_{BUS} (i.e. short circuit parameters) can also be obtained as follows:

The diagonal elements Y_{ii} (short circuit driving point admittances) is found by applying a unit voltage source between the i-th bus and the reference node which is ground normally and measuring or calculating its (i.e. ith bus) current while keeping all other buses *short circuited.* Similarly, the off diagonal elements Y_{ij} (short circuit transfer admittance) is found by applying a unit voltage source between the ith bus and the ground and measuring or calculating the current of jth bus while keeping all other buses short circuited (i.e. grounded).

We will show later using Graph theoretic formulation that the elements of Y_{BUS} matrix can also be found by graph theoretic approach i.e.

$$Y_{BUS} = A\,[Y]\,A^T$$

where A is the reduced (or bus) incidence matrix and Y is the primitive admittance matrix. We shall discuss this later in detail.

Thus we have shown that the Y_{BUS} matrix can be obtained directly *by inspection* from a given primitive network provided mutual coupling between the elements are neglacted which is usually the case in most of the power system studies such as load flow studies etc. In addition, the elements of Y_{BUS} matrix can also be obtained by *short circuit test* and also through a *graph theoretic approach*.

Though the Y_{BUS} matrix can be developed directly by inspection (i.e. there is a direct correspondence between the element of Y_{BUS} matrix and the corresponding power system network), this is not true in the case of Z_{BUS} matrix. There is no direct relation between the elements of Z_{BUS} matrix and the corresponding power system network. We can find the elements of Z_{BUS} matrix by the following method.

1. by finding *inverse of* Y_{BUS} matrix and thus we obtain

$$Z_{BUS} = Y_{BUS}^{-1}$$

2. by *open circuit test.* The performance equation of the power system network in the bus frame of reference using Z_{BUS} is:

$$\bar{E}_{BUS} = [Z_{BUS}]\, \bar{I}_{BUS} \qquad (2.19)$$

where

$$Z_{BUS} = \begin{bmatrix} Z_{11} & \cdots & & Z_{1n} \\ Z_{21} & Z_{22} & \cdots & Z_{2n} \\ & \vdots & & \\ Z_{n1} & Z_{n2} & \cdots & Z_{nn} \end{bmatrix}$$

The diagonal elements Z_{ii} are known as the *open circuit driving point impedance* and they are determined by applying a unit current source between the *i*th bus and the ground (i.e. reference node) and measuring the voltage of the same bus i.e. *i*th bus while keeping all other buses open circuited. The off-diagonal elements Z_{ij} are known as *open circuit transfer impedances* and they are found by applying a unit current source between the *i*th bus and the ground and measuring or calculating the voltage of *j*th bus while keeping other buses open circuited.

3. Z_{BUS} matrix can also be found *by building algorithm.* Several algorithms are available in the literature but all of them have nearly the common approach. In this case from the given primitive network (matrix), the elements of Z_{BUS} matrix are developed starting from the reference node known as the slack bus, in steps, by adding one element (either a branch or a link) at a time and in this way Z_{BUS} matrix is developed by building algorithm. We shall discuss the building algorithm for Z_{BUS} matrix later.

2.2.2 Development of Loop Impedance and Admittance Matrices

In this case, only the elements of loop impedance matrix i.e. Z_{loop} can be found *by inspection* and hence Z_{loop} matrix has direct correspondence with the given primitive network just like the Y_{BUS} matrix. To illustrate the method to develop Z_{loop} matrix, let us take the following example:

The elements of Z_{loop} matrix are developed by applying Kirchhoff's voltage laws and writing loop i.e. mesh equations of the given power system network. However, if there is any current source in the network, it must be transformed to the equivalent voltage source with series impedance. Let us take the example of Fig. 2.3 and write the loop i.e. mesh equations using Kirchhoff's voltage laws:

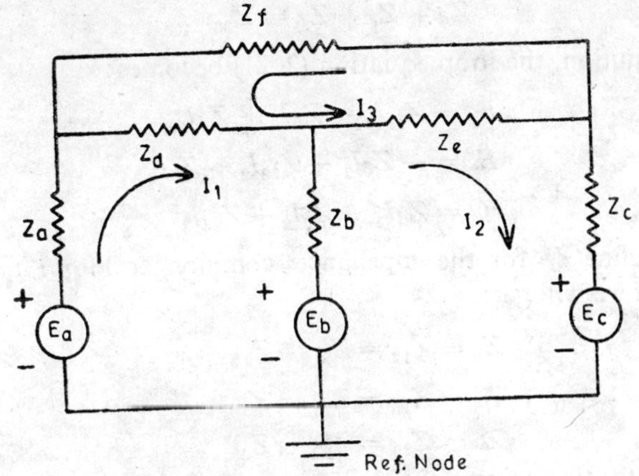

Fig. 2.3

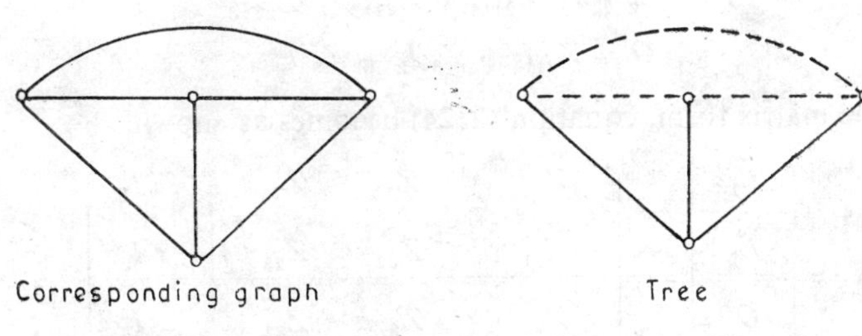

branches "————"; links "- - - - - -"

Fig. 2.4

$$E_a - E_b = I_1 Z_a + (I_1 - I_2) Z_b + (I_1 + I_3) Z_d$$
$$E_b - E_c = (I_2 - I_1) Z_b + I_2 Z_c + (I_2 + I_3) Z_e$$
$$O = (I_3 + I_1) Z_d + (I_3 + I_2) Z_e + I_3 Z_f \tag{2.20}$$

By collecting the coefficients of current (i.e. loop current I_1, I_2 and I_3 which are also called mesh currents), we obtain

$$E_a - E_b = (Z_a + Z_b + Z_d) I_1 - I_2 Z_b + I_3 Z_d$$
$$E_b - E_c = - I_1 Z_b + (Z_b + Z_c + Z_e) I_2 + I_3 Z_e$$
$$O = I_1 Z_d + I_2 Z_e + (Z_d + Z_e + Z_f) I_3 \tag{2.21}$$

Let, $E_a - E_b = E_1$ is the loop voltage of loop 1 i.e. resultant voltage source in the loop 1.

$E_b - E_c = E_2$ is the loop voltage of loop 2 i.e. resultant voltage source in the loop 2.

and let,

$$Z_a + Z_b + Z_d = Z_{11}$$
$$Z_b + Z_c + Z_e = Z_{22}$$

$$Z_d + Z_e + Z_f = Z_{33}$$

By this substitution, the loop equation (2.21) becomes:

$$E_1 = Z_{11}I_1 - Z_bI_2 + Z_dI_3$$
$$E_2 = -Z_bI_1 + Z_{22}I_2 + Z_eI_3$$
$$O = Z_dI_1 + Z_eI_2 + Z_{33}I_3 \quad (2.22)$$

Let us also define Z_{ij} for the impedance common to loop i and j. With this definition we will get:

$$Z_b = Z_{12} = Z_{21} = Z_{12}$$
$$Z_d = Z_{13} = Z_{31} = Z_{13}$$
$$Z_e = Z_{23} = Z_{32} = Z_{23} \quad (2.23)$$

With the substitution of eqn. (2.23) in the eqn. (2.22), we get

$$E_1 = Z_{11}I_1 - Z_{12}I_2 + Z_{13}I_3$$
$$E_2 = -Z_{21}I_1 + Z_{22}I_2 + Z_{23}I_3$$
$$O = Z_{31}I_1 + Z_{32}I_2 + Z_{33}I_3 \quad (2.24)$$

and in the matrix form, equation (2.24) becomes as shown,

$$\begin{bmatrix} E_1 \\ E_2 \\ O \end{bmatrix} = \begin{bmatrix} Z_{11} & -Z_{12} & Z_{13} \\ -Z_{21} & Z_{22} & Z_{23} \\ Z_{31} & Z_{32} & Z_{33} \end{bmatrix} \begin{bmatrix} I_1 \\ I_2 \\ I_3 \end{bmatrix} \quad (2.25)$$

The above equations (i.e. equation (2.25)) can be written in most general form as follows:

$$\begin{bmatrix} E_1 \\ E_2 \\ \cdot \\ \cdot \\ \cdot \\ E_n \end{bmatrix} = \begin{bmatrix} Z_{11} & Z_{12} & \cdots & Z_{1n} \\ Z_{21} & Z_{22} & \cdots & Z_{2n} \\ \cdot & \cdot & \cdots & \cdot \\ \cdot & \cdot & \cdots & \cdot \\ \cdot & \cdot & \cdots & \cdot \\ Z_{n1} & Z_{n2} & \cdots & Z_{nn} \end{bmatrix} \begin{bmatrix} I_1 \\ I_2 \\ \cdot \\ \cdot \\ \cdot \\ I_n \end{bmatrix} \quad (2.26)$$

and in the symbolic form the equation (2.26) becomes

$$\bar{E}_{\text{loop}} = [Z_{\text{loop}}]\,\bar{I}_{\text{loop}}$$

The diagonal elements of Z_{loop} matrix i.e. Z_{ii} is the *self loop impedance* and is equal to the sum of impedance of the elements in the loop "i". Similarly the off diagonal elements Z_{ij} which is known as the *mutual impedance*, is equal to the impedance of the elements common to loop i and j. The elements Z_{ij} is positive if loop i and j agrees and it is negative if loop i and j does not agree. This result is clear from the example shown above where $Z_{11} = Z_a + Z_b + Z_d$ = sum of the impedances of elements in loop 1 and such is the case with Z_{22} and Z_{33}.

And also $Z_{12} = -Z_b \rightarrow$ impedance of the element common to loop

1 and 2 where the loops do not agree and $Z_{13} = Z_d =$ imp. common to loop 1 and 3 where the loops agree. Note, in the case of Y_{BUS} matrix, all its off diagonal elements are negative.

From this it is clear, that the elements of Z_{loop} matrix can also be found directly by inspection from a given power system network (i.e. primitive network) and its order is $l \times l$ where l is the number of links, i.e. loops or meshes.

From graph theoretic point of view (which will be shown later in this chapter),

$$Z_{loop} = B[Z]\,B^T$$

where $Z =$ primitive impedance matrix

and $B =$ basic loop incidence matrix.

The inverse of Z_{loop} matrix is Y_{loop}, i.e.

$$Y_{loop} = Z^{-1}_{loop}$$

The Y_{loop} can only be found by inverting Z_{loop} matrix and thus there is no direct connection between Y_{loop} matrix and the actual network. We shall give later in this chapter, development of these network matrices from the graph theoretic approach.

2.2.3 *Primitive Network*

The data obtained from electricity boards or the power companies is in the form of primitive network (or primitive network matrix). Primitive network is a set of uncoupled elements which gives information regarding the characteristics of individual elements only. The primitive network can be represented in the impedance form as well as in the admittance form. The performance equation of any element $p-q$ in the impedance form will be (see Fig. 2.5).

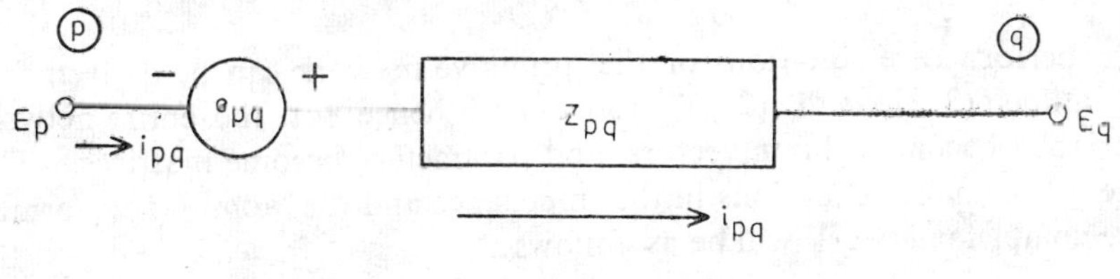

Fig. 2.5

Here v_{pq} voltage across the element $p-q$

i_{pq} current through the element $p-q$

e_{pq} voltage source in series with the element $p-q$

Z_{pq} self impedance of the element $p-q$

and E_p and E_q are the bus (i.e. nodal) voltages of the buses (i.e. nodes) p and q respectively. Assuming E_p at higher potential, we have

$$E_p + e_{pq} - Z_{pq} i_{pq} = E_q$$

i.e. $$E_p - E_q + e_{pq} = Z_{pq}\, i_{pq}$$

i.e. $$v_{pq} + e_{pq} = Z_{pq}\, i_{pq} \tag{2.27}$$

where $v_{pq} = E_p - E_q \rightarrow$ voltage across the element $p-q$. Similarly in the admittance form, the performance equation will be as in Fig. 2.6.

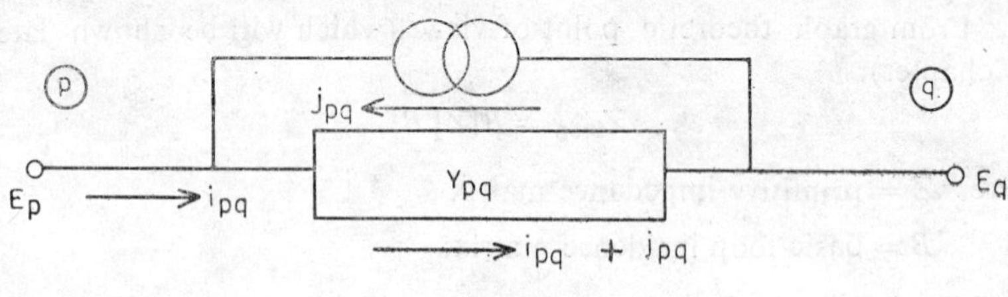

Fig. 2.6

Here j_{pq} is the current source in parallel with the element $p-q$ and Y_{pq} is the self admittance of the element $p-q$.

Now, (from Fig. 2.6)

$$i_{pq} + j_{pq} = Y_{pq}\, v_{pq} \tag{2.28}$$

where v_{pq} is the voltage drop across the element $p-q$.

From equation (2.27), we get,

$$i_{pq} = \frac{1}{Z_{pq}} v_{pq} + \frac{1}{Z_{pq}} e_{pq}$$

$$= Y_{pq}\, v_{pq} + Y_{pq}\, e_{pq}$$

where $$Y_{pq} = 1/Z_{pq}$$

Hence $$i_{pq} - Y_{pq}\, e_{pq} = Y_{pq}\, v_{pq} \tag{2.29}$$

Comparing the similar equations (2.29) and (2.28), we obtain,

$$j_{pq} = -\, Y_{pq}\, e_{pq} \tag{2.30}$$

The performance equations of the primitive network can be derived from equations (2.27) and (2.28) above and hence for the entire network, variables become column vectors and parameters become matrices. Thus the performance equations in the impedance and the admittance form for the complete network will be as follows:

$$\bar{v} + \bar{e} = [Z]\, \bar{i}$$

and $$\bar{i} + \bar{j} = [Y]\, \bar{v} \tag{2.31}$$

Here the diagonal elements of $[Z]$ and $[Y]$ viz. $Z_{pq\,pq}$ or $Y_{pq\,pq}$ are the self impedances/admittances for the element $p-q$ and the off diagonal elements $Z_{pq,rs}$ or $Y_{pq,rs}$ are mutual impedances/admittances, between elements $p-q$ and $r-s$. The primitive admittance matrix $[Y]$ is obtained by finding inverse of $[Z]$. However, if there is no mutual coupling between the elements, the matrices Y or Z will be diagonal.

2.3 Graph Theory

In order to describe the geometrical structure of the network, it is sufficient to replace the different power system components (of the corresponding power system network) such as generators, transformers and transmission lines etc. by a single line element irrespective of the characteristics of the power system components. The geometrical interconnection of these line elements (of the corresponding power system network) is known as a *graph* (rather *linear graph* as the graph means always a linear graph). Hence the graph 'G' $= (V, E)$, consists of a set of objects $V = (v_1, v_2, \ldots)$ called *nodes* or *vertices* and another set of objects $E = (e_1, e_2, \ldots)$ called *elements* or *edges.* A *subgraph* is a subset of the graph. A *path* is a subgraph of the connected elements with not more than two elements incident at a node. A node i.e. vertex and an edge i.e. element are *incident* if the node is the end point of the edge. A *loose-end* is the vertex with not more than one element incident to it. A *graph is connected* if and only if there exists a path between every pair of nodes. A single edge or a single node is a connected graph. If every edge of the graph is assigned a direction, the graph is termed as an *oriented graph.* The number of edges incident to a vertex with self loop counted twice is called the *degree of the vertex* i.e. node. A *tree* is a subgraph containing all the nodes of the original graph but no closed path and hence the *tree has the following properties*:

1. Tree is a subgraph containing all the vertex of the original graph.
2. Tree is connected.
3. Tree does not contain any closed path.

The *total number of trees* in a simply connected graph = det Y with every element assigned to an admittance of 1 mhos. Here Y is a bus admittance matrix.

To illustrate these points, let us take the following example:

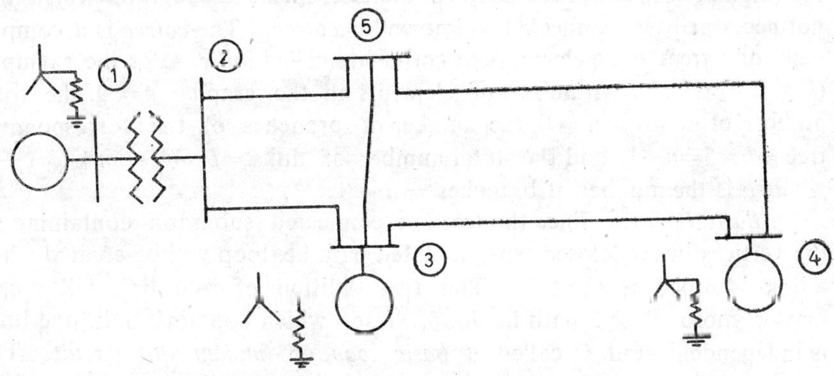

Fig. 2.7 Power system network

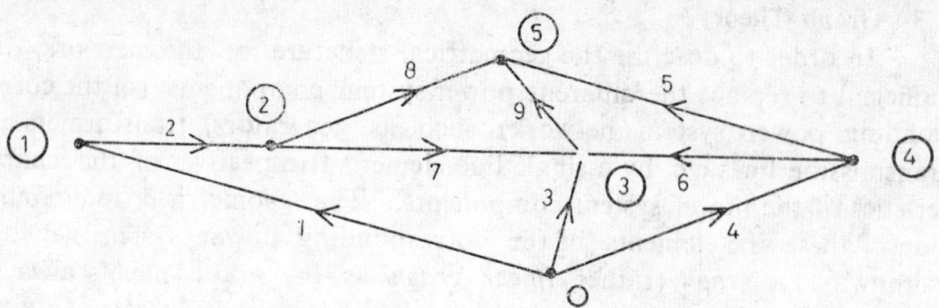

Fig. 2.8 (a) Oriented graph of the network of Fig. 2.7

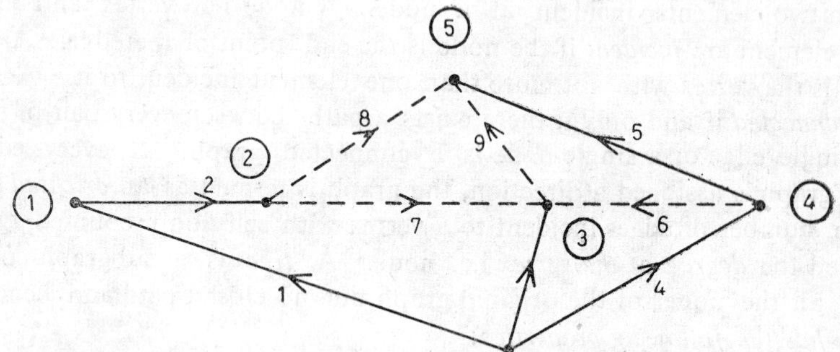

Fig. 2.8 (b) Tree and cotree of the corresponding graph of Fig. 2.8(a)

Note: 1. Thick lines indicate a tree and hence tree branches are shown as "————".

2. Dotted lines indicate cotree and hence the elements of cotree i.e. links are shown as "– – – – – –".

The *elements of the tree* are known as *branches.* The elements of the original graph not included in the tree, forms a subgraph which may not necessarily be connected, is known as *cotree.* The *cotree* is a complement of a tree. The elements of cotree are called *links.* For the example, (Fig. 2.8) the total number of elements in the graph $= e = 9$, the total number of nodes $= n = 6$, the number of branches of the corresponding tree $= b = 5 = n - 1$, and the total number of links $= l = 4 = e - n + 1 = e - b$, where b is the number of branches $= n - 1$.

Basic Loops: Since the tree is a connected subgraph containing all the vertices but no closed path, a closed path i.e. loop will be created when a link is added to the tree. Thus the addition of each link will create one or more closed path i.e. loop. Loop which contains only one link, is independent and is called a *basic loop* or *fundamental circuit.* The number of basic loops are equal to the number of links i.e. meshes " l " (where $l = e - n + 1$) and it has the same orientation as that of the corresponding link as shown in Fig. 2.9.

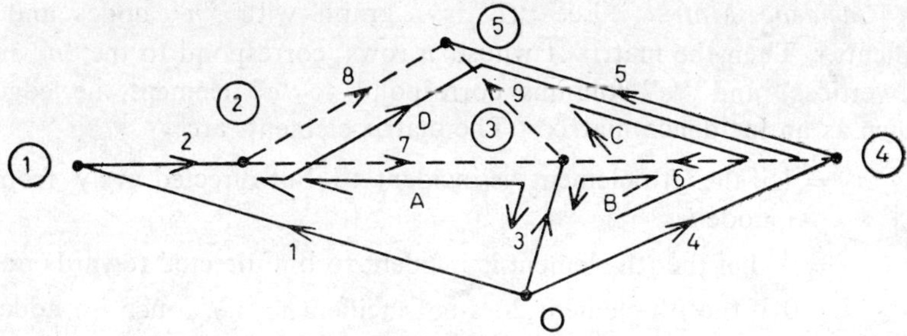

Fig. 2.9 Basic loops of the network shown in Fig. 2.7

Note: By adding the link 7 to the tree, a closed loop *A* is formed which has the same orientation as that of the corresponding link 7. Similarly by adding links 6, 8 and 9, the corresponding basic loops *B*, *D* and *C* having the same orientation as that of the links are formed. As the number of links are four, there are only four basic loops in this case.

Cut set: Cut-set is the minimal set of the elements in the connected graph which when removed, divide the graph in only two parts i.e. two connected subgraphs. *Basic cutsets* are those which contains only one branch and hence the number of basic cutsets are equal to the number of branches "*b*" ($b = n - 1$) and it has the same orientation as that of the corresponding branch as shown in the figure below (Fig. 2.10).

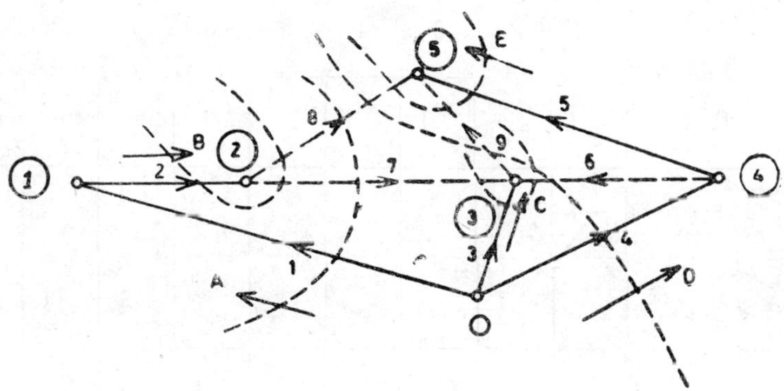

Fig. 2.10 Basic cutsets of the network shown in Fig. 2.7

Note: Since there are 5 branches, number of basic cutsets are also equal to five. For the above example, cutset A contains the branch 1 and also elements 7 and 8. Cutset B contains the branch 2 and also the elements 7 and 8 and so on. In every case the basic cutsets have the same orientation as that of the corresponding branch as shown in the Fig. 2.10.

2.3.1 *Matrix Representation of Graph*

Matrix representation of graphs is very important in the development of network equations. We shall discuss the following matrix representations of the graph.

Incidence Matrix: Let "*G*" is a graph with "*n*" nodes and "*e*" elements. Then the matrix $\bar{A}$ whose *n* rows correspond to the "*n*" nodes (i.e. vertices) and "*e*" columns correspond to "*e*" element, i.e. edges, is known as an incidence matrix. The matrix elements are:

$a_{ij} = 1$ if the *j*th element is incident to but directed away from the node *i*.

$= -1$ if the *j*th element is incident to but directed towards node *i*.

$= 0$ if the *j*th element does not incident to, i.e. touch *i*th node.

The dimension of this matrix is $n \times e$ and its rank is less than *n*.

Reduced incidence Matrix (i.e. Bus incidence matrix): Any node of the connected graph can be selected as the reference node and then the variables of the remaining $n - 1$ nodes which are termed as buses can be measured w.r.t. this assigned reference node. The matrix "*A*" obtained from the incidence matrix $\bar{A}$ by deleting the reference row (corresponding to the reference node) is termed as *reduced* or *bus incidence matrix* (the nnmber of buses in the connected graph is equal to $n - 1$ where *n* is the number of nodes). The order of this matrix is $(n - 1) \times e$ and its rows are linearly independent with rank equal to $n - 1$.

For Fig. 2.8, the incidence matrix $\bar{A}$ will be as shown below.

Element *e* / Nodes *n*	1	2	3	4	5	6	7	8	9
0	1		1	1					
1	−1	1							
$\bar{A}$ =2		−1					1	1	
3			−1			−1	−1		1
4				−1	1	1			
5					−1			−1	−1

(2.32)

The rows are linearly dependent and hence rank is less than *n* (i.e. 6).

Choosing node "0" as the reference node, the reduced incidence i.e. bus incidence matrix *A* for the above will be as shown below:

Bus $=(n-1)$ \ Element e	1	2	3	4	5	6	7	8	9
1	−1	1							
2		−1					1	1	
$A=$ 3			−1			−1	−1		1
4				−1	1	1			
5					−1			−1	−1

(2.33)

If the columns of "A" are arranged according to a particular tree (say as shown in Fig. 2.8(b)), the matrix A can be partitioned into two submatrices A_b of dimension $(n-1)(n-1) = (n-1)\,b$ (where b is the number of buses $= n-1$) and A_l of dimension $(n-1) \times l$ where l is the number of links $= e - n + 1$. The partitioned matrix of Fig. 2.8 is shown below:

Bus $=n-1$ \ e	Branches 1	2	3	4	5	Links 6	7	8	9
1	−1	1							
2		−1					1	1	
$A=$ 3			−1			−1	−1		1
4				−1	1	1			
5					−1			−1	−1

(2.34)

Bus $=n-1$ \ e	Branches	Links
Buses	A_b	A_l

(2.35)

i.e. $[A]_{(n-1)e} = [(A_b)_{(n-1)(n-1)}\ (A_l)_{(n-1)l}]$

A_b which corresponds to a tree, is a nonsingular matrix of rank $n-1=b$ $=5$ in this case.

Basic Loop i.e. Fundamental Circuit Matrix: The *basic loop* (i.e. *fundamental circuit matrix*) "B" $=(B_{ij})$ gives the incidence of elements with the basic loops of the connected graph. The elements of the matrix "B" is defined as

$b_{ij}=1$ if jth element is incident to (i.e. included in) ith basic loop and is oriented in the same direction.

$=-1$ -do- but oriented in the opposite direction.

$=0$ if ith basic loop does not include the jth element.

Basic loop incidence matrix "B" for the Fig. 2.9 will be as shown:

	Elements → 1	2	3	4	5	7	6	9	8
Basic loops A	1	1	−1			1			
$B=$ ↓ B			−1	1			1		
C			1	−1	−1			1	
D	1	1		−1	−1				1

(2.36)

The matrix "B" can also be partitioned into submatrices B_b and U_l where the columns of B_b correspond to branches and columns of U_l correspond to links. The partitioned matrix "B" is shown below.

l \ e	Branches 1	2	3	4	5	Links 7	6	9	8
$B=$ A	1	1	−1			1			
B			−1	1			1		
C			1	−1	−1			1	
D	1	1		−1	−1				1

	l \ e	Branch	Links
$=$ Basic loops	↓	B_b	U_l

(2.37)

$$= \left[\begin{array}{c|c} [B_b] & [U_l] \\ l \times (n-1) & l \times l \end{array} \right]$$

Here U_l is an identity matrix as shown.

Basic Cutset Incidence Matrix: The incidence of elements with the basic cutsets of a connected graph is given by basic cutset incidence matrix C. The elements of the matrix are:

$C_{ij} = 1$ if jth element is incident to (i.e. included in) ith basic cutset and oriented in the same direction.

$= -1$ -do- oriented in the opposite direction.

$= 0$ otherwise.

The basic cutset matrix C for Fig. 2.10 will be as shown:

	Basic cutsets "b" ↓ \ Elements "e" ⟶	1	2	3	4	5	6	7	8	9	
	A	1						−1	−1		
	B		1					−1	−1		
$C=$	C			1			1	1		−1	(2.38)
	D				1		−1		1	1	
	E					1			1	1	

The basic cutset incidence matrix C can be partitioned into submatrices U_b and C_l where columns of U_b (identity matrix) correspond to branches and columns of C_l correspond to links. Thus the matrix $[C]$ after partitioning becomes

b \ e	Branches	Links
Basic cutsets ↓	U_b	C_l

$$= \left[\begin{array}{c|c} [U_b] & [C_l] \\ (n-1)(n-1) & (n-1)_l \end{array} \right]$$

Branch Path Incidence Matrix: The incidence of branches to paths in a connected graph is shown by branch path incidence matrix "K", where the path is oriented from the bus to the reference node. *Path* is a *subgraph* of the connected elements where not more than two elements meet at a node. The elements of this matrix will be

$K_{ij} = 1$ if the jth branch is in path from ith bus to the reference and is oriented in the same direction (i.e. bus to the reference node).

$= -1$ -do- but oriented in the opposite direction.

$= 0$ otherwise.

For Fig. 2.8 (b), the branch path incidence matrix "K" will be:

Path ↓ $= n-1$ \ Branch $= n-1$ →	1	2	3	4	5
= 1	−1				
2	−1	−1			
3			−1		
4				−1	
5				−1	−1

(2.39)

The matrix K which relates branches with paths is a nonsingular matrix with rank $n-1 = b$ (i.e. buses) $= 5$ in this case.

2.3.2 *Relations*

We know that $[A] = [A_b \ A_l]$ where A is the reduced incidence matrix (from the equation (2.35)),

$$[A_b]_{(n-1)(n-1)} = [A_b]_{\text{bus*branches}}$$

and $[K]_{(n-1)(n-1)} = [K]_{\text{path*branches}}$ (from the equation (2.39)).

Then, $[A_b][K]^T = U$ (i.e. identity matrix)

$$\therefore \quad A_b = [K^T]^{-1} \text{ or } K^T = A_b^{-1} \tag{2.40}$$

it can also be shown that

$$[A_b][C_l] = [A_b]_{(n-1)(n-1)} * [C_l]_{(n-1)l}$$

$$= [A_l]_{(n-1)l} \tag{2.41}$$

i.e. $$[C_l] = A_b^{-1} A_l = K^T A_l \tag{2.42}$$

Here C_l is a submatrix of the basic cutset incidence matrix C.

Moreover it can easily be shown with the help of Kirchhoff's current law that

$$[A] \, \bar{i} = 0$$

where $\bar{i}$ is the column vector of element currents.

Partitioning the matrix $[A]$ and also the element current $\bar{i}$ in branches and links, we get,

$$[A_b \mid A_l] \begin{bmatrix} i_b \\ i_l \end{bmatrix} = 0 \tag{2.44}$$

where $\bar{i}$ is the element current and subscript "b" and "l" refers to branches and links respectively. Expanding the above Kirchhoff's current law equation (i.e. equation (2.44)) we get,

$$A_b\, i_b + A_l\, i_l = 0$$

i.e.

$$i_b = -\, A_b^{-1}\, A_l\, i_l = -\, C_l\, i_l \tag{2.45}$$

Similarly with the help of Kirchhoff's voltage law equation we can show that

$$[B]\, \bar{v} = 0 \tag{2.46}$$

where $\bar{v}$ is the voltage across the elements. And again partitioning for branches and links, we have

$$[B_b \mid U] \begin{bmatrix} v_b \\ v_l \end{bmatrix} = 0 \tag{2.47}$$

Here V_b and V_l are the column vectors of branch and link voltages respectively. Hence

$$B_b\, v_b + v_l = 0$$

i.e.

$$B_b\, v_b = -\, v_l$$

or

$$V_b = -\, B_b^{-1}\, V_l \tag{2.48}$$

Orthogonal Matrices: Let $\bar{X}$ and $\bar{Y}$ are column vectors. Then their inner (i.e. vector) product is defined as

$$\bar{X}^T\, \bar{Y} = [X_1\, X_2 \ldots] \begin{bmatrix} Y_1 \\ Y_2 \\ \vdots \end{bmatrix}$$

$$= X_1\, Y_1 + X_2\, Y_2 + \cdots \tag{2.49}$$

The vectors X and $\bar{Y}$ are orthogonal if $X^T Y = 0$, i.e.

$$X_1\, Y_1 + X_2\, Y_2 + \ldots = 0 \tag{2.50}$$

We can show that if the incidence matrix $[A]$ and basic loop incidence matrix $[B]$ of the connected graph G are arranged using the same order of edges, i.e., elements, then every row of B is orthogonal to every row of A i.e.

$$A \cdot B^T = B \cdot A^T = 0 \tag{2.51}$$

Similarly if edges in both B and C are arranged in the same order, then every row of B is orthogonal to every row of C i.e.

$$B \cdot C^T = C \cdot B^T = 0 \tag{2.52}$$

We can also show that equation (2.43) can be extended further and again with the help of Kirchhoff's current law we obtain

$$[C]\, \bar{i} = 0 \tag{2.53}$$

where $[C]$ is the basic cutset incidence matrix. The equation (2.53) indicates that the sum of the currents at the boundary of the cutset is zero. This equation is the cutset equation and corresponds to Kirchhoff's current law.

From the equation (2.53), we get,

$$[U_b \mid [C_l]] \begin{bmatrix} i_b \\ i_l \end{bmatrix} = 0$$

$$i_b + C_l\, i_l = 0$$

$$i_b = - C_l\, i_l \tag{2.54}$$

Thus the branch current can be expressed in terms of link currents.

2.4 Development of Network Matrices from Graph Theoretic Approach

2.4.1 Bus Admittance and Bus Impedance Matrices

The bus admittance matrix Y_{BUS} can be obtained with the help of bus incidence matrix $[A]$ which relates the variables and parameters of the primitive network to the bus quantities. The performance equation of the primitive network in the admittance form will be (refer to the equation (2.31).

$$\bar{i} + \bar{j} = [y]\, \bar{v}$$

Pre-multiplying the above by the bus incidence matrix A, we get

$$[A]\, \bar{i} + [A]\, \bar{j} = [A][Y]\, \bar{v} \tag{2.55}$$

We have already shown that

$$[A]\, \bar{i} = 0 \quad \text{(i.e. } KCL \text{ equation)}$$

Similarly. it can also be shown that

$$[A]\bar{j} = I_{BUS}$$

This is the algebraic sum of external current sources at each bus and is equal to the impressed bus current I_{BUS}. Thus the equation (2.55) becomes

$$I_{BUS} = A\, [Y]\, v \tag{2.56}$$

$$\text{Power in the bus frame of reference} = \bar{I}^{*T}_{BUS} \cdot \bar{E}_{BUS} \tag{2.57}$$

$$\text{Power in the primitive network} = \bar{j}^{*T}\, \bar{v} \tag{2.58}$$

As the power in the primitive network and the bus frame of reference must be equal i.e. transformation of variables must be power invariant, we have

$$\bar{I}^{*T}_{\text{BUS}}\,\bar{E}_{\text{BUS}}=\bar{j}^{*T}\,\bar{v} \tag{2.59}$$

Now $$[A]\,\bar{j}=\bar{I}_{\text{BUS}}$$

i.e. $$[A]\,j^{*T}=\bar{I}^{*T}_{\text{BUS}}$$

i.e. $$j^{*T}A^{*T}=I^{*T}_{\text{BUS}} \tag{2.60}$$

Since $[A]$ is a real matrix $\rightarrow A^*=A$ i.e.

$$\bar{j}^{*T}\,[A]^T=\bar{I}^{*T}_{\text{BUS}} \tag{2.61}$$

Substituting the equation (2.61) in the equation (2.59), we have,

$$j^{*T}\,[A]^T\,\bar{E}_{\text{BUS}}=\bar{j}^{*T}\,\bar{v} \tag{2.62}$$

Since the equation (2.62) is valid for all values of j, we have from the above

$$[A]^T\,\bar{E}_{\text{BUS}}=\bar{v} \tag{2.63}$$

Thus substituting eqn. (2.63) in the eqn. (2.56) we obtain

$$\begin{aligned} I_{\text{BUS}}=A\,[Y]\,v&=A\,[Y]\,A^T\,E_{\text{BUS}}\\ &=Y_{\text{BUS}}\,E_{\text{BUS}} \end{aligned} \tag{2.64}$$

Thus $$Y_{\text{BUS}}=A\,[Y]\,A^T \tag{2.65}$$

Since A is a singular matrix, $A\,[Y]\,A^T$ is a singular transformation of $[Y]$. And also,

$$Z_{\text{BUS}}=Y^{-1}_{\text{BUS}}=[A\,[Y]\,A^T]^{-1} \tag{2.66}$$

2.4.2 Branch Impedance and Branch Admittance Matrices

The branch admittance matrix Y_{BR} can be obtained with the help of basic cutset incidence matrix "C" which relates the variables and parameters of the primitive network to the branch quantities. The performance equation of the primitive network in the admittance form will be

$$\bar{i}+\bar{j}=[Y]\,v \tag{2.67}$$

By premultiplying by the basic cutset incidence matrix "C" the equation (2.67) becomes,

$$C\bar{i}+C\bar{j}=C\,[Y]\,v \tag{2.68}$$

We have already shown that $C\bar{i}=0$, i.e. sum of the currents at the boundary of the cutset is zero. However, it can easily be verified that $C\bar{j}$ is equal to I_{BR}, hence with these substitutions the equation (2.68) becomes,

$$\bar{I}_{BR}=C\,[Y]\,\bar{v} \tag{2.69}$$

Power in the branch frame of ref. $=I^{*T}_{BR}\,E_{BR}=$ power in the primitive network $=j^{*T}v$, (because transformation is power invariant) i.e.

$$\bar{I}^{*T}_{BR}\,\bar{E}_{BR} = \bar{j}^{*T}\,\bar{v} \tag{2.70}$$

However since $C\bar{j} = I_{BR}$, then $[C_j]^{*T} = I^{*T}_{BR}$

i.e. $\quad j^{*T}\,C^{*T} = I^{*T}_{BR}$ (2.71)

The matrix C is real i.e. $C^* = C$.

Thus the equation (2.71) becomes

$$j^{*T}\,C^T = I^{*T}_{BR} \tag{2.72}$$

Substituting the equation (2.72) for I_{BR} in equation (2.70) we get,

$$J^{*T}\,C^T\,\bar{E}_{BR} = j^{*T}\,\bar{v} \tag{2.73}$$

i.e. $$C^T\,\bar{E}_{BR} = \bar{v}$$

Because the above equation is true for all values of J.

Substituting the equation (2.73) in the equation (2.69), we get:

$$\bar{I}_{BR} = C[Y]\,\bar{v}$$

$$= C[Y]\,C^T\,\bar{E}_{BR}$$

But as $\quad I_{BR} = Y_{BR}\,E_{BR}$, we have

$$Y_{BR} = C[Y]\,C^T \tag{2.74}$$

Since "C" is singular, $C[Y]\,C^T$ is the singular transformation of "Y". And also

$$Z_{BR} = Y^{-1}_{BR} = [C[Y]\,C^T]^{-1} \tag{2.75}$$

2.4.3 Loop Impedance and Admittance Matrices

The loop impedance matrix Z_{loop} can be obtained with the help of basic loop incidence matrix "B" which relates the variables and parameters of the loop quantities with the primitive network matrix. Let us take the performance equation of the primitive network in the impedance form (refer to equation (2.31))

$$\bar{v} + \bar{e} = [Z]\,\bar{i} \tag{2.76}$$

Premultiplying the equation (2.76) by the basic loop incidence matrix "B", we obtain,

$$B\bar{v} + B\bar{e} = B[Z]\,\bar{i} \tag{2.77}$$

Now since $B\bar{v} = 0$ (according to the Kirchhoff's voltage law) and $B\bar{e}$ is the algebraic sum of the source voltages around each basic loop $= E_{\text{loop}}$. With these substitutions, eqn. (2.77) becomes:

$$\bar{E}_{\text{loop}} = B[Z]\,\bar{i} \tag{2.78}$$

Power in the primitive network $\quad = [\bar{i}^*]^T\,\bar{e}$

Power in the loop frame of ref. $\quad = \bar{I}^{*T}_{\text{loop}}\,\bar{E}_{\text{loop}}$

Power being invariant, we have

$$\bar{i}^{*T}\,\bar{e} = \bar{I}^{*T}_{\text{loop}}\,\bar{E}_{\text{loop}}$$

$$= \bar{I}^{*T}_{\text{loop}}\,[B]\,\bar{e} \tag{2.79}$$

As the equation (2.79) is true for all values of e, we get

$$\bar{i}^{*T} = \bar{I}^{*T}_{\text{loop}} [B] \tag{2.80}$$

Taking conjugate transpose of the equation (2.80), we get

$$\bar{i} = B^{*T} \bar{I}_{\text{loop}} \tag{2.81}$$

Since B is real matrix, $B^* = B$ i.e.

$$\bar{i} = B^T I_{\text{loop}} \tag{2.82}$$

Substituting the equation (2.82) in equation (2.78) we obtain

$$\bar{E}_{\text{loop}} = B[Z]\bar{i}$$

$$= B[Z]\, B^T \bar{I}_{\text{loop}} \tag{2.83}$$

But since $E_{\text{loop}} = Z_{\text{loop}}\, I_{\text{loop}}$, we get

$$Z_{\text{loop}} = B[Z]\, B^T \tag{2.84}$$

Since "B" is singular, $B[Z]\, B^T$ is a singular transformation of [Z] and also

$$Y_{\text{loop}} = Z^{-1}_{\text{loop}} = [B[Z]\, B^T]^{-1} \tag{2.85}$$

2.5 Augment cutset incidence matrix $\hat{C}$:

Fictitious cutsets (equal to the number of links) called *tie cutsets* can be introduced in order that the total number of cutsets are equal to the number of elements of the connected graph. Each tie cutset contains only one link and is oriented in the same direction as that of the link. The augmented cutset incidence matrix $\hat{C}$ can be formed by adding to the basic cutset incidence matrix C, an additional rows corresponding to these tie cutsets. Let us take the example shown on p. 28. (i.e. in Fig. 2.10a)

	e \ e	1	2	3	4	5	6
Basic cutsets							
$\hat{C}$ –	A	1				−1	1
	B		1		0	−1	1
	C		0	1		1	−1
	D				1	1	
Tie cutsets	E			0		1	
	F			0			1

(2.86)

Thus the augmented cutset incidence matrix $\hat{C}$ can be partitioned as shown below.

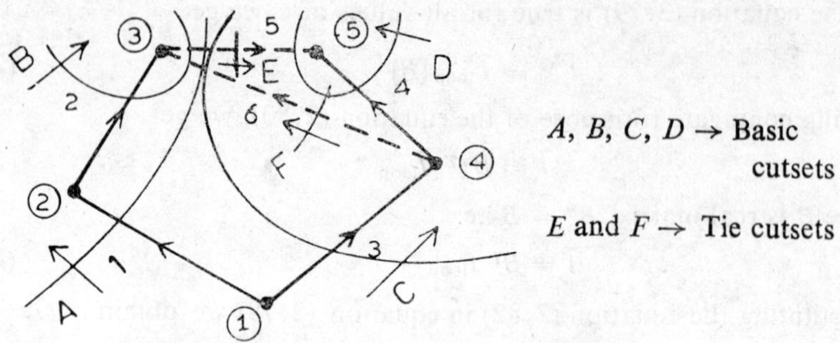

Fig. 2.10 (a) Augmented cutsets

$$\begin{array}{l} \\ \text{Basic cutsets} \\ \\ \text{Tie cutsets} \end{array} \begin{array}{c} \begin{array}{cc} \text{Branches} & \text{Links} \end{array} \\ \left[\begin{array}{c|c} U_b & C_l \\ \hline 0 & U_l \end{array}\right] \end{array} \tag{2.87}$$

The matrix $\hat{C}$ is a square nonsingular matrix of the dimension $e \times e$.

2.5.1 *Augmented Loop Incidence Matrix* $\hat{B}$

In order to have the total number of loops equal to the elements, fictitious loops called open loops which are equal to the number of branches, can be introduced. An open loop is defined as a path between adjacent nodes connected by a branch and is oriented in the same direction as that of the associated branch. The augmented loop incidence matrix is formed by adding to the basic loop incidence matrix "B" additional rows corresponding to the open loops (i.e. branches). Let us take an example (of Fig. 2.10 b).

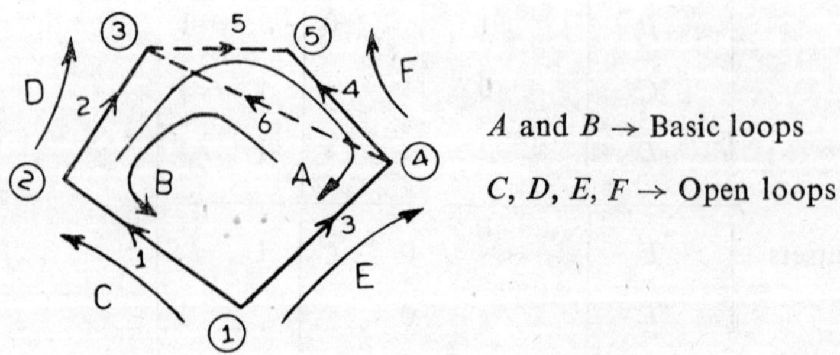

Fig. 2.10 (b) Augmented loops

	e \ e	1	2	3	4	5	6
Open loops	C	1			0		
	D		1				
	E			1			
$\hat{B}=$	F	0			1		
Basic loops	A	1	1	−1	−1	1	0
	B	−1	−1	1		0	1

(2.88)

Thus the matrix $\hat{B}$ can be partitioned as shown below,

	Branches	Links
Open loops	U_b	O
Basic loops	B_b	U_l

(2.89)

The matrix $\hat{B}$ is a square nonsingular matrix of the order exe.

2.6 Cutset and Circuit Equations

We have shown earlier that

$[C]\bar{i} = 0$ (according to Kirchhoff's current law)

and $[B]\bar{v} = 0$ (according to Kirchhoff's voltage law)

We have also shown that (see eqn. (2.73) for the calculation of Y_{BR} through graph theoretic approach)

$$\bar{v} = C^T \bar{E}_{BR}$$

Then partitioning this equation for branches and links, we have

$$\begin{bmatrix} \bar{v}_b \\ \bar{v}_l \end{bmatrix} = \begin{bmatrix} U_b \\ C_l^T \end{bmatrix} \bar{E}_{BR} \tag{2.90}$$

From the above (i.e. equation (2.90)), we get

$$\bar{v}_b = \bar{E}_{BR}$$

and

$$\bar{v}_l = C_l^T \bar{E}_{BR}$$

Hence

$$\begin{aligned} \bar{v} &= C^T \bar{E}_{BR} \\ &= C^T \bar{v}_b \end{aligned} \tag{2.91}$$

That is, voltage vector $\bar{v}$ can be expressed in terms of tree set of voltages v_b (i.e. branch voltages).

Similarly,

$\bar{i} = B^T \bar{I}_{loop}$ (see equation (2.82) for the calculation of Z_{loop} through graph theoretic approach).

Partitioning the above in terms of branches and links, we have

$$\begin{bmatrix} \bar{i}_b \\ \\ \bar{i}_l \end{bmatrix} = \begin{bmatrix} B_b^T \\ \\ U_l \end{bmatrix} \bar{I}_{loop} \tag{2.92}$$

From the equation (2.92), we get,

$$\bar{i}_b = B_b^T \bar{I}_{loop}$$

and

$$\bar{i}_l = \bar{I}_{loop}$$

Thus

$$\bar{i} = B^T \bar{I}_{loop} = B^T \bar{i}_l \tag{2.93}$$

i.e. element current is expressed in terms of cotree (i.e. link) currents.

2.7 Building Algorithm for the Bus Impedance Matrix Z_{BUS}

We shall discuss here an algorithm for developing bus impedance matrix Z_{BUS} from a given primitive network. Let us assume that Z_{BUS} matrix exists for a part of the primitive network known as partial network and the corresponding network equation for this partial network in the bus frame of reference is,

$$\bar{E}_{BUS_{m\times 1}} = [Z_{BUS}]_{m\times m} \bar{I}_{BUS_{m\times 1}} \tag{2.94}$$

From the above equation, it is clear that out of say n-buses of the primitive network, m buses are included in the partial network for which Z_{BUS} matrix exists. We shall take one element at a time from the remaining portion of the primitive network which are not included in the partial network for which the Z_{BUS} matrix is available and add it to the partial network and in this way, gradually Z_{BUS} matrix for the entire primitive network will be developed. However, the added element say $p-q$ between buses (p) and (q) may be a branch or a link. In case the partial network for which Z_{BUS} matrix is available includes only the bus p, then added element $p-q$ will naturally be a branch and in this way a new bus (q) will be added up to the partial network. Thus if $p-q$ is a branch, then a new bus q is added to the partial network and dimension of Z_{BUS} which was of the order $m \times m$ will now be of the order

$(m+1)(m+1)$. And hence corresponding to this new bus (q), a new row or column (say qth row/column) is added up in the Z_{BUS} matrix. In this way, the branch $p-q$ is simulated by calculating the elements in this new row or column. However, in case both the bus (p) and (q) were already included in the partial network but the element $p-q$ was not included, then the added element $p-q$ from the primitive network to the partial network will naturally be a link. Thus when the added element $p-q$ is a link, no new buses are added to the partial network as both the buses (p) and (q) are already included in the partial network for which Z_{BUS} matrix exists and hence dimension of the matrix Z_{BUS} will remain unchanged. However, the elements of the Z_{BUS} matrix will be modified in order to simulate the addition of this link.

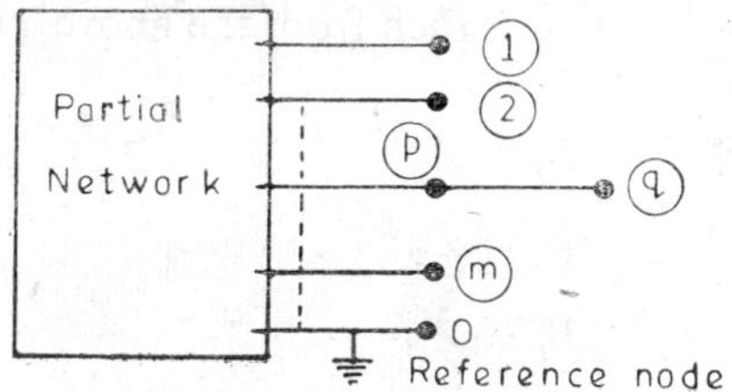

Addition of a branch $p-q$

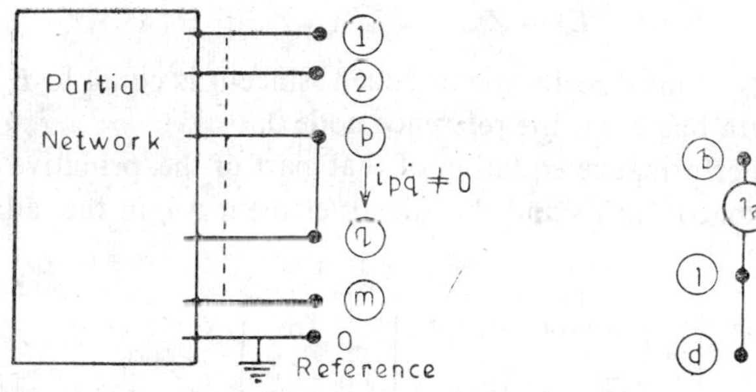

Addition of a link $p-q$

Connecting an ideal voltage source

Fig. 2.11

2.7.1 Addition of a Branch

As already mentioned, when the added element $p-q$ is a branch, a new bus q is created with the result, the dimension of Z_{BUS} matrix of the partial network will increase by one. The performance equation of the partial network with the addition of a branch $p-q$ will be as follows

$$\bar{E}_{BUS(m+1)\times 1} = [Z_{BUS}]_{(m+1)(m+1)} \bar{I}_{BUS(m+1)\times 1} \tag{2.95}$$

i.e.

$$\begin{bmatrix} E_1 \\ E_2 \\ \cdot \\ \cdot \\ E_m \\ E_q \end{bmatrix} = \begin{bmatrix} Z_{11} & Z_{12} & \dots & Z_{1m} & Z_{1q} \\ Z_{21} & Z_{22} & \dots & & Z_{2q} \\ & & \dots & & \cdot \\ & & \dots & & \cdot \\ & & \dots & & \cdot \\ Z_{m1} & & \dots & Z_{mm} & Z_{mq} \\ Z_{q1} & & \dots & & Z_{qq} \end{bmatrix} \begin{bmatrix} I_1 \\ I_2 \\ \cdot \\ \cdot \\ I_m \\ I_q \end{bmatrix} \quad (2.96)$$

Since the elements of the power system network are linear and bilateral, $Z_{qi} = Z_{iq}$, for $q = 1, \dots, n$. Therefore, it is clear from the above performance equation (i.e. eqn. (2.96)) that in this case a new bus q has been created which in turn creates a new row and a column (qth row and qth column). And hence, the addition of new branch $p-q$ needs the calculation of elements in the new row and column. Let us calculate first the elements Z_{qi} for $i = 1, \dots, n$ and $i \neq q$ (i.e. excluding diagonal elements Z_{qq}). To calculate these elements, we will apply a current source of 1 p.u. at the ith bus, i.e. $I_i = 1$ p.u. and keep remaining buses open-circuited, i.e. $I_k = 0$, $k = 1, \dots, m$ and $k \neq i$, then from the above performance equation (eqn. 2.96)), we have

$$\begin{aligned} E_1 &= Z_{1i} \\ E_2 &= Z_{2i} \\ &\vdots \\ E_p &= Z_{pi} \\ &\vdots \\ E_m &= Z_{mi} \\ &\vdots \\ E_q &= Z_{qi} \end{aligned} \quad (2.97)$$

In this way Z_{qi} can directly be obtained since it is equal to E_q, i.e. the voltage of the qth bus w.r.t. the reference node 0.

Now the performance equation of that part of the primitive network for which Z_{BUS} matrix exists and the added element p-q in the admittance form will be

$$\begin{bmatrix} i_{pq} \\ \bar{i}_{\rho\sigma} \end{bmatrix} = \begin{bmatrix} y_{pq\,pq} & \bar{y}_{pq\,\rho\sigma} \\ \bar{y}_{\rho\sigma\,pq} & [y_{\rho\sigma\,\rho\sigma}] \end{bmatrix} \begin{bmatrix} v_{pq} \\ \bar{v}_{\rho\sigma} \end{bmatrix} \quad (2.98)$$

Here i_{pq} and v_{pq} are the current through and the voltage drop accross the added element $p-q$. $\bar{i}_{\rho\sigma}$ and $\bar{v}_{\rho\sigma}$ are the column vectors of currents through and voltage drops across elements ρ–σ of the partial network. y_{pqpq} is the self admittance of the element $p-q$. $\bar{y}_{pq\rho\sigma}$ is the row vector of the mutual admittances between the added element $p-q$ and elements $\rho-\sigma$ of the partial network. $y_{\delta\sigma_{I}q}$ is the transpose of $y_{pq\,\rho\sigma} \cdot y_{\rho\sigma\,\rho\sigma}$ is the primitive admittance matrix of the partial network.

Expanding the equation (2.98), we get

$$i_{pq} = y_{pqpq}\, v_{pq} + \bar{y}_{pq\rho\sigma}\, \bar{v}_{\rho\sigma} \tag{2.99}$$

Since added element p–q is a branch, $i_{pq} = 0$ but v_{pq} may not be equal to zero as the element p–q may be mutually coupled to some elements ρ–σ of the partial network.

Thus the equation (2.99) becomes

$$y_{pqpq}\, v_{pq} + \bar{y}_{pq\rho\sigma}\, \bar{v}_{\rho\sigma} = 0$$

i.e.

$$v_{pq} = -\frac{\bar{y}_{pq\rho\sigma}\, \bar{v}_{\rho\sigma}}{y_{pqpq}} \tag{2.100}$$

But v_{pq} is the voltage drop across the element p–q, and hence $v_{pq} = E_p - E_q$ where E_p and E_q are the bus voltages at the buses p and q respectively i.e.

$$E_q = E_p - v_{pq} \tag{2.101}$$

Substituting in the equation (2.101), the equation (2.100), we get

$$E_q = E_p + \frac{\bar{y}_{pq\rho\sigma}\, \bar{v}_{\rho\sigma}}{y_{pqpq}}$$

$$= E_p + \frac{\bar{y}_{pq\rho\sigma}\, (\bar{E}_\rho - \bar{E}_\sigma)}{y_{pqpq}} \tag{2.102}$$

as $\bar{v}_{\rho\sigma} = \bar{E}_\rho - \bar{E}_\sigma$ are the voltage drops across elements ρ–σ of the partial network.

Substituting eqn. (2.97) in the eqn. (2.102) (i.e. $E_q = Z_{qi}$, $E_p = Z_{pi}$, $\bar{E}_\rho = \bar{Z}_{\rho i}$ and $\bar{E}_\sigma = \bar{Z}_{\sigma i}$ as $I_i = 1$ p.u. and other buses are open circuited), we get

$$Z_{qi} = Z_{pi} + \frac{\bar{y}_{pq\rho\sigma}\, (\bar{Z}_{\rho i} - \bar{Z}_{\sigma i})}{y_{pqpq}} \tag{2.103}$$

for $i = 1, \ldots, n$

$i \neq q$

Now if p is the reference bus, then $E_p = Z_{pi} = 0$, and the eqn. (2.103) becomes

$$Z_{qi} = \frac{\bar{y}_{pq\rho\sigma}\, (\bar{Z}_{pi} - \bar{Z}_{\sigma i})}{y_{pqpq}} \tag{2.104}$$

If p is not the reference bus, but the element p–q is not coupled to any elements ρ–σ of the partial network, then $y_{pq\rho\sigma} = 0$ and $Z_{pqpq} = 1/y_{pqpq}$ and thus equation (2.103) becomes

$$Z_{qi} = Z_{pi} \tag{2.105}$$

However if p is the reference bus and at the same time element p–q is also not mutually coupled to any elements ρ–σ of the partial network, i.e. $y_{pq\rho\sigma} = 0$, eqn. (2.103) becomes

$$Z_{qi} = 0. \tag{2.106}$$

To calculate the diagonal element Z_{qq}, we will inject a current source of 1 p.u. at the qth bus i.e. $I_q = 1$ p.u. and keep other buses open circuited, then eqn. (2.96) becomes

$$E_1 = Z_{1q}$$
$$E_2 = Z_{2q}$$
$$E_q = Z_{qq} \tag{2.107}$$

as in this case $I_q = 1$ p.u. and the remaining buses are open circuited.

Therefore, E_q gives directly Z_{qq}. Again taking eqn. (2.98) and expanding it we get,

$$i_{pq} = y_{pqpq}\, v_{pq} + \bar{y}_{pq\rho\sigma}\, \bar{v}_{\rho\sigma} \tag{2.108}$$

Since at the qth bus, the injected current I_q (= 1 p.u.) flows from the bus q towards the bus p and therefore, $i_{pq} = -I_q = -1$. Hence equation (2.108) becomes,

$$i_{pq} = y_{pqpq}\, v_{pq} + \bar{y}_{pq\rho\sigma}\, \bar{v}_{\rho\sigma}$$
$$= -I_q = -1$$

i.e.
$$v_{pq} = -\frac{1 + \bar{y}_{pq\rho\sigma}\, \bar{v}_{\rho\sigma}}{y_{pqpq}} \tag{2.109}$$

But,
$$v_{pq} = E_p - E_q$$

i.e.
$$E_q = E_p - v_{pq} \tag{2.110}$$

Substituting equation (2.109) in equation (2.110), we get

$$E_q = E_p + \frac{1 + \bar{y}_{pq\rho\sigma}\, \bar{v}_{\rho\sigma}}{y_{pqpq}}$$
$$= E_q + \frac{1 + \bar{y}_{pq\rho\sigma}\, [\bar{E}_\rho - \bar{E}_\sigma]}{y_{pqpq}} \tag{2.111}$$

Substituting eqn. (2.107) in eqn. (2.111), we get

$$Z_{qq} = Z_{pq} + \frac{1 + \bar{y}_{pq\rho\sigma}\, [\bar{Z}_{\rho q} - \bar{Z}_{\sigma q}]}{y_{pqpq}} \tag{2.112}$$

If p is a reference bus, then $E_p = Z_{pq} = 0$ and then the equation (2.112) becomes

$$Z_{qq} = \frac{1 + \bar{y}_{pq\rho\sigma}[\bar{Z}_{\rho q} - \bar{Z}_{\sigma q}]}{y_{pqpq}} \tag{2.113}$$

If p is not the reference bus, but the element $p-q$ is not mutually coupled to any elements $\rho-\sigma$ of the partial network, then

$$\bar{y}_{pq\rho\sigma} = 0 \quad \text{and} \quad 1/y_{pqpq} = z_{pqpq}$$

and thus the equation (2.112) becomes

$$Z_{p\rho} = Z_{pq} + Z_{pqpq} \tag{2.114}$$

However, if p is the reference bus and at the same time the element $p-q$ is also not mutually coupled to any elements $p-\sigma$ of the partial network, eqn. (2.112) becomes,

$$Z_{qq} = Z_{pqpq} \tag{2.115}$$

2.7.2 *Addition of Link*

As discussed earlier, if the added elements $p--q$ is a link, no new bus is created, i.e. the dimension of the Z_{BUS} matrix remains the same but its elements are recalculated. However, so as to use the same algorithm as that of the branch, we will connect an ideal voltage source e_l in series with the element $p-q$ such that the current through the element $p-q$ is zero. In this way (as shown in Fig. 2.12) a fictitious bus (l) will be created and the element $p-l$ in which no current is flowing because of the choice of the voltage source e_l, will be treated as a branch. The fictitious bus l is removed later.

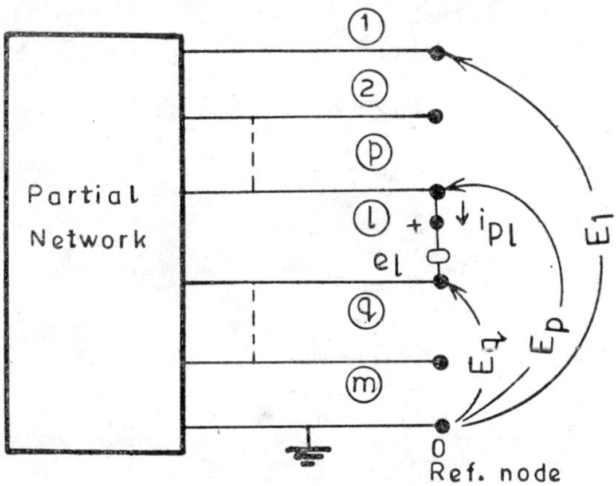

Fig, 2.12 Connecting an ideal voltage source e_l

Since voltage source e_l is connected between the bus (q) and the fictitious bus (l) we have

$$E_l = E_q + e_l$$

where E_l is the fictitious bus voltage (i.e. the voltage of the fictitious bus l).

No doubt all the bus voltages say E_1, E_2, E_p and E_q etc. are measured with respect to the reference node 0, let us assume that only for the lth bus, reference bus is q, then naturally with the definition, we have

$$E_l = e_l \tag{2.116}$$

Since $p-l$ is a branch, fictitious bus l has been created with the result the dimension of Z_{BUS} matrix will be increased to $(m+1)(m+1)$. The performance equation in the bus frame of reference now becomes

$$\begin{bmatrix} E_1 \\ E_2 \\ \vdots \\ E_m \\ e_l \end{bmatrix} = \begin{bmatrix} Z_{11} \cdots & Z_{1m} Z_{1l} \\ Z_{21} \cdots & Z_{2m} Z_{2l} \\ \vdots & \vdots \\ Z_{m1} \cdots & Z_{mm} Z_{ml} \\ Z_{l1} & Z_{lm} Z_{ll} \end{bmatrix} \begin{bmatrix} I_1 \\ I_2 \\ \vdots \\ I_m \\ I_l \end{bmatrix} \tag{2.117}$$

Thus now because of the creation of a fictitious bus l, a new row and a new column has been created whose elements $Z_{li} = Z_{il}$ will be calculated by using the same algorithm as that of the branch.

To find the elements Z_{li} for $i = 1, ..., n$ and $i \neq l$ (i.e. excluding the diagonal element), we will apply a unit current source at the ith bus, i.e. $I_i = 1$ p.u. and keep other buses open circuited, i.e. $I_k = 0$ for $k = 1. ..., n$ and $k \neq i$, then equation (2.117) becomes,

$$\begin{aligned} E_1 &= Z_{1i} \\ E_2 &= Z_{2i} \\ \vdots &\quad \vdots \\ E_p &= Z_{pi} \\ E_q &= Z_{qi} \\ E_m &= Z_{mi} \\ e_l &= Z_{li} \end{aligned} \tag{2.118}$$

i.e. $E_k = Z_{ki}$ for $k = 1, ..., n$

$k \neq l$

and $e_l = Z_{li}$

Since bus voltage at the pth bus is higher than that of the qth bus, we have

$$E_q + e_l + v_{pl} = E_p$$

i.e.

$$e_l = E_p - E_q - v_{pl} \tag{2.119}$$

where v_{pl} is the voltage drop across the fictitious branch $p-l$ which may not be equal to zero as the branch $p-l$ may be coupled to some elements $\rho-\sigma$ of the partial network but we know that $i_{pl} = 0$, since $p-l$ is a branch.

The performance equation of the partial network (i.e. the part of the primitive network for which Z_{BUS} exists) along with the added branch $p-l$ in the admittance form will be,

$$\begin{bmatrix} i_{pl} \\ \bar{i}_{\rho\sigma} \end{bmatrix} = \begin{bmatrix} y_{plpl} & \bar{y}_{pl\rho\sigma} \\ \bar{y}_{\rho\sigma pl} & [y_{\rho\sigma\rho\sigma}] \end{bmatrix} \begin{bmatrix} v_{pl} \\ \bar{v}_{\rho\sigma} \end{bmatrix} \tag{2.120}$$

where i_{pl} and v_{pl} are the current and voltage for the fictitious branch $p-l$ and $\bar{i}_{\rho\sigma}$ and $\bar{v}_{\rho\sigma}$ for the elements $\rho-\sigma$ of the partial network, and

y_{plpl} is the self admittance of the element $p-l$

$\bar{y}_{pl\rho\sigma}$ is the row vector of mutual admittance between the element $p-l$ and the elements $\rho-\sigma$ of the partial network,

$\bar{y}_{\rho\sigma pl}$ is the transpose of $\bar{y}_{pl\rho\sigma}$,

and $[y_{\rho\sigma\rho\sigma}]$ is the primitive admittance matrix of the partial network.

Expanding the eqn. (2.120), we get

$$i_{pl} = y_{plpl}\, v_{pl} + \bar{y}_{pl\rho\sigma}\, \bar{v}_{\rho\sigma} = 0 \tag{2.121}$$

i.e.

$$v_{pl} = -\frac{\bar{y}_{pl\rho\sigma}\, \bar{v}_{\rho\sigma}}{y_{plpl}} \tag{2.122}$$

Substituting the equation (2.122) in the equation (2.119), we get,

$$e_l = E_p - E_q - v_{pl}$$

$$= E_p - E_q + \frac{\bar{y}_{pl\rho\sigma}\, \bar{v}_{\rho\sigma}}{y_{plpl}} \tag{2.123}$$

Since e_l is an ideal voltage source, we have

$$y_{plpl} = y_{pqpq}$$

and

$$y_{pl\rho\sigma} = \bar{y}_{pq\rho\sigma}$$

and also

$$\bar{v}_{\rho\sigma} = \bar{E}_\rho - \bar{E}_\sigma \tag{2.124}$$

Substituting these in the equation (2.123), we get

$$e_l = E_p - E_q + \frac{\bar{y}_{pq\rho\sigma}[\bar{E}_\rho - \bar{E}_\sigma]}{y_{pqpq}} \tag{2.125}$$

Substituting the relations (2.118) in the equation (2.125), we get

$$Z_{li} = Z_{pi} - Z_{qi} + \frac{\bar{y}_{pq\rho\sigma}[\bar{Z}_{\rho i} - \bar{Z}_{\sigma i}]}{y_{pqpq}} \tag{2.126}$$

$$i = 1, 2, \ldots, n$$
$$i \neq l$$

If (p) is a reference bus, $Z_{pi} = 0$ and the equation (2.126) becomes

$$Z_{li} = -Z_{qi} + \frac{\bar{y}_{pq\rho\sigma}[\bar{Z}_{\rho i} - Z_{\sigma i}]}{y_{pqpq}} \tag{2.127}$$

If p is not the reference bus but the element $p-q$ is not mutually coupled to any elements $\rho-\sigma$ of the partial network, then

$$\bar{y}_{pq\rho\sigma} = 0$$

and $$1/y_{pqpq} = Z_{pqpq} \tag{2.128}$$

Substituting equation (2.128) in equation (2.126), we get

$$Z_{li} = Z_{pi} - Z_{qi} \tag{2.129}$$

However, if p is the reference bus and at the same time element $p—q$ is not mutually coupled, eqn. (2.126) becomes

$$Z_{li} = -Z_{qi} \tag{2.130}$$

To calculate the diagonal element Z_{ll}, we will apply a current source of 1 p.u., at the lth bus with respect to the bus q i.e. $I_l = 1$ p.u. and is measured with respect to the qth bus keeping other buses open circuited and then eqn. (2.117) becomes

$$E_1 = Z_{1l}$$

$$E_2 = Z_{2l}$$

$$\vdots$$

$$E_m = Z_{ml}$$

$$e_l = Z_{ll} \tag{2.131}$$

i.e. $$E_k = Z_{kl} \quad \text{for} \quad k = 1, \ldots, n \quad k \neq l$$

and $$e_l = Z_{ll}$$

Since $I_l = 1$ p.u. and i_{pl} is the current from the bus p towards l, we have

$$i_{pl} = -I_l = -1 \tag{2.132}$$

Let us take eqn. (2.120) and write it in the expanded form as given below

$$i_{pl} = y_{plpl}\, v_{pl} + \bar{y}_{pl\rho\sigma}\, \bar{v}_{\rho\sigma} \tag{2.133}$$

$$= -I_l = -1$$

i.e. $$v_{pl} = -\frac{1 + \bar{y}_{pl\rho\sigma}\bar{v}_{\rho\sigma}}{y_{plpl}} \tag{2.134}$$

Substituting equation (2.134) in equation (2.119), we get

$$e_l = E_p - E_q + \frac{1 + \bar{y}_{pl\rho\sigma}\, \bar{v}_{\rho\sigma}}{y_{plpl}} \tag{2.135}$$

Now $$\bar{v}_{\rho\sigma} = \bar{E}_\rho - \bar{E}_\sigma$$

and $$y_{pl\rho\sigma} = y_{pq\rho\sigma};\; y_{plpl} = y_{pqpq}$$

and finally substituting eqn. (2.131) in eqn. (2.135), we get

$$Z_{ll} = Z_{pl} - Z_{ql} + \frac{1 + \bar{y}_{pq\rho\sigma}\,[\bar{Z}_{\rho l} - \bar{Z}_{\sigma l}]}{y_{pqpq}} \tag{2.135a}$$

If p is a reference bus, $Z_{pl} = 0$, then eqn. (2.135a) becomes

$$Z_{ll} = -Z_{ql} + \frac{1 + \bar{y}_{pq\rho\sigma}\,[\bar{Z}_{\rho l} - \bar{Z}_{\sigma l}]}{y_{pqpq}} \tag{2.136}$$

Further if p is not the reference bus and p–q is not mutually coupled to any elements $\rho-\sigma$ of the partial network, the eqn. (2.135a) becomes

$$Z_{ll} = Z_{pl} - Z_{ql} + Z_{pqpq} \tag{2.137}$$

Finally, if p is the reference but $p-q$ is also not mutually coupled, we get

$$Z_{ll} = -Z_{ql} + Z_{pqpq} \tag{2.138}$$

In the end we shall delete the fictitious bus (l) by short circuiting the voltage source e_l which is in series with the link $p-q$ and thus will modify the elements of Z_{BUS}.

Let us take eqn. (2.117) we write it in the partitioned form as shown modify below,

$$\begin{bmatrix} \bar{E}_{BUS} \\ \\ e_l \end{bmatrix} = \left[\begin{array}{c|c} [Z_{BUS}] & \bar{Z}_{il} \\ \hline \bar{Z}_{lj} & Z_{ll} \end{array}\right] \begin{bmatrix} \bar{I}_{BUS} \\ \\ I_l \end{bmatrix} \tag{2.139}$$

Expanding the eqn. (2.139), we get

$$\bar{E}_{BUS} = [Z_{BUS}]\, \bar{I}_{BUS} + \bar{Z}_{il}\, I_l \tag{2.140}$$

$$e_l = \bar{Z}_{lj}\, \bar{I}_{BUS} + Z_{ll}\, I_l \tag{2.141}$$

By short circuiting voltage source e_l, i.e. $e_l = 0$, the fictitious bus l will automatically be eliminated. And therefore,

$$e_l = \bar{Z}_{lj}\, \bar{I}_{BUS} + Z_{ll}\, I_l = 0$$

i.e.

$$I_l = -\frac{\bar{Z}_{lj}\, \bar{I}_{BUS}}{Z_{ll}} \tag{2.142}$$

Substituting eqn. (2.142) in eqn. (2.140), we get

$$\bar{E}_{BUS} = [Z_{BUS}]\, \bar{I}_{BUS} - \frac{\bar{Z}_{il}\, Z_{lj}}{Z_{ll}}\, \bar{I}_{BUS}$$

$$= \left([Z_{BUS}] - \frac{\bar{Z}_{il}\, Z_{lj}}{Z_{ll}} \right) \bar{I}_{BUS}$$

$$= [Z_{BUS} \text{ (after elimination)}]\, \bar{I}_{BUS} \tag{2.143}$$

i.e. Z_{BUS} (after elimination, i.e. modified)

$$= Z_{BUS\,(\text{original})} - \frac{Z_{il}\, Z_{lj}}{Z_{ll}} \tag{2.144}$$

Therefore any element Z_{ij} of the modified Z_{BUS} matrix will be

$$Z_{ij\,(\text{modified})} = Z_{ij\,(\text{original})} - \frac{Z_{il}\, Z_{lj}}{Z_{ll}} \tag{2.145}$$

$$i = 1, \ldots, n; \quad j = 1, \ldots, n$$

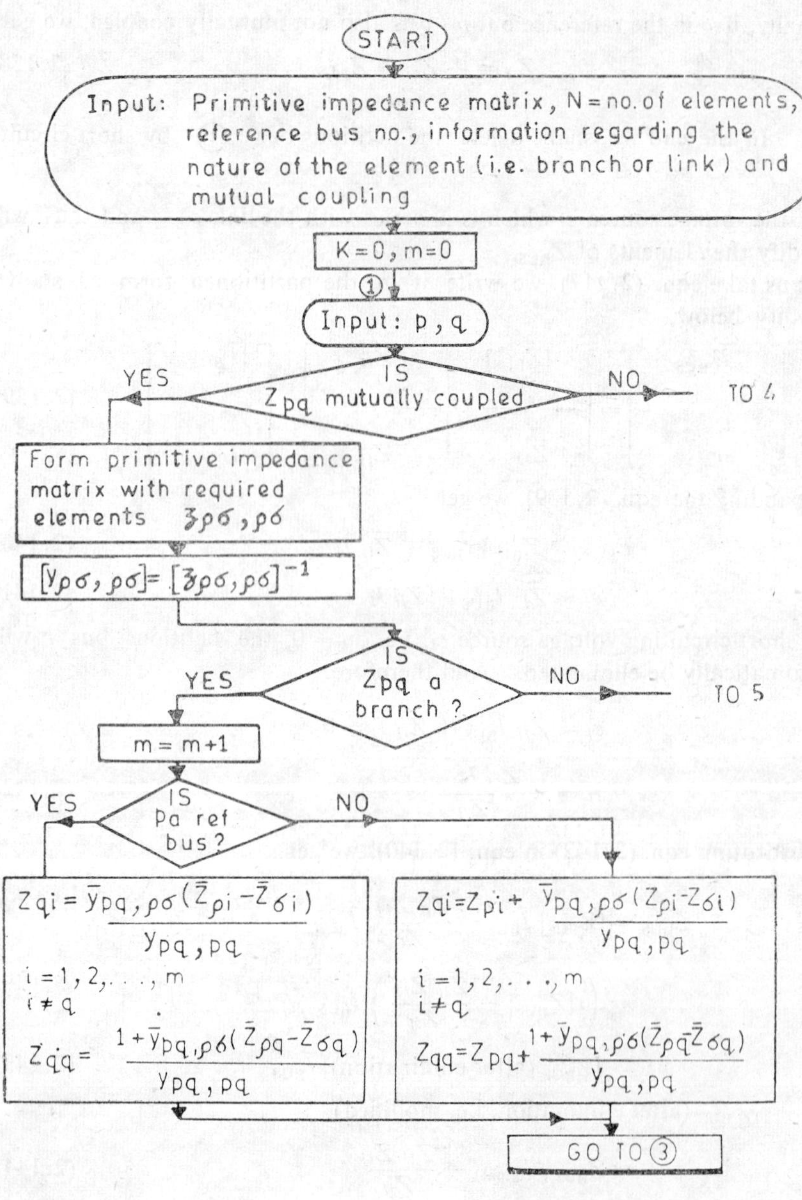
START
Input: Primitive impedance matrix, N = no. of elements, reference bus no., information regarding the nature of the element (i.e. branch or link) and mutual coupling
K = 0, m = 0
1
Input: p, q
IS Zpq mutually coupled
YES
NO
TO 4
Form primitive impedance matrix with required elements zρσ, ρσ
[yρσ, ρσ] = [zρσ, ρσ]^-1
IS Zpq a branch ?
YES
NO
TO 5
m = m+1
IS Pa ref bus?
YES
NO
Zqi = ȳpq, ρσ (Z̄ρi − Z̄σi) / ypq, pq
i = 1, 2, . . ., m
i ≠ q
Zqq = 1 + ȳpq, ρσ (Z̄ρq − Z̄σq) / ypq, pq
Zqi = Zρi + ȳpq, ρσ (Zρi − Zσi) / ypq, pq
i = 1, 2, . . ., m
i ≠ q
Zqq = Zpq + 1 + ȳpq, ρσ (Z̄ρq − Z̄σq) / ypq, pq
GO TO 3

(a)

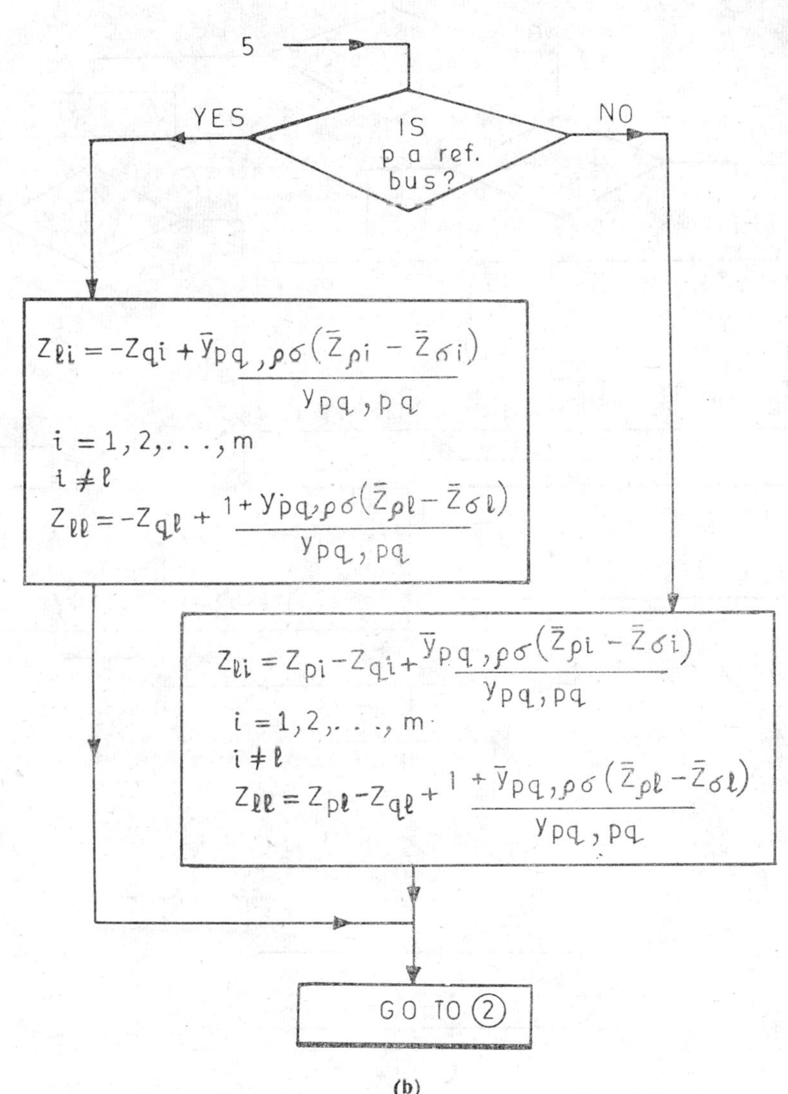
5
YES
IS
p a ref.
bus?
NO
$Z_{\ell i} = -Z_{qi} + \frac{\bar{y}_{pq,\rho\sigma}(\bar{Z}_{\rho i} - \bar{Z}_{\sigma i})}{y_{pq,pq}}$
$i = 1, 2, \ldots, m$
$i \neq \ell$
$Z_{\ell\ell} = -Z_{q\ell} + \frac{1 + \bar{y}_{pq,\rho\sigma}(\bar{Z}_{\rho\ell} - \bar{Z}_{\sigma\ell})}{y_{pq,pq}}$
$Z_{\ell i} = Z_{pi} - Z_{qi} + \frac{\bar{y}_{pq,\rho\sigma}(\bar{Z}_{\rho i} - \bar{Z}_{\sigma i})}{y_{pq,pq}}$
$i = 1, 2, \ldots, m$
$i \neq \ell$
$Z_{\ell\ell} = Z_{p\ell} - Z_{q\ell} + \frac{1 + \bar{y}_{pq,\rho\sigma}(\bar{Z}_{\rho\ell} - \bar{Z}_{\sigma\ell})}{y_{pq,pq}}$
GO TO (2)

(b)

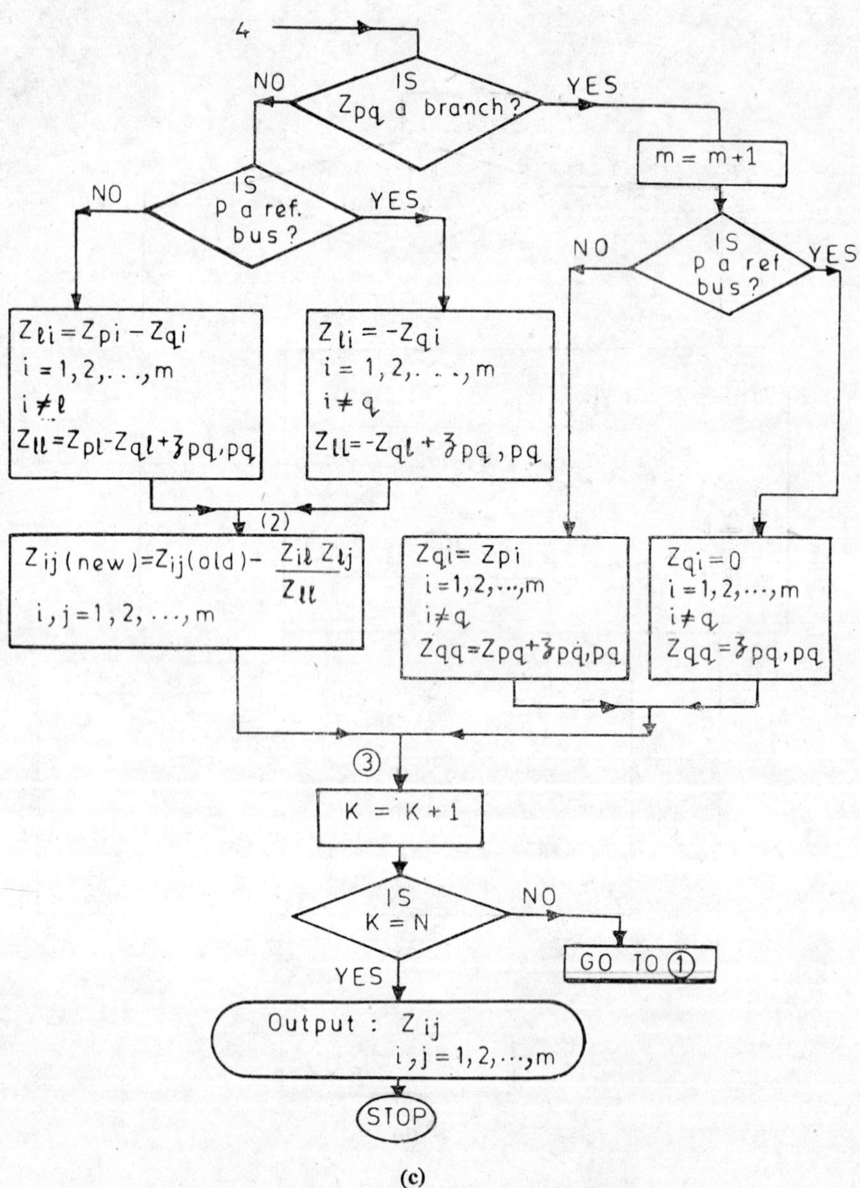

(c)

Fig. 2.13 Flow diagram for building Z_{BUS} algorithm

In this way the elements of Z_{BUS} matrix will get modified due to addition of a link.

2.8 Modification of Z_{BUS} Matrix Due to Changes in the Primitive Network

The changes in the primitive network can be the addition of an element, removal of an element or changes in the impedance of an element. Due to these changes, the corresponding Z_{BUS} matrix will also change.

If an element is added to the primitive network, then Z_{BUS} matrix of the primitive network may be taken as Z_{BUS} matrix at that stage and the same building algorithm may be used to calculate the elements of new Z_{BUS} matrix depending upon whether the new element is a branch or a link.

The removal of an element or changing its impedance is the same thing. If the element to be deleted is not mutually coupled to any other elements, then the effect of its deletion is taken into account by connecting an element in parallel to this element (i.e. connecting a link) whose impedance is the negative of the impedance of the element to be deleted. If the impedance of the element is to be changed then connect a link in parallel with this element so that resultant impedance of this combination is the desired value.

However, if the element to be removed or the element whose impedance is to be changed, is mutually coupled to some other elements then we follow the procedure given below.

Let the performance equation of the original system be

$$E_{BUS} = Z_{BUS}\, I_{BUS} \tag{2.146}$$

Now because of changes in the primitive network, which may be deletion of an element or the change of impedance of an element, the bus impedance matrix becomes Z'_{BUS} and hence the performance equation now becomes

$$E'_{BUS} = Z'_{BUS}\, I_{BUS} \tag{2.147}$$

where E'_{BUS} are the new bus voltages. Now suppose, we keep the original Z_{BUS} but make appropriate changes in the bus currents of the effected elements (i.e. element to be deleted and the elements to which it is mutually coupled) such that bus voltages are new bus voltages, i.e. E'_{BUS}, then the performance equation (2.147) becomes,

$$E'_{BUS} = Z_{BUS}\,(I_{BUS} + \Delta I_{BUS}) \tag{2.148}$$

Let an element $p-q$ which is mutually coupled to an element $\rho-\sigma$ is to be deleted or its impedance is to be changed, then in accordance with equation (2.148), we get

$$\Delta I_k = \Delta I_{pq} \quad \text{for } k = p$$

$$\Delta I_k = -\Delta I_{pq} \quad \text{for } k = q$$

$$\Delta I_k = \Delta I_{\rho\sigma} \text{ for } k = \rho$$

$$\Delta I_k = -\Delta I_{\rho\sigma} \text{ for } k = \sigma \tag{2.149}$$

and $\Delta I_k = 0$ for remaining buses.

Similarly in order to calculate the element Z'_{ij} for the modified bus impedance matrix Z'_{BUS}, we connect a current source of 1 p.u. at the jth bus i.e. $I_j = 1$ p.u. and keep other buses open-circuited. Then we have,

$$I_j = 1 \text{ p.u.}$$

$$I_k = 0 \text{ for } k = 1, \ldots, n \quad k \neq j \tag{2.150}$$

Then from the equation (2.148) we have

$$E'_i = \sum_{k=1}^{n} Z_{ik}\,[I_k + \Delta I_k] \tag{2.151}$$

Substituting eqns. (2.149) and (2.150) in eqn. (2.151), we get

$$E'_i = Z_{ij} + Z_{ip}\,\Delta I_{pq} - Z_{iq}\,\Delta I_{pq} + Z_{i\rho}\,\Delta I_{\rho\sigma} - Z_{i\sigma}\,\Delta I_{\rho\sigma}$$

$$= Z_{ij} + (Z_{ip} - Z_{iq})\,\Delta I_{pq} + (Z_{i\rho} - Z_{i\sigma})\,\Delta I_{\rho\sigma} \tag{2.152}$$

Let $\alpha = \begin{bmatrix} p \\ \rho \end{bmatrix}$ and $\beta = \begin{bmatrix} q \\ \sigma \end{bmatrix}$

Then by defining these new variables, equation (2.152) takes the form

$$E'_i = Z_{ij} + (Z_{i\alpha} - Z_{i\beta})\,\Delta I_{\alpha\beta} \tag{2.153}$$

Let $[Y_s]$ is the primitive admittance matrix of the original network and $[Y_s']$ is that of the modified network, then the performance equation of the primitive network in the admittance form due to this change will be,

$$\Delta I_{\alpha\beta} = [[Y_s] - [Y_s']]\, v'_{\gamma\delta} \tag{2.154}$$

But

$$v'_{\gamma\delta} = E'_\gamma - E'_\delta \tag{2.155}$$

and

$$E_\gamma' = Z_{\gamma j} + (Z_{\gamma\alpha} - Z_{\gamma\beta})\,\Delta I_{\alpha\beta} \tag{2.156}$$

and also

$$E_\delta' = Z_{\delta j} + (Z_{\delta\alpha} - Z_{\delta\beta})\,\Delta I_{\alpha\beta} \tag{2.157}$$

Substituting equations (2.156) and (2.157) in equation (2.155), we get,

$$v'_{\gamma\delta} = Z_{\gamma j} - Z_{\delta j} + (Z_{\gamma\alpha} - Z_{\delta\alpha} - Z_{\gamma\beta} + Z_{\delta\beta})\,\Delta I_{\alpha\beta} \tag{2.158}$$

Substituting eqn. (2.158) in eqn. (2.154), we get

$$\Delta I_{\alpha\beta} = [[Y_s] - [Y_s']]\,[Z_{\gamma j} - Z_{\delta j}] + [Z_{\gamma\alpha} - Z_{\delta\alpha} - Z_{\gamma\beta} + Z_{\delta\beta}]\,\Delta I_{\alpha\beta}$$

i.e.

$$\Delta I_{\alpha\beta} - [[Y_s] - [Y_s']]\,[Z_{\gamma\alpha} - Z_{\delta\alpha} - Z_{\gamma\beta} + Z_{\delta\beta}]\,\Delta I_{\alpha\beta}$$

$$= [[Y_s] - [Y_s']]\,[Z_{\gamma j} - [Z_{\delta j}]] \tag{2.159}$$

$$[U - [[Y_s] - [Y_s']]]\,[Z_{\gamma\alpha} - Z_{\delta\alpha} - Z_{\gamma\beta} + Z_{\delta\beta}]\,\Delta I_{\alpha\beta}$$

$$= [[Y_s] - [Y_s']]\,[Z_{\gamma j} - Z_{\delta j}] \tag{2.160}$$

i.e.

$$[\Delta I_{\alpha\beta}] = [U - [[Y_s] - [Y_s']]\,[Z_{\gamma\alpha} - Z_{\delta\alpha} - Z_{\gamma\beta} + Z_{\delta\beta}]]^{-1}\,[[Y_s] - [Y_s']]\,[Z_{\gamma j} - Z_{\delta j}]$$

$$= M^{-1} \Delta Y_s [Z_{\gamma j} - Z_{\delta j}] \tag{2.161}$$

where $\quad M = [U - [[Y_s] - [Y_s']] [Z_{\gamma\alpha} - Z_{\delta\alpha} - Z_{\gamma\beta} + Z_{\delta\beta}]$

and $\quad \Delta Y_s = [Y_s] - [Y_s']$

Substituting equation (2.161) in equation (2.153), we get

$$E_i' = Z_{ij} + [Z_{i\alpha} - Z_{i\beta}] [M^{-1}] \Delta Y_s [Z_{\gamma j} - Z_{\delta j}] \tag{2.162}$$

Since $I_j = 1$ p.u. and the rest of the buses are open circuited, the modified bus impedance matrix Z'_{BUS} will be found as follows,

$$Z_{ij\,(\text{modified})} = Z_{ij} + [Z_{i\alpha} - Z_{i\beta}] (M^{-1}) (\Delta Y_s][Z_{\gamma j} - Z_{\delta j}] \tag{2.163}$$

for $\quad i = 1, \ldots, n$

and $\quad j = 1, \ldots, n.$

2.9 Distinction between Standard Network Formulation and Graph Theoretic Formulation

The network matrices in bus, loop and branch frame of reference have been developed using both standard network formulation as well as graph theoretic formulation. Out of these different types of network matrices, the formulation using Y_{BUS} and Z_{BUS} is quite common for various kinds of power system studies such as load flow studies and short circuit studies etc.

The Z_{BUS} and Y_{BUS} matrices have been devoloped using standard network technique as well as using graph theoretic technique. If the mutual coupling between the elements of the power system network is neglected which is normally true in most of the cases, the application of standard network technique yields Y_{BUS} matrix directly by inspection. Z_{BUS} matrix is obtained by finding inverse of Y_{BUS}. Z_{BUS} matrix can also be obtained by conducting open circuit test and Y_{BUS} matrix by conducting short circuit test. As a matter of fact, the development of Z_{BUS} matrix by building algorithm is actually the outcome of applying open circuit test at different buses in sequence. In this case, both Y_{BUS} and Z_{BUS} matrices are developed by conducting appropriate tests even though mutual coupling between elements exist.

Obtaining Y_{BUS} and Z_{BUS} matrices by graph theoretic approach is valid even though mutual coupling exists between the elements of primitive network. Graph theoretic formulation takes into account the structure of the network. Here appopriate incidence matrices are developed which transform the primitive network (i.e. set of uncoupled elements) to network matrices such as Y_{BUS} or Z_{BUS}. This method of formulation is difficult for a large scale network because of the size and complications involved.

2.10 Numerical Example (Building Algorithm)

1. (a) Form the bus impedance matrix Z_{BUS} by building algorithm for the following power system network whose parameters (i.e. per unit

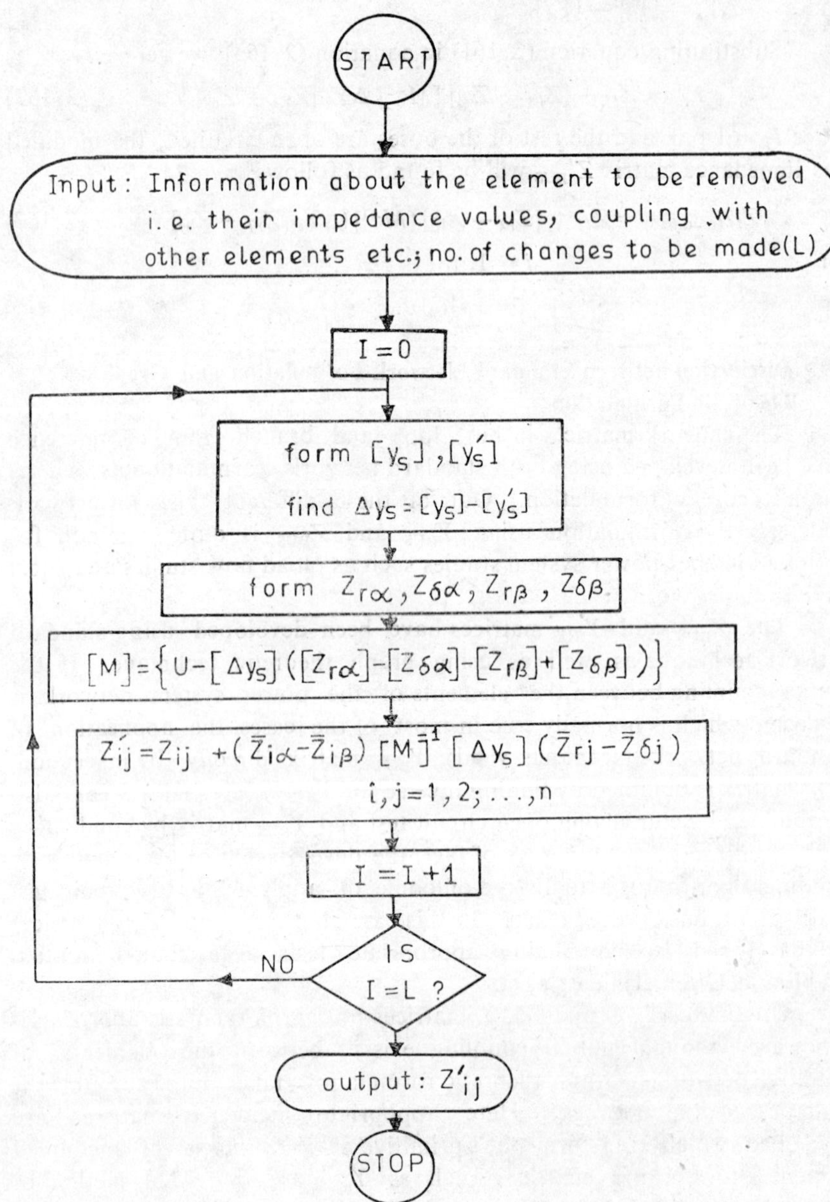

Fig. 2.14 Modification of Z_{BUS} for elimination of network elements

impedances) are given below. Modify Z_{BUS} matrix thus obtained to include the addition of an element 2–5 with an impedance of 0.4.

(b) Modify the original Z_{BUS} matrix of part (a) when the element 3–4 is removed. Modify also the Z_{BUS} matrix thus obtained if the element 1–2 mutually coupled to element 1–5 is deleted.

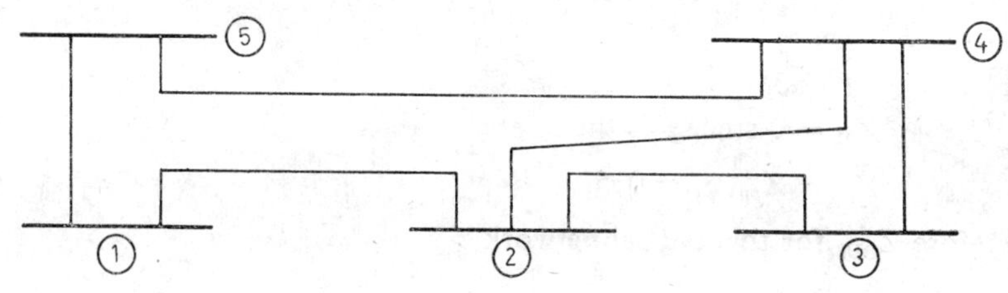

Element No.	Self impedances		Mutual impedances	
	Bus No. p–q	$Z_{pq\ pq}$	Bus No m—n	$Z_{pq\ mn}$
1	1-2	0.5	1–5	0.1
2	2–3	0.5	2–4	0.2
3	3–4	0.25		
4	2–4	0.6	4–5	0·2
5	4–5	0.75		
6	1–5	0.4		

Note : Impedances are in per unit.

Solution Part (a) For this problem, the given network is built up by adding the elements one by one, starting from the bus (1) which is taken as the reference. The elements are added in the order of their numbering, viz. 1, 2, 3, 4, 5 and 6. Thus the 4 and 6 become links and the remaining, branches. The graph of the complete network is shown below.

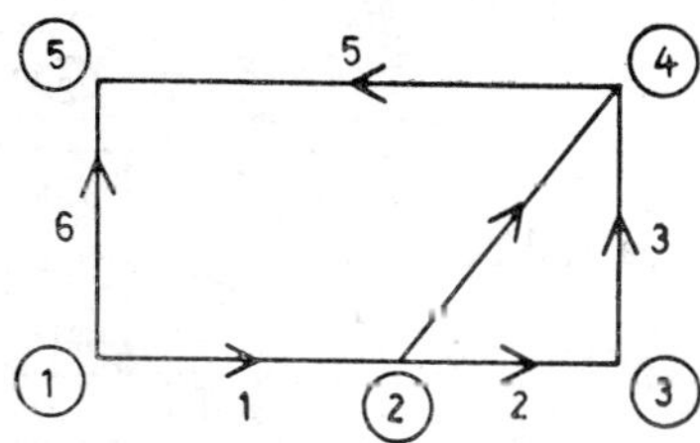

The building up of the network is started with bus (or node) (1).

Step 1: Element 1 is added to the bus (1). This creates a new bus (2).

So the added element is a branch. The question of mutual coupling does not arise.

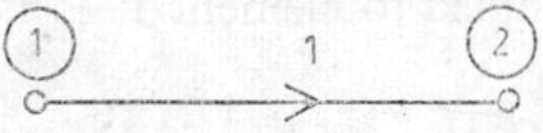

Now $p = 1, q = 2$

$Z_{21} = 0$ since p is the reference

$Z_{22} = Z_{pq, pq} = 0.5.$

Therefore Z_{BUS} for this partial network

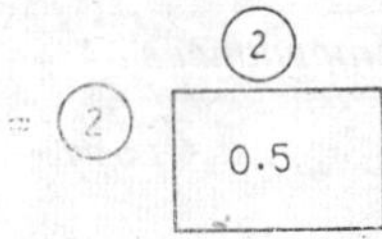

Step 2: Next the element 2 is added. Addition of this element results in the creation of the new bus (3). So it will be a branch. It is not mutually coupled to the element, 1, of the partial network.

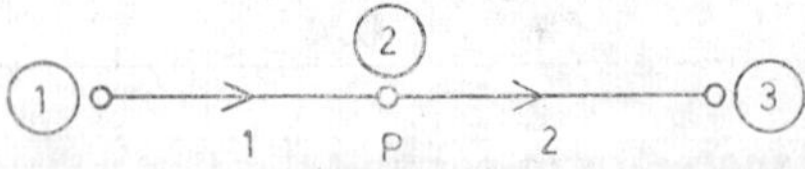

Now $p = 2, q = 3$

and $Z_{23, 23} = 0.5$

$Z_{qi} = Z_{pi}$. Therefore $Z_{32} = Z_{22} = 0.5 = Z_{23}$

$Z_{qq} = Z_{pq} + z_{pq, pq} = Z_{23} + 0.5 = 0.5 + 0.5 = 1.0$

Therefore Z_{BUS} for this partial network =

	(2)	(3)
(2)	0.5	0.5
(3)	0.5	1.0

Step 3 : Next the element 3 is added. It gets added to the bus (3) and creates a new bus (4). Also, it is not coupled to either element, 1 or element 2 of the partial network.

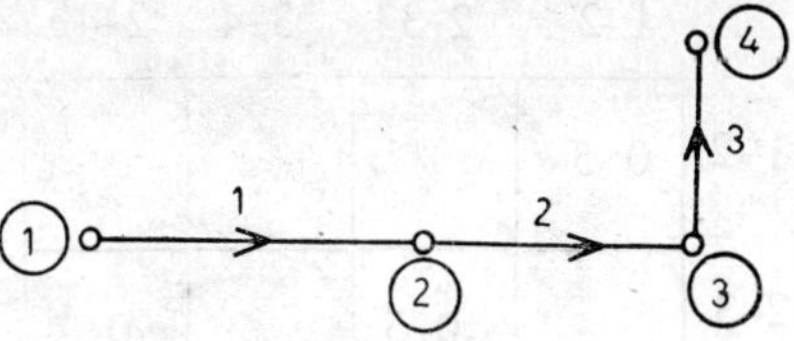

Therefore, $p = 3, q = 4, z_{34,34} = 0.25$

$$Z_{qi} = Z_{pi}, \text{ i.e., } Z_{4i} = Z_{3i} \text{ for } i = 2, 3$$

Therefore, $Z_{42} = Z_{32} = 0.5 = Z_{24}$, and $Z_{43} = Z_{33} = 1.0 = Z_{34}$

$$Z_{qq} = Z_{pq} + z_{pq,pq}$$

Therefore, $Z_{44} = Z_{34} + z_{34,34} = 1.0 + 0.25 = 1.25.$

Therefore, for this new partial network,

$Z_{BUS} =$

	(2)	(3)	(4)
(2)	0.5	0.5	0.5
(3)	0.5	1.0	1.0
(4)	0.5	1.0	1.25

Step 4: The element 4 is added. It gets added between the buses (2) and (4), both of which are already present in the partial network. So, it is a link. It has mutual coupling to the element No. 2 connected between $\rho = 2$ and $\sigma = 3$.

Therefore, $p = 2, q = 4, \rho = 2$ and $\sigma = 3$.

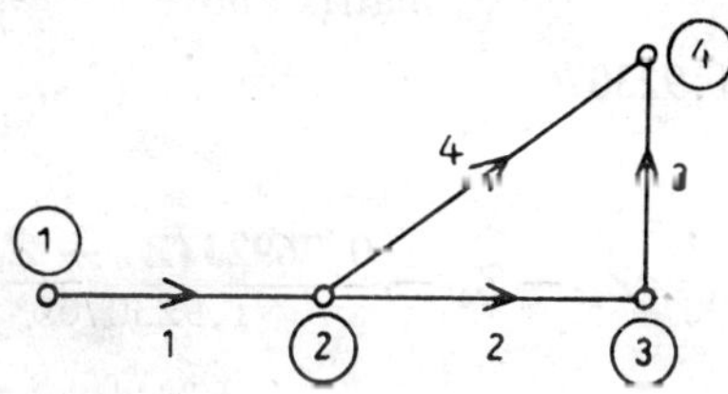

The primitive impedance matrix for this partial network is

$[z] =$

	1–2	2–3	3–4	2–4
1–2	0.5			
2–3		0.5		0.2
3–4			0.25	
2–4		0.		0.6

Therefore, the primitive admittance matrix of this partial network,

$y = [z]^{-1} =$

	1–2	2–3	3–4	2–4
1–2	2.0			
2–3		2.307692		−0.76923
3–4			4.0	
2–4		−0.76923		1.923076

Now, with the fictitious node 'l', the following analysis is carried out.

$$Z_{li} = Z_{pi} - Z_{qi} + \frac{\bar{y}_{pq,\,p\sigma}[\bar{Z}_{\rho i} - \bar{Z}_{\sigma i}]}{y_{pq,\,pq}}$$

$$i = 2,\ 3,\ 4$$

and $\qquad i \neq l$

Here $\quad \bar{y}_{pq,\,p\sigma} = -\ 0.76923,\ $ taken from the primitive admittance matrix above.

and $\quad y_{pq,\,pq} = 1.923076 \qquad$,, $\qquad$,, $\qquad$,,

Therefore,

$$Z_{l1} = Z_{1i} - Z_{4i} - \frac{0.76923\,(Z_{2i} - Z_{3i})}{1.923076}$$

for $i = 2, 3, 4$

$$Z_{l2} = Z_{22} - Z_{42} - \frac{.76923\,(Z_{22} - Z_{32})}{1.923076}$$

$$= (0.5 - 0.5) - \frac{0.76923\,(0.5 - 0.5)}{1.923076} = 0 = Z_{2l}$$

$$Z_{l3} = Z_{23} - Z_{43} - \frac{0.76923\,(0.5 - 1.0)}{1.923076}$$

$$= (0.5 - 1.0) - \frac{0.76923\,(0.5\ \cdot}{1.923076} = -0.3 = Z_{2l}$$

$$Z_{l4} = Z_{24} - Z_{44} - \frac{0.76923\,(Z_{24} - Z_{34})}{1.923076}$$

$$= (0.5 - 1.25) - \frac{0.76923\,(0.5 - 1.0)}{1.923076}\,\bar{y} = -0.55 = Z_{4l}$$

Now,

$$Z_{ll} = Z_{pl} - Z_{ql} + \frac{1 + \bar{y}_{pq,\rho\sigma}\,(\bar{Z}_{\rho l} - \bar{Z}_{\sigma l})}{pq,pq}$$

$$= Z_{2l} - Z_{4l} + \frac{[1 - 0.76923\,(Z_{2l} - Z_{3l})]}{1.923076}$$

$$= 0 - (-0.55) + \frac{[1 - 0.76923\,(0 - (-0.3))]}{1.923076}$$

$$= 0.95$$

Next, the fictitious node is eliminated as follows:

Let $$[X] = \frac{Z_{il}\,\bar{Z}_j}{Z_{ll}}$$

Then, $Z_{BUS\,(new)} = Z_{BUS\,(old)} - [X]$

$$\text{Now } [X] = \frac{1}{0.95}\begin{bmatrix} 0 \\ -0.3 \\ -0.55 \end{bmatrix}\begin{bmatrix} 0 & -0.3 & -0.55 \end{bmatrix}$$

$$= \begin{bmatrix} 0 & 0 & 0 \\ 0 & 0.0947368 & 0.1736842 \\ 0 & 0.1736842 & 3.318421 \end{bmatrix}$$

Therefore,

$Z_{\text{BUS (new)}} =$

0.5	0.5	0.5
0.5	1.0	1.0
0.5	1.0	1.25

$- X$

	②	③	④
②	0.5	0.5	0.5
③	0.5	0.9052632	0.8263158
④	0.5	0.8263158	0.931579

Step 5: The element 5 is added. It gets added to the bus (4) of the partial network and creates the new bus (5). So, it is a branch. It is mutually coupled to the element number 4, connected between $\rho = 2$ and $\sigma = 4$.

Therefore, $p = 4$, $q = 5$, $\rho = 2$ and $\sigma = 4$.

The primitive impedance matrix of the partial network is

$Z =$

	1–2	2–3	3–4	2–4	4–5
1–2	0.5				
2–3		0.5		0.2	
3–4			0.25		
2–4		0.2		0.6	0.2
4–5				0.2	0.75

Since we are interested in elements 4 and 5, we take away z_{11} and z_{33}, the inpedances of the uncoupled elements and invert the remaining submatrix. Then we get

$$\begin{array}{c} \\ 2\text{-}3 \\ 2\text{-}4 \\ 4\text{-}5 \end{array} \begin{array}{c} \begin{array}{ccc} 2\text{-}3 & 2\text{-}4 & 4\text{-}5 \end{array} \\ \left[\begin{array}{ccc} 0.5 & 0.2 & 0 \\ 0.2 & 0.6 & 0.2 \\ 0 & 0.2 & 0.75 \end{array}\right]^{-1} \end{array} =$$

	2–3	2–4	4–5
2–3	⊗	⊗	⊗
2–4	⊗	⊗	-0.5714286
4–5	⊗	-0.5714286	1.4857142

×–Not required here.

Therefore, $y_{pq,pq} = 1.4857142$, and

$y_{pq,p\sigma} = -0.5714286.$

Now $$Z_{qi} = Z_{pi} + \frac{\bar{y}_{pq,p\sigma}(\bar{Z}_{\rho i} - \bar{Z}_{\sigma i})}{y_{pq,pq}} \quad \text{for } i = 2, 3, 4$$

Therefore, $$Z_{5i} = Z_{4i} - \frac{0.5714286\,(Z_{2i} - Z_{4i})}{1.4857142} \quad \text{for } i = 2, 3, 4$$

Therefore, $$Z_{52} = Z_{42} - \frac{0.5714286\,(Z_{22} - Z_{42})}{1.4857142} \quad \text{for } i = 2$$

$$= 0.5 - \frac{0.5714286\,(0.5 - 0.5)}{1.4857142} = 0.5$$

$$Z_{53} = Z_{43} - \frac{0.5714286\,(Z_{23} - Z_{43})}{1.4857142} \quad \text{for } i = 3$$

$$= 0.8263158 - \frac{0.5714286\,(0.5 - 0.8263158)}{1.4857142}$$

$$= 0.9518217$$

$$Z_{54} = Z_{44} - \frac{0.5714286\,(Z_{24} - Z_{44})}{1.485742} \quad \text{for } i = 4$$

$$= 0.931579 - \frac{0.5714286\,(0.5 - 0.931579)}{1.4857142}$$

$$= 1.0975709$$

Now $$Z_{qq} = Z_{pq} + \frac{1 + \bar{y}_{pq,p\sigma}(\bar{Z}_{pq} - \bar{Z}_{\sigma q})}{y_{pq,pq}}$$

Therefore, $$Z_{55} = Z_{45} + \frac{1 - 0.5714286\,(Z_{25} - Z_{45})}{1.4857142}$$

$$= 1.0975709 + \frac{1 - 0.5714286\,(0.5 - 1.0975709)}{1.4857142}$$

$$= 2.0004828$$

Therefore, Z_{BUS} for this partial network is therefore equal to

$Z_{BUS} =$

	(2)	(3)	(4)	(5)
(2)	0.5	0.5	0.5	0.5
(3)	0.5	0.9052632	0.8262158	0.9518217
(4)	0.5	0.8263158	0.931579	1.0975709
(5)	0.5	0.9518217	1.0975709	2.0004828

Step 6: Finally the element number 6 is added. It gets added between buses (5) and (1) of the partial network. It is a link. It is mutually coupled to the element 1 which is between the buses $\rho = 1$ and $\sigma = 2$. Therefore, $p = 1$, $q = 5$, $\rho = 1$ and $\sigma = 2$.

From the primitive admittance matrix obtained in earlier problem we have

$$y_{pq,pq} = y_{15,15} = 2.6315789 \quad \text{and}$$

$$y_{pq,\rho\sigma} = y_{15,12} = -\,0.5263157$$

Therefore,

$$Z_{li} = Z_{pi} - Z_{qi} + \frac{y_{pq,\rho\sigma}(\bar{Z}_{\rho i} - \bar{Z}_{\sigma i})}{y_{pq,pq}} \quad \text{for} \quad i = 2, 3, 4, 5$$

$$= 0 - Z_{5i} + \frac{(-0.5263157)(0 - Z_{2i})}{2.6315789}$$

$$(Z_{1i} = Z_{1l} = 0).$$

$$Z_{li} = -\,Z_{5i} + 0.2Z_{2i} \quad \text{for} \quad i = 2, 3, 4 \text{ and } 5.$$

$$Z_{l2} = -Z_{52} + 0.2Z_{22} = -0.5 + 0.2 \times 0.5 = -0.4.$$

$$Z_{l3} = -Z_{53} + 0.2Z_{23} = -0.9518217 + 0.2 \times 0.5 = -0.8518217$$

$$Z_{l4} = -Z_{54} + 0.2Z_{24} = -1.0975709 + 0.2 \times 0.5 = -0.9975709$$

$$Z_{l5} = -Z_{55} + 0.2Z_{25} = -2.0004828 + 0.2 \times 0.5 = -1.9004828$$

And $$Z_{ll} = -Z_{51} + \frac{1 - 0.5263157(-Z_{2l})}{2.6315789}$$

$$= 1.9004828 + \frac{1 - 0.5263157(+0.4)}{2.6315789}$$

$$= 2.2004828.$$

$$[X] = \frac{Z_{il} Z_{lj}}{Z_{ll}}.$$

$$= \frac{1}{2.2004828} \begin{bmatrix} -0.4 \\ -0.8518217 \\ -0.9\ 95709 \\ -1.9004828 \end{bmatrix} \begin{bmatrix} -0.4 & -0.8518217 & -0.9975709 & -1.9004828 \end{bmatrix}$$

$$= \begin{bmatrix} 0.072711 & 0.15484 & 0.18133 & 0.34546 \\ 0.15484 & 0.32974 & 0.38616 & 0.73568 \\ 0.18133 & 0.38616 & 0.45225 & 0.86156 \\ 0.38616 & 0.73568 & 0.86156 & 1.6413828 \end{bmatrix}$$

Therefore

$Z_{BUS\,(new)} = \{Z_{BUS\,(old)} - X\}$ is the final solution, which is as follows.

Z_{BUS}(final) =

	②	③	④	⑤
②	0.42789	0.34516	0.31867	0.15454
③	0.34516	0.58255	0.44015	0.21614
④	0.31867	0.44015	0.47932	0.23601
⑤	0.15454	0.21614	0.23601	0.3591

Modification of Z_{BUS} when an element of only self impedance of 0.4 is added between the buses (2) and (5).

The added element is a link. Here $p=2$ and $q=5$. It is not mutually coupled to any of other elements.

Therefore

$$Z_{li} = Z_{pi} - Z_{qi} \quad \text{for} \quad i = 2, 3, 4 \text{ and } 5$$

and

$$Z_{lt} = Z_{pl} - Z_{ql} + Z_{pq,pq}$$

Now $\quad Z_{li} = (Z_{2i} - Z_{5i})$; Therefore, $\quad Z_{l2} = Z_{22} - Z_{52} = 0.27335$

$$Z_{l3} = Z_{23} - Z_{53} = 0.12902$$

$$Z_{l4} = Z_{24} - Z_{54} = 0.08266$$

$$Z_{l5} = Z_{25} - Z_{55} = -0.20456$$

$$Z_{ll} = Z_{pl} - Z_{ql} + 0.4$$

$$= Z_{2l} - Z_{5l} + 0.4 = 0.87791.$$

Therefore, $\quad [X] = \dfrac{\bar{Z}_{il}\bar{Z}_{lj}}{Z_{ll}}$

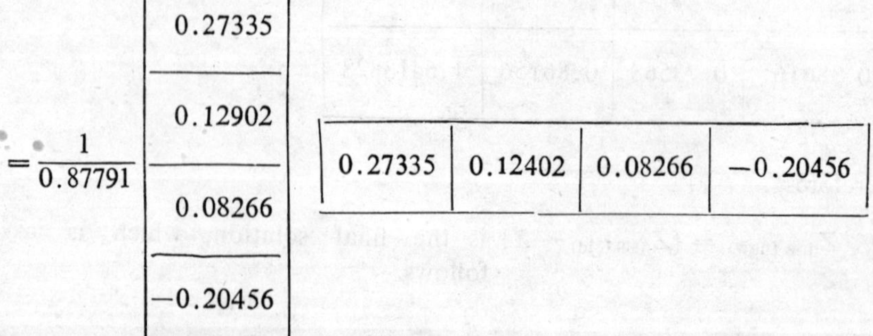

$$= \frac{1}{0.87791} \begin{bmatrix} 0.27335 \\ 0.12902 \\ 0.08266 \\ -0.20456 \end{bmatrix} \begin{bmatrix} 0.27335 & 0.12402 & 0.08266 & -0.20456 \end{bmatrix}$$

0.08511	0.040172	0.025737	−0.06392
0.040172	0.018961	0.012147	−0.030062
0.025737	0.012147	0.0077828	−0.01926
−0.06392	−0.0300062	−0.01926	0.047664

Therefore, Modified $Z_{BUS} = Z_{BUS(old)} - [X]$

0.34277	0.30498	0.29293	0.21823
0.30498	0.56358	0.42800	0.24620
0.29293	0.42800	0.47854	0.25527
0.21823	0.24620	0.25527	0.31143

2. (a) The element 3–4 is removed from the network of Problem 2.3. It is not coupled to any of other elements of the network. Removal of this element is, therefore, simulated by adding an element as a link, having an impedance equal to the *negative* of that of the element 3–4 and in parallel with the element 3–4. So, the formulae for the addition of a link are used.

Here $p = 3$, $q = 4$ and $z_{34,34} = -0.25$.

$$Z_{li} = Z_{pi} - Z_{qi} - (Z_{3i} - Z_{4i}) \qquad i - 2, 3, 4, 5$$

Therefore

$$Z_{l2} = (Z_{32} - Z_{42}) = 0.02649$$

$$Z_{l3} = (Z_{33} - Z_{43}) = 0.14235$$

$$Z_{l4} = (Z_{34} - Z_{44}) = -0.03917$$

and

$$Z_{l5} = (Z_{35} - Z_{45}) = -0.01987$$

And,

$$Z_{ll} = Z_{pl} - Z_{ql} + z_{pq,pq}$$

$$= Z_{3l} - Z_{4l} - 0.25 = -0.06848.$$

(Values for Z_{32}, Z_{42}, Z_{33}, Z_{43} are taken from the Z_{BUS} already determined in *I* part of Numerical Example 1.)

$$[X] - \frac{\bar{Z}_{li}\,\bar{Z}_{lj}}{Z_{ll}}$$

$= \frac{(-)\,1}{0.06848}$

0.02649
0.14235
−0.03917
−0.01987

0.02649	0.14235	−0.03917	−0.01987

= (−)

0.010247	0.055065	−0.015152	−0.0076882
0.055065	0.2959	−0.081423	−0.041303
−0.015152	−0.081423	0.022404	0.011365
−0.0076882	−0.041303	0.011365	0.0057654

Therefore, Modified Z_{BUS} is equal to

=

0.42789	0.34516	0.31867	0.15454
0.34516	0.58255	0.44015	0.21614
0.31867	0.44015	0.47932	0.23601
0.15454	0.21614	0.23601	0.3591

— [X] given above

=

	②	③	④	⑤
②	0.43813	0.40022	0.30351	0.14685
③	0.40022	0.87845	0.35872	0.17483
④	0.30351	0.35872	0.50172	0.24737
⑤	0.14685	0.17483	0.24737	0.36486

(b) Modification of Z_{BUS} if the element 1—2 mutually coupled to the element 1—5 is deleted.

Here $u-v = 1-2$ and $\rho-\sigma = 1-5$

The elements of the new Z_{BUS} are given by

$$Z'_{ij} = Z_{ij} + (\bar{Z}_{i\alpha} - \bar{Z}_{i\beta})\,[M]^{-1}\,[\Delta y_s]\,[\bar{Z}_{\gamma i} - \bar{Z}_{\delta i}] \quad -\ (1)$$

$i = 2, 3, 4, 5$

and $j = 2, 3, 4, 5$

With the element 3—4 removed (in part (a)), the primitive admittance matrix will be the old one minus 1st row and 1st column. From this new $[y]$, the submatrix $[y_s]$ corresponding to the elements 1—5 and 1—2 can be written down as

$y_s =$

	1—5	1—2
1—5	2.631578	−0.5263157
1—2	−0.5263157	2.1052631

and the submatrix, y'_s with the element 1—2 removed will be

$y'_s =$

	1—5	1—2
1—5	$\frac{1}{0.4} = 2.5$	0
1—2	0	0

Therefore

$[\Delta y'_s] = y_s - y'_s =$

	1—5	1—2
1—5	0.131578	−0.5263157
1—2	−0.5263157	2.1052631

Since for the solution of the problem indices are taken as 1—2 and 1—5, the $[\Delta y_s]$ to be taken as follows:

$$[\Delta y_s] = \begin{array}{c|c|c|} & 1-2 & 1-5 \\ \hline 1-2 & 2.1052631 & -0.5263157 \\ \hline 1-5 & -0.5263157 & 0.131578 \\ \hline \end{array}$$

Now the matrix M, is given by

$$M = \{U - [\Delta y_s]\,([Z_{\gamma\alpha}] - [Z_{\delta\alpha}] - [Z_{\gamma\beta}] + [Z_{\delta\beta}])\}$$

where the indices $\alpha\beta$ and $\gamma\delta$ are

$$\alpha\beta = 12, 15 \quad \text{and} \quad \gamma\delta = 12, 15$$

i.e., $\alpha = 1.1, \quad \beta = 2.5, \quad \gamma = 1.1 \quad \text{and} \quad \delta = 2.5$

Therefore,

$$\left.\begin{array}{l} Z_{\gamma\alpha} = \begin{bmatrix} Z_{11} & Z_{11} \\ Z_{11} & Z_{11} \end{bmatrix} = 0 \\ Z_{\delta\alpha} = \begin{bmatrix} Z_{21} & Z_{21} \\ Z_{51} & Z_{51} \end{bmatrix} = 0 \\ Z_{\gamma\beta} = \begin{bmatrix} Z_{12} & Z_{15} \\ Z_{12} & Z_{15} \end{bmatrix} = 0 \end{array}\right\} \text{since bus (1) is taken as reference.}$$

and $$Z_{\beta\delta} = \begin{bmatrix} Z_{22} & Z_{25} \\ Z_{52} & Z_{55} \end{bmatrix} = \begin{bmatrix} 0.43813 & 0.14685 \\ 0.14685 & 0.36486 \end{bmatrix}$$

$$M = \left[\begin{bmatrix} 1 & 0 \\ 0 & 1 \end{bmatrix} - \begin{bmatrix} 2.1052631 & -0.5263157 \\ -0.5263157 & 0.131578 \end{bmatrix} \begin{bmatrix} 0.43813 & .4685 \\ 0.14685 & .36486 \end{bmatrix}\right]$$

$$= \begin{bmatrix} 0.1549106 & -0.1171263 \\ 0.2112724 & 1.029819 \end{bmatrix}$$

Let $M^{-1} = 0.1842754$

Therefore

$$M^{-1} = \frac{1}{0.1842754} \begin{bmatrix} 1.029819 & 0.1171263 \\ -0.2112724 & 0.1549106 \end{bmatrix}$$

$$= \begin{bmatrix} 5.5884778 & 0.6356046 \\ -1.1465035 & 0.8406472 \end{bmatrix}$$

Therefore

$$M^{-1}\Delta ys = \begin{bmatrix} 11.430687 & -2.857672 \\ -2.8561373 & 0.7140334 \end{bmatrix}$$

For the elements of the first row, $i = 2$.

Therefore

$$Z'_{2j} = [Z_{2j}] + [Z_{22}\ Z_{25}] \begin{bmatrix} 11.430687 & -2.857672 \\ -2.8561373 & 0.7140334 \end{bmatrix} \begin{bmatrix} Z_{2j} \\ Z_{5j} \end{bmatrix}$$

$$= [Z_{2j}] + [0.43813\ \ 0.14685] \begin{bmatrix} 11.430687 & -2.857672 \\ -2.8561273 & 0.7140334 \end{bmatrix} \begin{bmatrix} Z_{2j} \\ Z_{5j} \end{bmatrix}$$

$$= [Z_{2j}] + [4.5887031 \quad -1.147176] \begin{bmatrix} Z_{2j} \\ Z_{5j} \end{bmatrix}$$

For $j = 2$,

$$Z'_{22} = Z_{22} + [4.5887031 \quad -1.147176] \begin{bmatrix} Z_{22} \\ Z_{52} \end{bmatrix}$$

Therefore

$$Z'_{22} = 0.43813 + [4.5887031 \quad -1.147176] \begin{bmatrix} 0.43813 \\ 0.14685 \end{bmatrix}$$

$$= 0.43813 + 1.8419856 = 2.2801156$$

For $j = 3$,

$$Z'_{23} = Z_{23} + [4.5887031 \quad -1.147176] \begin{bmatrix} Z_{23} \\ Z_{53} \end{bmatrix}$$

$$= 0.40022 + [4.5887031 \quad -1.147176] \begin{bmatrix} 0.40022 \\ 0.17483 \end{bmatrix}$$

$$= 0.40022 + 1.6359411 = 2.0361611$$

For $j = 4$,

$$Z'_{24} = Z_{24} + [4.5887031 \quad -1.147176] \begin{bmatrix} Z_{24} \\ Z_{54} \end{bmatrix}$$

$$= 0.30351 + [4.5887031 \quad -1.147176] \begin{bmatrix} 0.30351 \\ 0,24737 \end{bmatrix}$$

$$= 0.30351 + 1.1089403 = 1.4124503$$

For $j = 5$,

$$Z'_{25} = Z_{25} + [4.5887031 \quad -1.147176] \begin{bmatrix} Z_{25} \\ Z_{55} \end{bmatrix}$$

$$= 0.14685 + [4.5887031 \quad -1.147176]\begin{bmatrix} 0.14685 \\ 0.36486 \end{bmatrix}$$

$$= 0.14685 + 0.2552924 = 0.4021424$$

For the elements of second row, $i = 3$

Therefore,

$$Z'_{3j} = Z_{3j} + [Z_{32}\ Z_{35}]\begin{bmatrix} 11.430687 & -2.857672 \\ -2.8561373 & 0.7140334 \end{bmatrix}\begin{bmatrix} Z_{2j} \\ Z_{5j} \end{bmatrix}$$

$$= Z_{3j} + [0.40022 \ 0.17483]\begin{bmatrix} 11.430687 & -2.857672 \\ -2.8561373 & 0.7140334 \end{bmatrix}\begin{bmatrix} Z_{2j} \\ Z_{5j} \end{bmatrix}$$

$$= Z_{3j} + [4.075451 \quad -1.018863]\begin{bmatrix} Z_{2j} \\ Z_{5j} \end{bmatrix}$$

Now, for $j = 3$,

$$Z'_{33} = Z_{33} + [4.075451 \quad -1.018863]\begin{bmatrix} Z_{23} \\ Z_{53} \end{bmatrix}$$

$$= 0.87845 + [4.075451 \quad -1.018863]\begin{bmatrix} 0.40022 \\ 0.17483 \end{bmatrix}$$

$$= 0.87845 + 1.4529491 = 2.331399$$

For $j = 4$

$$Z'_{34} = Z_{34} + [4.075451 \quad -1.018863]\begin{bmatrix} Z_{24} \\ Z_{54} \end{bmatrix}$$

$$= 0.35872 + [4.075451 \quad -1.018863]\begin{bmatrix} 0.30351 \\ 0.24737 \end{bmatrix}$$

$$= 0.35872 + 0.9849039 = 1.3436239$$

For $j = 5$,

$$Z'_{35} = Z_{35} + [4.075451 \quad -1.018863]\begin{bmatrix} Z_{25} \\ Z_{55} \end{bmatrix}$$

$$= 0.17483 + [4.075451 \quad -1.018863]\begin{bmatrix} 0.14685 \\ 0.36486 \end{bmatrix}$$

$$= 0.17483 + 0.2267376 = 0.4015676.$$

For the elements of third row, $i = 4$:

$$Z'_{4j} = Z_{4j} + [Z_{42} \quad Z_{45}]\begin{bmatrix} 11.430687 & -2.857672 \\ -2.8561373 & 0.7140334 \end{bmatrix}\begin{bmatrix} Z_{2j} \\ Z_{5j} \end{bmatrix}$$

$$=Z_{4j}+[0.30351\ \ 0.24737]\begin{bmatrix}11.430687 & -2.857672\\ -2.8561373 & 0.7140334\end{bmatrix}\begin{bmatrix}Z_{2j}\\ Z_{5j}\end{bmatrix}$$

$$=Z_{4j}+[2.7624254\quad -0.6907015]\begin{bmatrix}Z_{2j}\\ Z_{5j}\end{bmatrix}$$

So, for $j=4$

$$Z'_{44}-Z_{44}+[2.7624254\quad -0.6907015]\begin{bmatrix}Z_{24}\\ Z_{54}\end{bmatrix}$$

$$=0.50172+[2.762454\quad -0.6907015]\begin{bmatrix}0.30351\\ 0.24737\end{bmatrix}$$

$$=0.50172+0.6675648$$

$$=1.1692848$$

For $j=5$,

$$Z'_{45}=Z_{45}+[2.762454\quad -0.6907015]\begin{bmatrix}Z_{25}\\ Z_{55}\end{bmatrix}$$

$$=0.24737+[2.7624254\quad -0.6907015]\begin{bmatrix}0.14685\\ -0.36486\end{bmatrix}$$

$$=0.24737+0.1536528$$

$$=0.4010228$$

The diagonal element in the last row is given by

$$Z'_{55}=Z_{55}+[Z_{52}\quad Z_{55}]\begin{bmatrix}11.430687 & -2.857672\\ -2.8561373 & 0.7140334\end{bmatrix}\begin{bmatrix}Z_{25}\\ Z_{55}\end{bmatrix}$$

$$=0.36486+[0.14685\ \ 0.36486]\begin{bmatrix}11.430687 & -2.857672\\ -2.8561373 & 0.7140334\end{bmatrix}\begin{bmatrix}0.14685\\ 0.36486\end{bmatrix}$$

$$=0.36486+0.0354118$$

$$=0.4002718$$

For the remaining elements, symmetry is made use of.

Therefore, the modified Z'_{BUS} is given by

Z'_{BUS} =

	②	③	④	⑤
②	2.2801156	2.0361611	1.4124503	0.4021424
③	2 0361611	2.331399	1.3436239	0.4015676
④	1.4124503	1.3436239	1.1692848	0.4010228
⑤	0.4021424	0.4015676	0.4010228	0.4002718

The above result can be compared to the Z_{BUS} obtained directly for the remaining network, which is given below.

Actual Z_{BUS} =

	②	③	④	⑤
②	2.15	1.95	1.35	0.40
③	1.95	2.25	1.35	0.40
④	1.35	1.35	1.15	0.40
⑤	0.40	0.40	0.40	0.40

Problems

2.1 For the network shown below (i.e. Fig. 1) draw its graph and mark also a tree. Give the total number of edges (i.e. elements), nodes, buses and branches

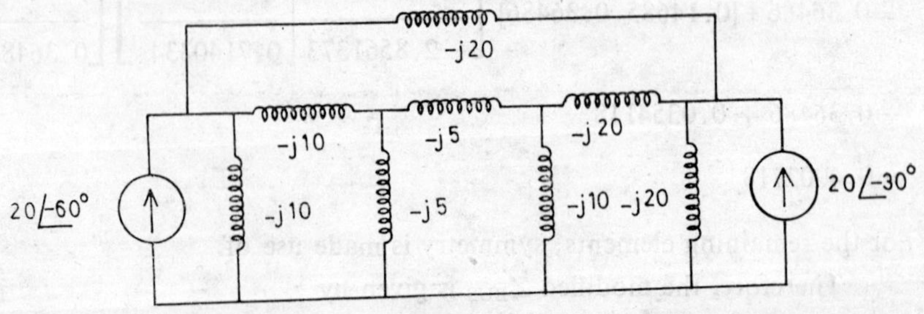

Fig. 1

for this graph. Write also its nodal equations and determine the elements of Y_{BUS} matrix directly by inspection.
Values shown are currents and admittances in per unit.

2.2 For the network shown in Fig. 1, write the loop equations and also determine the elements of Z_{loop} matrix directly by inspection.

2.3 For the network shown in Fig. 2, draw its graph and mark a tree. How many trees this graph will have. Mark the basic cutsets and basic loops (i.e. fundamental circuits) and form the incidence matrices $\hat{A}$, A, B, C and K.

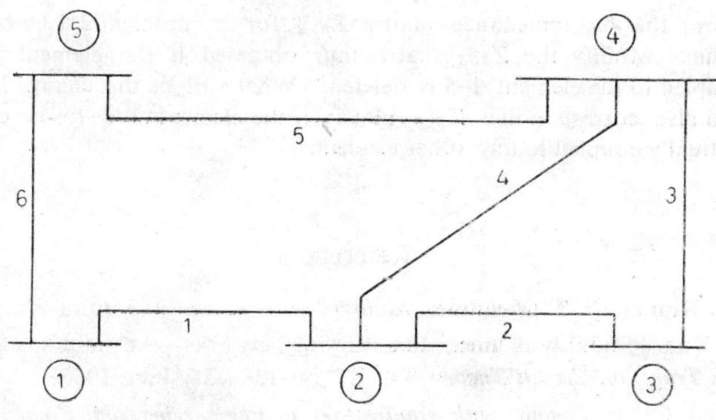

Fig. 2

Impedance for this network in p.u. is given below:

Element Number	Self impedances		Mutual impedances	
	Bus No. $p-q$	imp. $Z_{pq\ pq}$	Bus No. $m-n$	imp. $Z_{pq\ mn}$
1	1–2	0.6	1–5	0.15
2	2–3	0.55	3–4	0.2
3	3–4	0.25		
4	2–4	0.65	4–5	0.25
5	4 5	0.45		
6	5–1	0.5		

2.4 For the previous problem (i.e. problem 2.3), determine Y_{BUS}, Z_{BUS}, Y_{BR}, Z_{BR}, Y_{loop} and Z_{loop} by graph theoretic techniques.

2.5 For the network shown in Fig. 3, draw its graph. Mark basic cutsets, basic loops and open loops and also form the incidence matrices A, $\hat{A}$, B, $\hat{B}$, C, $\hat{C}$ and K.

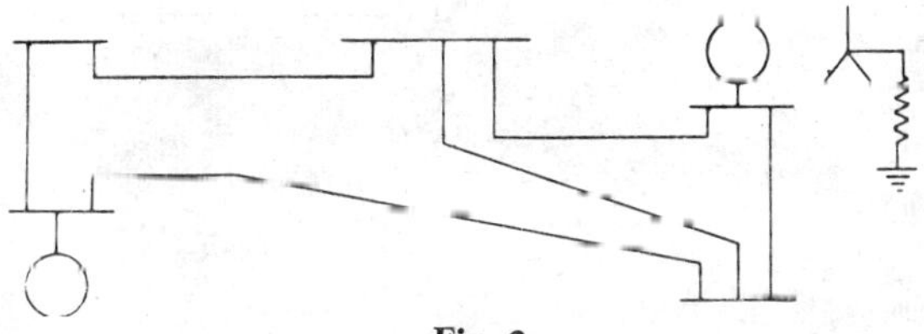

Fig. 3

2.6 For the previous problem (i.e. problem 2.5), verify the following relations:

(i) $A_b K^T = U$

(ii) $A_b C_l = A_l$

(iii) $A B^T = BA^T = 0$

(iv) $B C^T = C B^T = 0$

(v) $C \bar{i} = 0$

(vi) $B \bar{v} = 0$

2.7 Form the bus impedance matrix Z_{BUS} for the problem 2.3 by building algorithm. Modify the Z_{BUS} matrix thus obtained if the element 1–2 mutually coupled to the element 1–5 is deleted. What will be the change in procedure and also corresponding Z_{BUS} matrix if the element 1–2 to be deleted is not mutually coupled to any other element.

References

1. W.H. Kim and H.E. Meadows, *Modern Network Analysis*, John Wiley, New York.
2. E.V. Kuh, "Stability of linear time varying networks—the state space approach", *IEEE Trans. on Circuit Theory*, Vol. 12, pp 150–156, June 1956.
3. N. Deo, *Graph Theory with Application to Engineering and Computer Science*, Prentice Hall, N.J.
4. H.W. Hale and J.B. Ward, "Digital computation of driving point and transfer impedances", *Trans. AIEE*, Vol. 76, Pt. III, pp 476–481.
5. Ahmed, H. El-Abiad, "Algorithm for direct computation and modification of solution matrices of notworks including mutual impedances", *IEEE Proc. PICA Conf.*, pp 150–156, 1963.
6. G.W. Staggand Ahmed, H. El-Abiad, *Computer Methods in Power System Analysis*, McGraw-Hill, N.Y.

Chapter 3

Power System Components and Their Representation

3.1 Introduction

For the purpose of power system studies such as load flow studies, short circuit studies or transient stability studies, we need to develop the mathematical model of the power system network. In order to develop fairly accurate models of the power system network, it is necessary that the model should reflect correctly the terminal behaviour of each component of the network for the purpose (i.e. study) for which the model has been developed. This is because the terminal behaviour of the power system components differs from normal i.e. steady state conditions to abnormal, i.e. transient conditions. The different components which constitute the power system are; synchronous machines transmission network and transformers, distribution network including static loads, composite loads and dynamic loads such as induction machines etc.

In this chapter, we have discussed in detail the representation i.e. model of transmission and distribution network. For this purpose we have also discussed the different types of transformations available in the power system, their properties, uses and limitations. The modelling i.e. representations of synchronous machines is discussed only briefly in this chapter and its detailed modelling including the modelling of automatic voltage regulator and exciter and also that of governor and prime mover is given in the Chapter 9. The modelling of static loads and induction machines are also given in the Chapter 9 and that of transformer in the Chapter 6.

In addition to this, the analysis of multiphase systems which may become a reality in the near future because of restrictions due to rights of way and other advantages of multiphase systems especially its higher efficiency and lesser cost is also discussed briefly in this chapter.

3.2 Power System Network

Power system can be subdivided into three major parts viz:

(i) Power generation

(ii) Power transmission, and

(iii) Power distribution.

Today, mostly, power is generated by 3-phase synchronous generators, transmitted by 3-phase transmission lines and again distributed through 3-phase distribution networks. Three phase synchronous generators are always designed to generate 3-phase balanced excitation i.e. to generate three phase voltages which are equal in magnitude and displaced from each other by an equal angle of 120 electrical degrees. Thus the excitation (i. e. generation) is balanced. The three phase transmission line and the three phase distribution network are either arranged symmetrically or if they are unsymmetric in arrangement, they are transposed at regular intervals to balance their electrical characteristics with the result, the three phase network comprising of transmission and distribution system are also balanced. And hence during the normal, i. e. steady state operations, both, the three phase excitation, i. e. three phase generation or supply as well as the three phase transmission and distribution networks are balanced and hence, so far as the calculations are concerned, they, can be treated as a single phase system for the analysis.

However, if there is a fault on the system, the excitation becomes unbalanced (i. e. the three phase voltages may not be equal in magnitude or may not be displaced by an equal angle of 120 electrical degrees) but the three phase network consisting of transmission and distribution network are balanced. Thus it becomes a case of unbalanced excitation connected to the three phase balanced network. Such situations are usually tackled by transforming the actual phasor quantities to component quantities by linear time dependent power invariant transformtions because after the transformation, these phasor quantities, which were coupled earlier, becomes uncoupled and the differential equations describing the dynamic behaviour of the system which have time dependent i. e. variable coefficients, after transformation, become differential equations with constant coefficients. In practice, a large number of such transformations are available but all of them have the same common features which will be discussed later.

If in addition to the unbalanced excitation, the three phase network is also unbalanced, there is no advantage of transforming, them into component quantities because even after transformation, the phasor quantities remain coupled. But, such situations are not very common in the power system.

Let us take a sample power system network as shown in Fig. 3.1

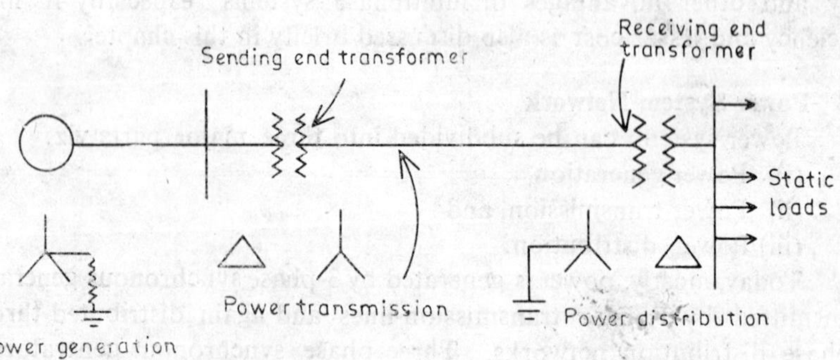

Fig. 3.1 One line diagram of a typical power system network

where the figure represents the one line diagram of a typical power system network. The impedance diagram of this sample network will be as shown in Fig. 3.2.

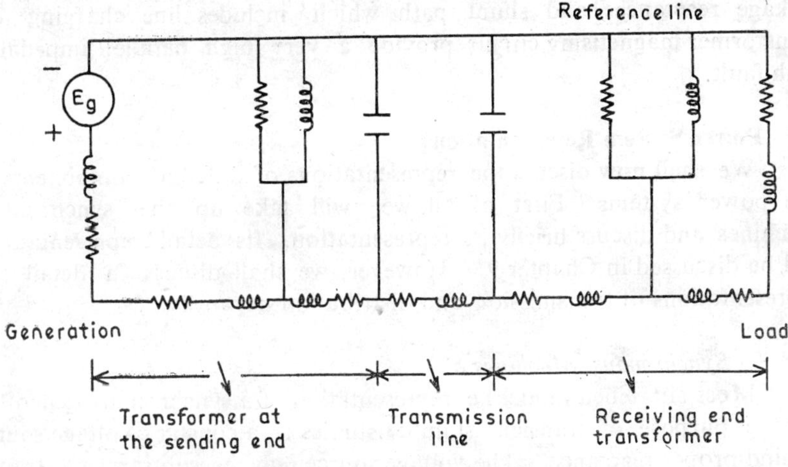

Fig. 3.2 Impedance diagram

In the impedance diagram as shown in Fig. 3.2, the different components of the power system as shown by the single line diagram (see Fig. 3.1) are replaced by their equivalent circuits viz. the synchronous generator at the generating station by a constant voltage source (or an appropriate voltage source) behind proper impedance, the transformers at the sending or at the receiving end by their equivalent circuits and the transmission line by normal π-equivalent circuit. The above diagrams are only valid for normal operations (i.e. for steady state analysis such as load flow or planning studies) when, both the excitation and also the three phase network are balanced and thus they can be analysed as a single phase system.

In many studies, the synchronous generator armature effective resistance, transformer winding resistance, transmission line resistance, line charging and the magnetising circuit of the transformers are neglected and thus we obtain the reactance diagram as shown in Fig. 3.3.

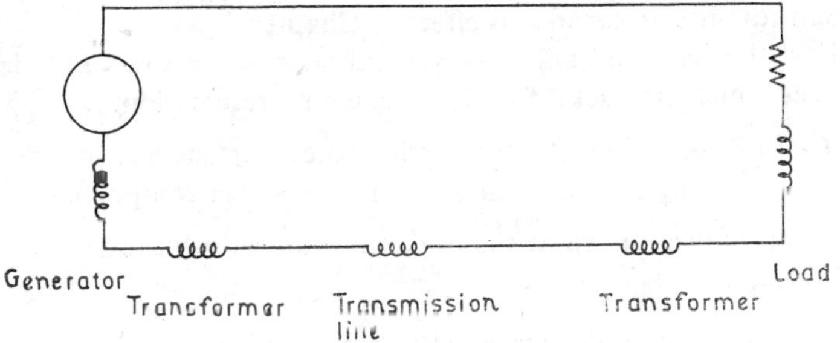

Fig. 3.3 Reactance diagram

Assumptions in drawing the above diagram is quite accurate for many power system studies, such as short circuit studies etc. as winding resistances including line resistances are quite small compared to the leakage reactances and shunt path which includes line charging and transformer magnetising circuit provide a very high parallel impedance with fault.

3.3 Power System Representations

We shall now discuss the representations of different components of the power systems. First of all, we will take up the synchronous machines and discuss briefly its representation. Its detail representations will be discussed in Chapter 9. However, we shall discuss in detail the representations of transmission and distribution networks.

3.3.1 Synchronous Machines

Most simplified model i.e. representation of a synchronous generator for the purpose of transient stability studies is a constant voltage source behind proper reactance. The voltage source may be subtransient, transient or steadystate voltages and the reactance may be the corresponding reactances. In this model, as we shall see in Chapter 9, saliency and changes in the flux linkages are neglected. However, to understand this model, let us take a synchronous generator operating at no load before a 3-phase short circuit is applied at its terminals. The current flowing in the synchronous generator just after the three phase short circuit occurs at its terminals is similar to the current that flows in a R-L circuit upon which suddenly an a. c. voltage is applied. Hence the current will have both the a. c. (i. e. steady state) component as well as the d.c. (i.e. transient) component which decays exponentially with the time constant L/R. However, if this d. c. component is neglected the oscillograph of the a. c. component of the current that flows in the synchronous generator just after the fault occurs, will have the shape as shown in Fig. 3.4.

Just after the fault, the current is maximum as the air gap flux which generates voltage is maximum at the instant the fault occurs than a few cycles later as the armature reaction flux produced due to a very large lagging current in the armature provides nearly a demagnetising effect. We shall discuss in detail this effect in Chapter 9.

For this case of unloaded synchronous generator, let us define few reactances which are useful for the machine representation.

Oa = Peak value of symmetrical a.c. current i.e. peak value of symmetrical current which includes d.c. component.

This current is also known as the peak value of the subtransient current.

Hence,

$$\text{r.m.s. value of } \textit{subtransient current} = \bar{I}'' = Oa/\sqrt{2} \tag{3.1}$$

Now if the first few cycles where the current decrements is very fast

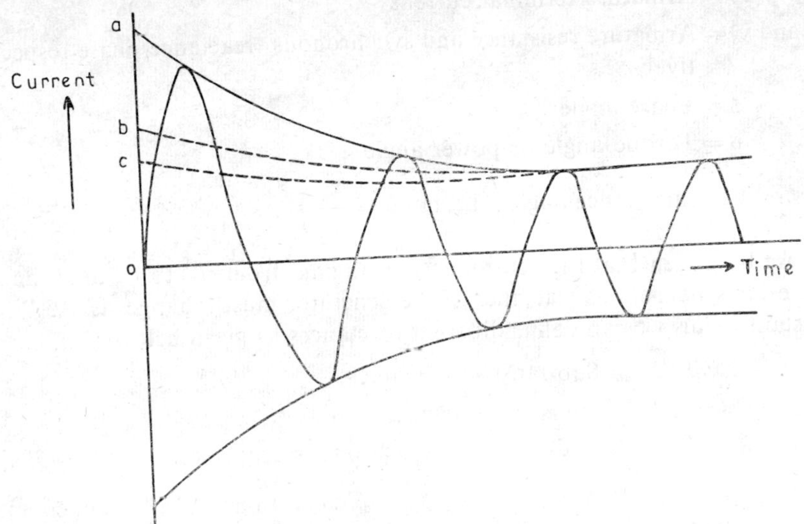

Fig. 3.4 Oscilogram of the current that flows in the unloaded synchronous generator just after a 3-ϕ fault at its terminals excluding d.c. component

are neglected, and the current envelop is extended upto the zero time we get the intercept *Ob*. *Ob* is the peak value of the transient current, and hence,

$$\text{r.m.s. value of } \textit{transient current} = I' = Ob/\sqrt{2} \tag{3.2}$$

However the steady state value of the short circuit current (i.e. sustained value of the short circuit current)

$$= Oc/\sqrt{2} = I \tag{3.3}$$

Since the excitation is constant from no-load to the instant when the three phase short circuit occurs, the excitation voltage in the synchronous generator will remain constant. This excitation voltage "E_g" is known as an open circuit voltage or the no-load induced e.m.f. (see phasor diagram given in Fig. 3.5).

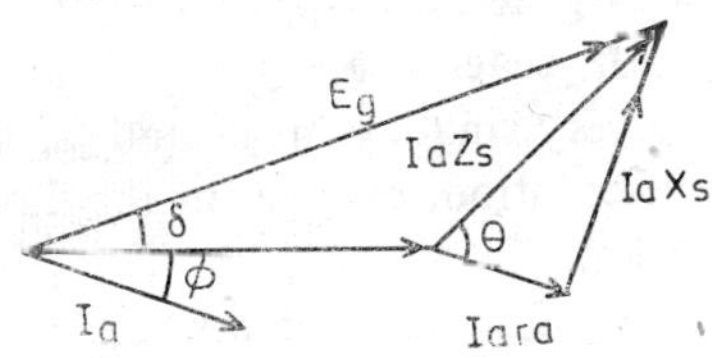

Fig 3.5 Phasor diagram of a nonsalient pole synchronous generator for steady state analysis

E_g = Excitation voltage, i.e. open circuit voltage (or no-load induced e.m.f.)

V = Full load terminal voltage

I_a = Armature terminal current

r_a and x_s = Armature resistance and synchronous reactance/phase respectively

ϕ = Phase angle

δ = Torque angle or power angle

θ = Impedance angle $\left(\text{i.e. } \tan\theta = \dfrac{X_s}{r_a} \right)$

But we have seen that the current is changing from I'' to I' to I and hence the synchronous reactance of the generator must change as "E_g" is constant. Thus we can define different reactances as given below :

$X'' = E_g/I''$ = Subtransient reactance,

$X' = E_g/I'$ = Transient reactance, and

$X = E_g/I$ = Steady state, i.e synchronous reactance. (3.4)

Note : These reactances will be discussed in more detail from circuit point of view in the Chapter 9.

However, since armature reaction flux produced at this instant (which is produced by a very large lagging armature current as this current is limited only by armature impedance where winding resistance is negligible compared to the synchronous reactance) is nearly demagnetising since it acts along the direct axis of the machine, it is more appropriate to call these reactances as direct axis reactances viz. X_d'' or X_d' or X_d. Then with this approximation, we can express the earlier equation (3.4) by the following equation.

$X_d'' = E_g/I''$ = direct axis subtransient reactance (3.5)

$X_d' = E_g/I'$ = direct axis transient reactance (3.6)

$X_d = E_g/I$ = direct axis component of the steady state i.e. synchronous reactance. (3.7)

Here I'', I' and I are the subtransient, transient and steady state i.e. sustained value of short circuit currents and E_g is the excitation i.e. open circuit voltage (no-load induced emf) in the armature. Hence the simplest model of the synchronous generator is a constant voltage "E_g" in series with the proper imp. or reactance, i.e. X_d'' or X_d' or X_d (armature winding resistance being very small, is neglected) as shown in the Fig. 3.6.

In the above representation, changes in the flux linkages and saliency have been neglected.

However, in the case of power systemwhen the synchronous machine connected to the power system is operating at load before the fault occurs, the synchronous generator is represented by an appropriate voltage source behind proper rectances as shown in Fig. 3.7.

This representation can easily be obtained for any fault in the power system with the help of Thevenin's equivalent circuit, from which it is clear that the flux linkages and hence the internal voltage of the

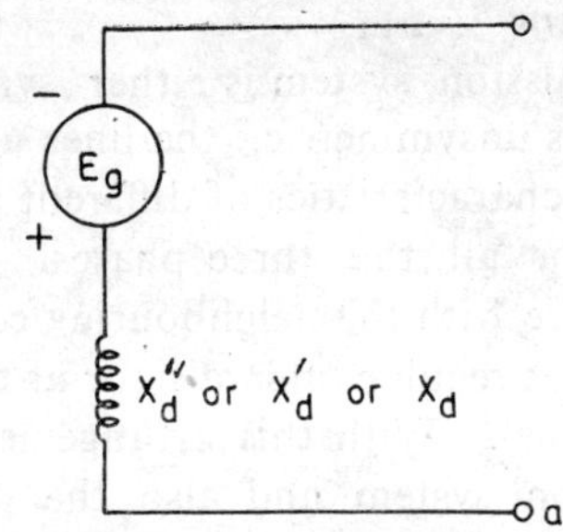

Fig. 3.6 Simplified model of sy. gen.

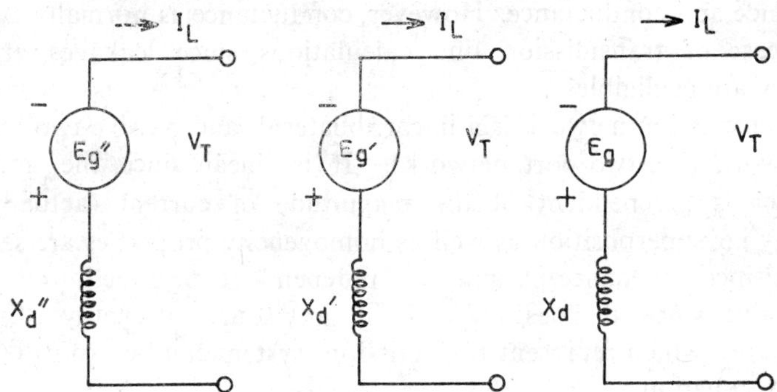

Fig. 3.7 Simplest model of sy. gen.

machine remains constant; only its phase angle changes. Now, the machine equations which are based upon the model of Fig. 3.7, are as shown below,

$$E_g'' = V_T + jI_L X_d'' \tag{3.8}$$

$$E_g' = V_T + jI_L X_d' \tag{3.9}$$

$$E_g = V_T + jI_L X_d \tag{3.10}$$

where E_g'', E_g' and E_g are subtransient, transient and steady state i.e. excitation voltage respectively of the synchronous generator and V_T is the synchronous generator terminal voltage and I_L is the synchronous generator current.

However, for the synchronous motor connected to the power system, equations (3.8)-(3.10) take the following form:

$$E_g'' = V_T - jI_L X_d'' \tag{3.11}$$

$$E_g' = V_T - jI_L X_d' \tag{3.12}$$

and

$$E_g = V_T - jI_L X_d \tag{3.13}$$

Equations (3.8) and (3.13) represent most simplified model of synchronous machines for the purposes of transient stability studies. We shall deal the detailed analysis and modelling of synchronous machines in Chapter 9.

3.3.2 Transmission System

Three-phase transmission system is either symmetric (i.e. balanced) in arrangement or if it is unsymmetric, the lines are transposed in order to balance the electrical characteristics of different phases so as to obtain the same parameters for all the three phases. This also helps in the elimination of interference with the neighbouring communication circuits. The lines are transposed at regular intervals (or at the generating stations and the main substations). With this arrangement the 3-phase network consisting of transmission system and also the distribution system are assumed to be symmetric i.e. balanced.

The transmission line has four parameters viz. resistance, inductance, capacitance and conductance. However, conductance, is normally neglected in the case of transmission line calculations since leakages at normal frequency are negligible.

Transmission network is a linear, bilateral and passive two terminals pair network (i.e. two port network). It is linear since the impedance parameter is independent of the magnitude of current (actually both additivity i.e. superposition as well as homogeneity properties are satisfied), bilateral since impedance parameter is independent of direction of current and passive since it does not contain any source of energy. Such two port network which represent transmission system can be expressed by the following equations :

$$V_S = AV_R + BI_R \tag{3.14}$$

$$I_S = CV_R + DI_R \tag{3.15}$$

i.e.

$$\begin{bmatrix} V_S \\ I_S \end{bmatrix} = \begin{bmatrix} A & B \\ C & D \end{bmatrix} \begin{bmatrix} V_R \\ I_R \end{bmatrix} \tag{3.16}$$

The *ABCD* constants are known as a generalized circuit constants. The matrix

$$\begin{vmatrix} A & B \\ C & D \end{vmatrix} = 1 \tag{3.17}$$

i.e. $AD - BC = 1$. From this we conclude that the 3-phase network is reciprocal and at the same time it is also symmetric since $A = D$. The transmission system can be represented by the following block diagram.

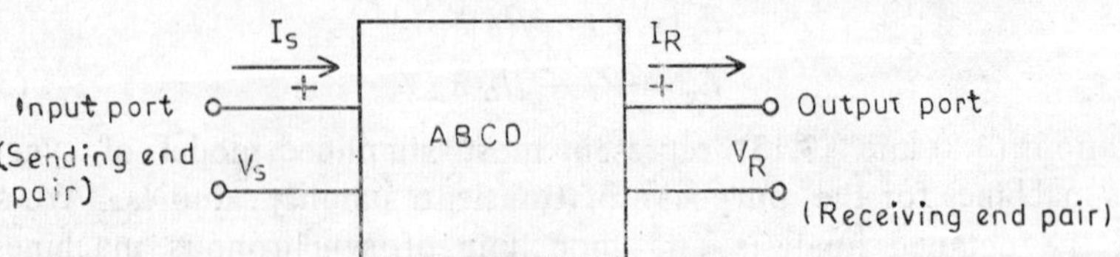

Fig. 3.8 Transmission network

Here V_S and I_S are sending end (i.e. input) voltage and current respectively and V_R and I_R are the receiving and (i.e. output) voltage and current respectively.

3.4 Three-Phase Power System Network

In the case of normal operation (i.e during steady state condition), the 3-phase system is solved as an equivalent single phase system since both the generation i.e. excitation as well as the transmission and distribution networks are symmetric, i.e. balanced. This is because, the three phase synchronous generator is designed to generate balanced voltage, i.e. to provide balanced excitation and the three phase transmission or distribution network are transposed to balance the electrical characteristics of different phases and thus the 3-phase power system network as a whole is symmetric and balanced. This is true only for normal i.e. steady state analysis.

However, during an abnormal operation such as during faults etc., even though the three-phase network is balanced but the excitation becomes unbalanced. There may also be situations when both the three phase excitation and 3-phase (transmission and distribution) networks are unbalanced. However, this latest situation is quite uncommon in the power system operation.

The three-phase network can be represented both in the impedance form (Fig. 3.9) and in the admittance form (Fig. 3.11) as shown.

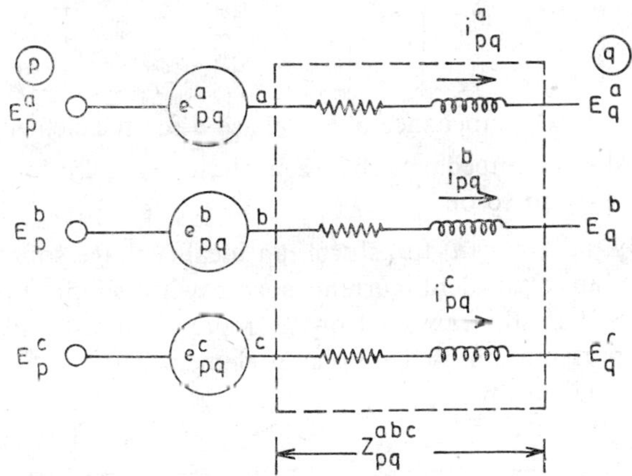

Fig. 3.9 3-phase network representation in impedance form

The single line diagram representation of Fig. 3.9 will be as shown.

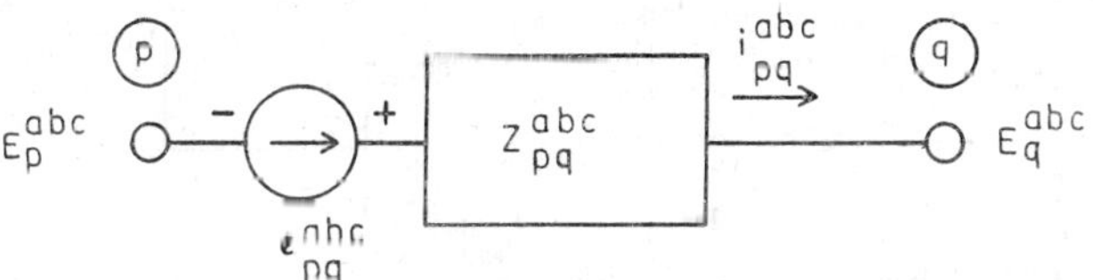

Fig. 3.10

Here $\bar{e}_{pq}^{a,b,c} = \begin{bmatrix} e_{pq}^{a} \\ e_{pq}^{b} \\ e_{pq} \end{bmatrix}$ is the column vector of voltage sources in phases *a*, *b* and *c*. (3.18)

$\bar{i}_{pq}^{a,b,c} = \begin{bmatrix} i_{pq}^{(a)} \\ i_{pq}^{(b)} \\ i_{pq}^{(c)} \end{bmatrix}$ is the column vector of currents through phases *a*, *b*, *c*. (3.19)

$[z_{pq}^{a,b,c}]$ is the 3-phase impedance matrix (3 × 3) of the element $p-q$.

$\bar{E}_{p}^{a,b,c}$ is the bus voltage vector at the bus *p*.

and $E_{q}^{a,b,c}$ is the bus voltage vector at the bus *q*.

Thus the performance equation of the above network will be

$$\bar{E}_{p}^{a,b,c} + \bar{e}_{pq}^{a,b,c} - [Z_{pq}^{a,b,c}]\,\bar{i}_{pq}^{a,b,c} = \bar{E}_{q}^{a,b,c} \tag{3.20}$$

or

$$\bar{E}_{p}^{a,b,c} - \bar{E}_{q}^{a,b,c} + \bar{e}_{pq}^{a,b,c} = [Z_{pq}^{a,b,c}]\,\bar{i}_{pq}^{a,b,c} \tag{3.21}$$

Let $\bar{E}_{p}^{a,b,c} - \bar{E}_{q}^{a,b,c}$ is voltage drop across the element between buses *p* and $q = \bar{v}_{pq}^{abc}$; then we have

$$\bar{v}_{pq}^{abc} + \bar{e}_{pq}^{abc} = [z_{pq}^{abc}]\,\bar{i}_{pq}^{abc} \tag{3.22}$$

i.e.

$$\begin{bmatrix} v_{pq}^{a} \\ v_{pq}^{b} \\ v_{pq}^{c} \end{bmatrix} + \begin{bmatrix} e_{pq}^{a} \\ e_{pq}^{b} \\ e_{pq}^{c} \end{bmatrix} = \begin{bmatrix} z_{pq}^{aa} & z_{pq}^{ab} & z_{pq}^{ac} \\ z_{pq}^{ba} & z_{pq}^{bb} & z_{pq}^{bc} \\ z_{pq}^{ca} & z_{pq}^{cb} & z_{pq}^{cc} \end{bmatrix} \begin{bmatrix} i_{pq}^{a} \\ i_{pq}^{b} \\ i_{pq}^{c} \end{bmatrix} \tag{3.23}$$

Here z_{pq}^{aa} is the self impedance of the phase *a* for the element $p-q$.

and Z_{pq}^{ab} = Mutual impedance between phases *a* and *b* for the element $p-q$ and so on.

Similarly as we can transform an ideal voltage source with series impedance to an equivalent current source with shunt admittance (as already discussed in the case of 1-phase representation), the following is admittance representation of the 3-phase element between the buses *p* and *q* in the admittance form,

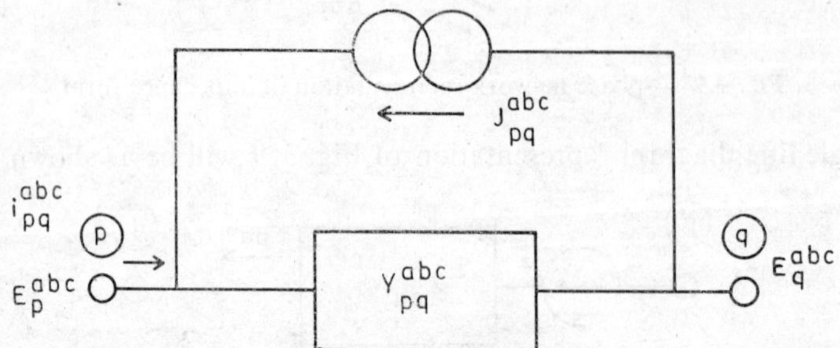

Fig. 3.11 Three-phase network representation in the admittance form

From Fig. 3.11, we obtain directly,

$$\bar{i}^{abc}_{pq} + \bar{j}^{abc}_{pq} = [y^{abc}_{pq}]\, \bar{v}^{abc}_{pq} \tag{3.24}$$

where $\bar{j}^{abc}_{pq}$ is the current sources in phases a, b and c for the element p–q,

and $[y^{abc}_{pq}] = [z^{abc}_{pq}]^{-1}$ is the 3-phase admittance matrix of the dimension 3×3, and,

$$\bar{j}^{abc}_{pq} = -[y^{abc}_{pq}]\, \bar{e}^{abc}_{pq} = -[Z^{abc}_{pq}]^{-1}\, \bar{e}^{abc}_{pq} \tag{3.25}$$

Thus we have the performance equation in the admittance form,

$$\begin{bmatrix} j^a_{pq} \\ j^b_{pq} \\ j^c_{pq} \end{bmatrix} + \begin{bmatrix} i^a_{pq} \\ i^b_{pq} \\ i^c_{pq} \end{bmatrix} = \begin{bmatrix} y^{aa}_{pq} & y^{ab}_{pq} & y^{ac}_{pq} \\ y^{ba}_{pq} & y^{bb}_{pq} & y^{bc}_{pq} \\ y^{ca}_{pq} & y^{cb}_{pq} & y^{cc}_{pq} \end{bmatrix} \begin{bmatrix} v^a_{pq} \\ v^b_{pq} \\ v^c_{pq} \end{bmatrix} \tag{3.26}$$

3.5 Classification of Three-phase Power System Network

A power system network, in general consisting of three balanced sub-networks—generation, transmission and distribution networks—possesses certain special kinds of symmetries. These are rotational symmetries in plane, reflection symmetries in space and translational symmetries in time. Based upon symmetry considerations alone, following are the two broad classifications of three phase network:

1. *Rotating network elements:* These network elements possess only rotational symmetries i.e. symmetries in this case are the symmetry operations of proper rotations in the plane of paper carrying the network elements, about an axis passing through the centroid of the network structure and at right angle to the plane of paper.

2. *Stationary elements:* These network elements possess rotational symmetries in plane and also reflection symmetries in space. The reflection symmetries correspond to the reversal of the network elements about its axes of symmetries.

Before we proceed further in the analysis of 3-phase network, we shall discuss symmetries and its consequences and show how based solely on these symmetries, transformation matrices to diagonalize the coefficient matrix of the network can be developed.

3.5.1 Symmetries

Symmetric network has its elements so arranged that certain permutations of the elements result in a network configuration which is indistinguishable from the original. For an n-port network, symmetry means that if the port voltages are permuted amongst themselves in certain manner then the corresponding port currents are also similarly permuted. This implies that the coefficint matrix of the network is invariant to these permutations.

In power system networks, the permutations of port variables are essentially equivalent to the familiar symmetry operation, such as, rotation about an axis and/or rcflection about the axes of symmetries of the network.

Symmetry operations of permutations on the port variables of a symmetric network have four basic properties: (i) the symmetry operations of this set possess closure property, (ii) the symmetry operations are associative, (iii) there is an identity operation which leaves the network in its original state, and (iv) every symmetry operation has a corresponding inverse operation which undoes the effect of symmetry operation. These properties serve as the defining postulates of a group.

Proposition: The symmetry operations, i.e., permutations of port variables, constitue a group.

3.5.2 Group Theoretic Formulation (*Appendix*-2)

Since the symmetry operations constitute a group, these can be represented formally by a group of permutation matrices, each of which commutes with the coefficient matrix of the symmetric network.

The n-port symmetric network equation in the impedance form under steady state condition is

$$\bar{Y} = [A]\bar{X} \tag{3.27}$$

where $\bar{X}$ is a n-vector input (excitation), $\bar{Y}$ an n-vector output (response) and, $[A]$ the coefficient matrix of the network. If the symmetries of the network structure, and thus of the network equation (3.27), be represented by symmetry operations, $R = A, B, \ldots$, etc., the effect of each of these symmetry operations is to permute the port variables in some manner. For rotational symmetries, for example, a vector $\bar{X} = [X_1, X_2 \ldots X_n]^T$ is replaced after permutation by a vector, $X' = [X_n X_1 X_2 \ldots X_{n-1}]^T$. The relationship between $\bar{X}$ and $\bar{X}'$ in the matrix form is then,

$$\begin{bmatrix} X_n \\ X_1 \\ X_2 \\ \vdots \\ X_{n-1} \end{bmatrix} = \begin{bmatrix} 0 & 0 & \ldots & 1 \\ 1 & 0 & \ldots & 0 \\ 0 & 1 & \ldots & \\ & & & \\ 0 & \ldots & 1 & 0 \end{bmatrix} \begin{bmatrix} X_1 \\ X_2 \\ \cdot \\ \vdots \\ X_n \end{bmatrix}$$

The $n \times n$ matrix of 1 and 0 in the above equation is known as a permutation matrix. In general, the effect of the symmetry operations $R = A, \ldots$ etc. may be formally expressed using permutation matrices, $D(R)$ in the form

$$X' = D[R]\bar{X} \tag{3.28}$$

where $\bar{X}'$ is the vector resulting from the permutation on the vector $\bar{X}$. Evidently, these matrices are such that for any two operations R_1 and R_2,

$D(R_1R_2) = D(R_1)D(R_2)$. In particular, if A is the identity operation then $D(A) = I$, and accordingly for any operation R, $I = D(RR^{-1}) = D(R)D(R^{-1})$. Thus, $D(R^{-1}) = D^{-1}(R)$. In all the matrices, $D(R)$ then constitutes under multiplication, a representation of the symmetry operations.

Symmetry of the network equation (3.27) implies that the output vector $\bar{Y}$ can be similarly permuted

$$\bar{Y}' = D[R]\bar{Y} \tag{3.29}$$

where $\bar{Y}'$ is the output corresponding to the input $\bar{X}'$, that is,

$$\bar{Y}' = [A]\bar{X}' \tag{3.30}$$

From equations (3.28) to (3.30),

$$D(R)\bar{Y} = AD(R)\bar{X} \tag{3.31}$$

or,

$$\bar{Y} = D^{-1}(R)AD(R)\bar{X} \tag{3.32}$$

Comparing with equation (3.27)

$$A = D^{-1}(R)\,A\,D(R) \tag{3.33}$$

or

$$D(R)\,A = A\,D(R) \tag{3.34}$$

Proposition: Each of the permutation matrices $D(R)$ representing symmetry operation $R = A, B, \ldots$, etc., commutes with the coefficient matrix A of the network equation.

3.5.3 *Network with Rotational Symmetries*

The network elements with rotation symmetries alone are referred to as rotating elements. The symmetry operations in case of networks with rotational symmetries constitute a cyclic (and hence commutative) group, that is, all the operations are generated by a single operation when successively applied. In this case, the symmetry operations are $R = p^0, p^1, p^2, \ldots p^k$ denoting p applied k times. These facts can be demonstrated for n-phase symmetric networks also.

The cyclic nature of the symmetry operations evidently means that the permutation matrices also make a cyclic group. Then writing $D(P)$ simply as D,

$$D[P^k] = D^k$$

It can be shown that the coefficient matrix A is a linear combination of permutation matrices D^k for $k = 1, 2, \ldots$ etc.

Consider the relationship between D^k and matrices that commute with them. Let δ_k denote a $n \times 1$ column vector with a 1 in the kth row and zero elsewhere. Then A_k, the kth column vector of matrix A,

$$A = \begin{bmatrix} A_{11} & A_{12} & \cdots & A_{1n} \\ A_{21} & A_{22} & \cdots & A_{2n} \\ A_{n1} & A_{n2} & \cdots & A_{nn} \end{bmatrix} = [A_1\ A_2 \ldots A_k \ldots A_n]$$

is given by,

$$A_k = \begin{bmatrix} A_{1k} \\ A_{2k} \\ \vdots \\ A_{nk} \end{bmatrix} = A\,\delta_k = [A_{1k}\,\delta_1 + A_{2k}\,\delta_2 + \ldots + A_{kk}\,\delta_k + \ldots + A_{nk}\cdot\delta_n] \tag{3.35}$$

In particular, the first column vector of the coefficient matrix A_1

$$A_1 = \begin{bmatrix} A_{11} \\ A_{21} \\ \vdots \\ A_{n_1} \end{bmatrix} = A\,\delta_1 = [A_{11}\,\delta_1 + A_{21}\,\delta_2 + \ldots + A_{nl}\,\delta_n] \tag{3.36}$$

Also $$D^{k-1}\,A\,\delta_1 = A\,D^{k-1}\,\delta_1 \tag{3.37}$$

It can be verified that, $D^{k-1}\,\delta_1 = \delta_k$.

Substituting $A_k = A\,\delta_k$ in equation (3.37),

$$D^{k-1}\,A\,\delta_1 = A\,D^{k-1}\,\delta_1 = A\,\delta_k = A_k \tag{3.38}$$

From equations (3.36) and (3.38)

$$A\,\delta_k = D^{k-1}\,[A]\,\delta_1 = D^{k-1}\,A_1 = [A]\,\delta_k$$

for $k = 0, 1, \ldots, n-1$. Since this is true for every δ_k, it is true for any arbitrary vector $\bar{X}$ and therefore,

$$[A] = [A_{11}\,D^0 + A_{21}\,D^1 + \ldots A_{n_1}\,D^{n-1}] \tag{3.39}$$

Proposition: The coefficient matrix A which commutes with the permutation matrices D^k, is a linear combination of the permutation matrices D^k, $k = 0, 1, \ldots, n-1$.

The above propositions and results help in constructing similarity transformation matrix α for diagonalizing the permutation matrices D^k belonging to the cyclic group of rotation.

Construction of Similarity Transformation: Similarity transformation matrix α can be constructed for diagonalizing permutation matrices D^k for networks displaying only rotational symmetries. It is evident that these permutation matrices D^k which represent symmetry operations corresponding to rotations, are periodic and therefore, all the eigenvalues of D^k are distinct and equal to the roots of unity. Moreover, the permutation matrices D^k commute in pair and accordingly they can be shown to have in common the mutually orthogonal columns of the following matrix as their eigenvectors:

$$\alpha = \frac{1}{\sqrt{n}} \begin{bmatrix} 1 & 1 & 1 & 1 \\ 1 & a & a^2 & a^3 \\ 1 & a^2 & a^4 & a^6 \\ \vdots & \vdots & \vdots & \vdots \\ 1 & a^{n-1} & & \end{bmatrix}$$

where $$a = (1)^{1/n} = e^{j2\pi/n} \tag{3.40}$$

The similarity transformation matrix α then diagonalizes each one of the permutation matrices D^k, $k = 0, 1, 2, \ldots, n-1$, that is

$$\alpha^{*T} \; D^k \; \alpha = \text{Diag } D^k = \text{eigenvalues of } D^k \tag{3.41}$$

for $k = 0, 1, \ldots, n-1$.

Since the coefficient matrix A of the network equation (3.27) commutes with the permutation matrices D^k and is also a linear combination of permutation matrices D^k, it will have the same set of n independent eigenvectors as the permutation matrices D^k. This results in a very important proposition:

Proposition: The unitary matrix α which diagonalizes the permutation matrices D^k and whose columns are eigenvectors of D^k, also diagonalizes the coefficient matrix A of the network equation. It can then be concluded that,

$$\alpha^{*T} [A] \alpha = \text{diag } [A] = \text{eigenvalues of } [A]$$

Example: Steady state analysis of 3-phase symmetric power system networks displaying rotational symmetries:

Consider a 3-phase symmetric subnetwork between buses p and q as shown in Fig. 3.12. The network equation in the impedance form for this subnetwork is,

$$\bar{E}_p^{abc} - \bar{E}_q^{abc} = [Z_{pq}^{abc}] \, \bar{i}_{pq}^{abc} \tag{3.42}$$

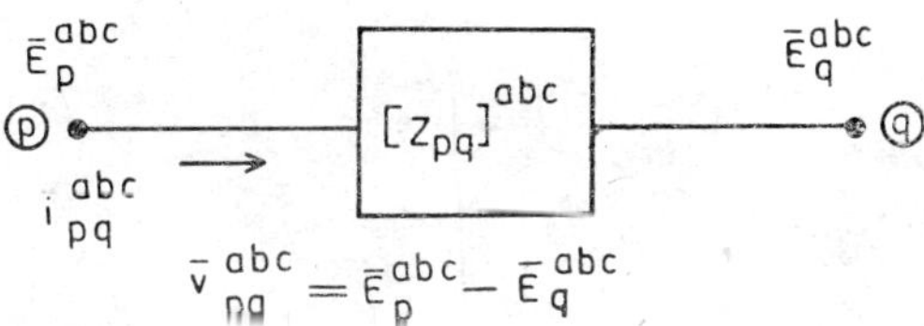

Fig. 3.12. Three-phase power system network

Here, $\bar{E}_p^{abc} - \bar{E}_q^{abc} = \bar{v}_{pq}^{abc}$ is the column vector of voltage drop across the 3-phase element $p-q$; $\bar{E}_p^{abc}$ and $\bar{E}_q^{abc}$ are the column vectors of bus voltages at the buses p and q respectively, and $[Z_{pq}^{abc}]$ is the impedance matrix of the 3-phase element $p-q$. Equation (3.42) can be reproduced as

$$\bar{v}_{pq}^{abc} = [Z_{pq}^{abc}]\bar{i}_{pq}^{abc}$$

or,

$$\begin{bmatrix} v_{pq}^{a} \\ v_{pq}^{b} \\ v_{pq}^{c} \end{bmatrix} = \begin{bmatrix} Z_{pq}^{aa} & Z_{pq}^{ab} & Z_{pq}^{ac} \\ Z_{pq}^{ba} & Z_{pq}^{bb} & Z_{pq}^{bc} \\ Z_{pq}^{ca} & Z_{pq}^{cb} & Z_{pq}^{cc} \end{bmatrix} \begin{bmatrix} i_{pq}^{a} \\ i_{pq}^{b} \\ i_{pq}^{c} \end{bmatrix} \qquad (3.43)$$

The symmetry operations for the 3-phase element $p-q$ are rotations through 120° (elec) represented by the symmetry element C_3^1, through 240° (elec) represented by the symmetry element C_3^2 and, through 360° (elec), that is, zero rotation represented by the identity element E, about an axis perpendicular to the plane of paper. These operations constitute a group known as a permutation group of order 3, as there are three distinct elements in the group (Fig. 3.14). They are represented by the following permutation matrices $D[R]$ which are proper orthogonal matrices

$$D(E) = \begin{bmatrix} 1 & 0 & 0 \\ 0 & 1 & 0 \\ 0 & 0 & 1 \end{bmatrix} \qquad (3.44)$$

$$D[C_3^1] = \begin{bmatrix} 0 & 1 & 0 \\ 0 & 0 & 1 \\ 1 & 0 & 0 \end{bmatrix}$$

$$D[C_3^2] = \begin{bmatrix} 0 & 0 & 1 \\ 1 & 0 & 0 \\ 0 & 1 & 0 \end{bmatrix} \qquad (3.45)$$

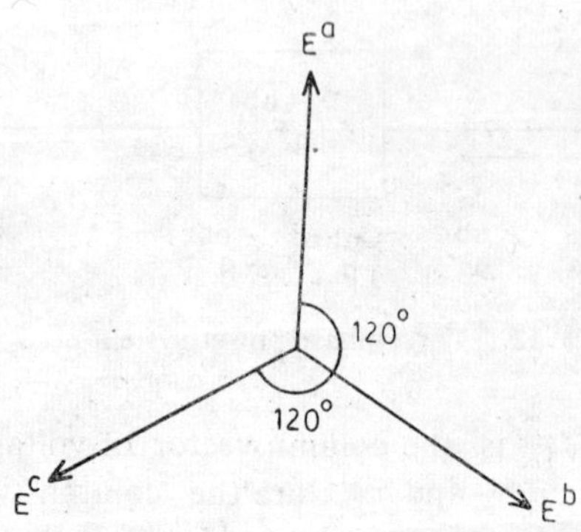

Fig. 3.13 Voltage phaser diagram

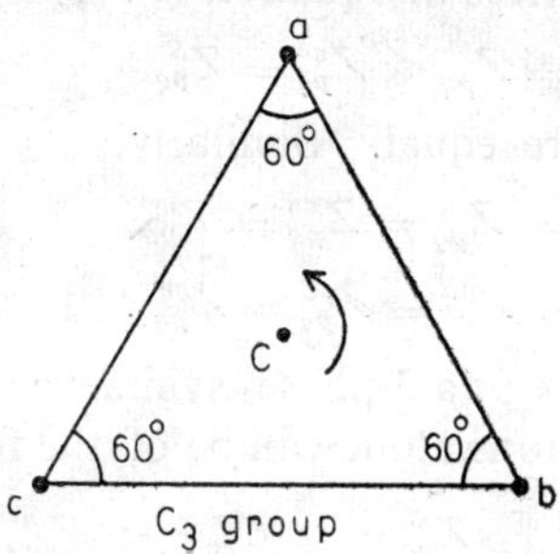

Fig. 3.14 Three-ϕ rotating elements

The network structure, and hence the network equation (3.43), remains invariant with the permutations of port voltages $\bar{v}_{pq}^{abc}$ which cause similar permutations in the port currents $\bar{i}_{pq}^{abc}$, i.e.,

$$\begin{aligned} D[E]\,\bar{v}_{pq}^{abc} &= \bar{v}_{pq}^{abc} \\ D[E]\,\bar{i}_{pq}^{abc} &= \bar{i}_{pq}^{abc} \\ D[C_3^1]\,\bar{v}_{pq}^{abc} &= \bar{v}_{pq}^{bca} \\ D[C_3^1]\,\bar{i}_{pq}^{abc} &= \bar{i}_{pq}^{bca} \\ D[C_3^2]\,\bar{v}_{pq}^{abc} &= \bar{v}_{pq}^{cab} \\ D[C_3^2]\,\bar{i}_{pq}^{abc} &= \bar{i}_{pq}^{cab} \end{aligned} \tag{3.46}$$

It can be easily verified that,

$$D[C_3^1]\,D[C_3^1] = D[C_3^2],$$

and

$$D[C_3^1]\,D[C_3^1]\,D_3^1[C_3^1] = D[C_3^2]\,D[C_3^1] = D[E]$$

Since all the three group elements $D[C_3^1]$, $D[C_3^2]$ and $D[E]$ are generated by a single element $D[C_3^1]$, it is known as cyclic group. Also,

$$D[C_3^1]\;D[C_3^2] = D[C_3^2]\,[D\,C_3^1]$$

i.e., the group is commutative (abelian group). Substituting coefficient matrix $A = Z_{pq}^{abc}$ in equation (3.33)

$$[Z_{pq}^{abc}] = D[R]\,[Z_{pq}^{abc}]D^{-1}[R]$$

Now for $R = C_3^1$, $D[R]$ becomes $D[C_3^1]$, and $D^{-1}[R]$ becomes $D[C_3^2]$ because $D^{-1}[C_3^1] = D[C_3^2]$.

Hence

$$[Z_{pq}^{abc}] = D[C_3^1]\,[Z_{pq}^{abc}]\,D[C_3^2] \tag{3.47}$$

or, (after appropriate substitutions).

$$\begin{bmatrix} Z_{pq}^{aa} & Z_{pq}^{ab} & Z_{pq}^{ac} \\ Z_{pq}^{ba} & Z_{pq}^{bb} & Z_{pq}^{bc} \\ Z_{pq}^{ca} & Z_{pq}^{cb} & Z_{pq}^{cc} \end{bmatrix} = \begin{bmatrix} Z_{pq}^{bb} & Z_{pq}^{bc} & Z_{pq}^{ba} \\ Z_{pq}^{cb} & Z_{pq}^{cc} & Z_{pq}^{ca} \\ Z_{pq}^{ab} & Z_{pq}^{ac} & Z_{pq}^{aa} \end{bmatrix} \tag{3.48}$$

From the equality of matrices in equation (3.48)

$$Z_{pq}^{aa} = Z_{pq}^{bb} = Z_{pq}^{cc} = Z_{pq}^{S} \tag{3.48a}$$

that is, self impedances are equal. Similarly,

$$Z_{pq}^{ab} = Z_{pq}^{bc} = Z_{pq}^{ca} = Z_{pq}^{m1}$$

and,

$$Z_{pq}^{ac} = Z_{pq}^{ba} = Z_{pq}^{cb} = Z_{pq}^{m2}$$

Thus the coefficient matrix of a 3-phase symmetric element with the symmetry operations of rotations alone will be of the form,

$$[Z_{pq}^{abc}] = \begin{bmatrix} Z_{pq}^{S} & Z_{pq}^{m1} & Z_{pq}^{m2} \\ Z_{pq}^{m2} & Z_{pq}^{S} & Z_{pq}^{m1} \\ Z_{pq}^{m1} & Z_{pq}^{m2} & Z_{pq}^{S} \end{bmatrix} \tag{3.49}$$

Similar results can be obtained for $R = C_3^2, E$. The permutation matrices $D[R]$ for $R = C_3^1, C_3^2, E$ commute in pair and therefore these permutation matrices, which form an abelian group, will have common linear independent eigenvectors of the form,

$$\begin{bmatrix} 1 & 1 & 1 \\ 1 & a^2 & a \\ 1 & a^4 & a^2 \end{bmatrix} \tag{3.50}$$

where $a = 3\sqrt{1} = e^{j2\pi/3}$.

Thus the unitary matrix α whose columns are eigenvectors of $D[R]$, simultaneously diagonalize each of the permutation matrices $D[R]$ for $R = C_3^1, C_3^2, E$, that is

$$\alpha^{*T} D[R]\alpha = \text{diag } D[R] = \text{eigenvalues of } D[R]$$

The columns of the unitary matrix α are orthonormal, that is columns are orthogonal to each other and are also normalized to unity by dividing the column vector by its norm $\sqrt{3}$. Thus the matrix α, which can be found by inspection, is of the form,

$$\frac{1}{\sqrt{3}}\begin{bmatrix} 1 & 1 & 1 \\ 1 & a^2 & a \\ 1 & a^4 & a^2 \end{bmatrix} = \frac{1}{\sqrt{3}}\begin{bmatrix} 1 & 1 & 1 \\ 1 & a^2 & a \\ 1 & a & a^2 \end{bmatrix} \tag{3.51}$$

because $a^4 = a$.

The coefficient matrix Z_{pq}^{abc} commutes with the permutation matrices $D[R]$ and is also a linear combination of $D[R]$. Therefore, the same unitary matrix α will also diagonalize the coefficient matrix Z_{pq}^{abc}. It is thus evident that the unitary matrix α of form given by equation (3.51) is the familiar symmetrical component transformation matrix.

This shows the derivation of diagonalizing transformation for networks having cyclic symmetry using properties of regular commutative permutation groups. An alternative method to develop the same transformation matrix α using representation theory of finite groups is also available. This method is specially suited if the network has, in addition to rotational symmetries, reflection symmetries and can be considered together.

Representation Theory Approach. A group of matrices is said to form a representation of group, $G=(E, A, B...)$ if there exists a correspondence between the matrices $D[R]$ and group elements, $R = A, B...$ etc such that if the group element R_1 is being represented by the permutation matrix $D[R_1]$ and R_2 by $D[R_2]$, then R_1, R_2 will be represented by $D[R_1]\,D[R_2]$.

From any representation $D(R)$, another representation may be obtained by similarity transformation to each matrix of the group. A representation is said to be reducible if it can be converted to the block diagonal form called 'reduced-out representation' through a similarity transformation, the submatrix block on the diagonal of the 'reduced out representation' are the irreducible representation of the group. If α is some similarity transformation, the reduced-out representation is,

$$\overline{D}R = \alpha^{*T}\, D[R]\,\alpha = \begin{bmatrix} D^1R & & 0 \\ & D^2R & \\ 0 & & \ddots \\ & & \cdot \end{bmatrix}$$

where $D^1[R]$, etc are the irreducible representations of the group.

Several important properties of group representations and their characters are derived from a basic theorem concerning elements of matrices which constitute irreducible representations of the group. This theorem is known as 'orthogonality theorem' and is stated as follows:

$$\sum_R [D^i(R)_{mn}]\,[D^j(R)_{m'n'}]^* = \left[\frac{h^2}{\sqrt{l_i\,l_j}}\,\delta_{ij}\,\delta'_{mm}\,\delta'_{nn}\right] \tag{3.52}$$

Here, h is the order of the group, l_i the dimension of ith representation which is the order of each of the matrices constituting the representation and $D^i[R_{mn}]$ the element in the mth row and nth column of a matrix corresponding to a symmetry operation, R in the ith irreducible representation.

Given below are some important rules about irreducible representation and their characters which are the consequences of the above orthogonality theorem:

(i) The sum of the squares of the dimensions of the irreducible representation is equal to the order of group,

$$\sum l_i^2 = l_i^2 + l_2^2 + \ldots = h \tag{3.53}$$

where l_1 is the dimension of the ith irreducible representation and h is the order of the group.

(ii) The sum of the squares of characters in any irreducible representation is equal to the order of the group, that is,

$$\sum_R [X^i(R)]^2 = h \tag{3.54}$$

where $X^i(R)$ is the character of the ith irreducible representation.

(iii) The vectors whose components are the characters of two different irreducible representations are orthogonal, that is,

$$\sum_R X^i(R)\, X^j(R)^* = 0 \quad \text{for } i \neq j$$

(iv) In a given representation (reducible or irreducible), characters of all matrices belonging to the symmetry operation in the same class are identical.

(v) The number of irreducible representations in a group is equal to the number of classes in the group.

For any reducible representation, it is always possible to find some similarity transformation matrix α which will reduce each matrix of the reducible representations to one consisting of blocks on the diagonal, each of which belongs to an irreducible representation of the group. Since characters (trace) of a matrix are not changed by similarity transformations,

$$X[R] = \sum_j a_j\, x_j[R]$$

Where $X[R]$ is the character of the reducible representation and $X^j[R]$ is that of jth irreducible representation. Hence,

$$a_l = \left(\frac{1}{h}\right) X[R]\, X^i[R] \tag{3.55}$$

Here, $X^i[R]$ is the character of ith irreducible representation and a_l is the number of times the irreducible-representation $D^i[R]$ appears in the block diagonal of the reduced-out representation of $D[R]$, that is,

$$\bar{D}R = \alpha^{*T} D[R]\,\alpha = \begin{bmatrix} D^1R & & & & & & 0 \\ & D^2R & & & & & \\ & & \cdot & & & & \\ & & & \cdot & & & \\ & & & & \cdot & & \\ & & & & & \cdot & \\ & 0 & & & \left\{\begin{matrix} D^iR & & \\ & \cdot & \\ & & \cdot \\ & & & D^iR \end{matrix}\right. & \end{bmatrix}$$

(a_l times: the block D^iR repeated)

Consider the example of 3-phase symmetric network displaying only rotational symmetries as shown in Fig. 3.12. The permutation matrices $D[R]$ for $R = C_3^1, C_3^2, E$ (equations (3.44) and (3.45)) form the reducible representation of the group elements $R = C_3^1, C_3^2, E$. It has already been shown that they form a cyclic abelian group of order 3, and since each element in a cyclic group is in a separate class, the number of classes are also 3. Therefore, the number of irreducible representations will also be equal to 3. If l_1, l_2 and l_3 are the dimensions of these irreducible representations, then from equation (3.53),

$$l_2^2 + l_2^2 + l_3^2 = h = 3$$

or,
$$l_1 = l_2 = l_3 = 1.$$

Therefore, all the three irreducible representations will have the dimension of l. Since the vectors, whose components are the characters of two different irreducible representations, are orthogonal and the characters of all the matrices belonging to the operation in the same class are identical, the character table of the irreducible representations will be as in Table 3.1.

Table 3.1 Character Table of Irreducible Representations

	$R = E$	$R = C_3^1$	$R = C_3^2$
$X^1 [R]$	1	1	1
$X^2 [R]$	1	a	a^2
$X^3 [R]$	1	a^2	a

where, $a = e^{j2\pi/3}$

Since the dimensions of all the irreducible representations $D^i [R]$, $i = 1, 2, 3$ are 1, it is evident that these irreducible representations will be of the same form as that of their characters. The irreducible representation of the group is given in Table 3.2.

Table 3.2 Irreducible Representation

	E	C_3^1	C_3^2
$D^1 [R]$	1	1	1
$D^2 [R]$	1	a	a^2
$D^3 [R]$	1	a^2	a

The number of times each of these irreducible representations $D^i [R]$ appears at the diagonal of the reduced-out representations $\bar{D} [R]$ can be obtained from equation (3.55) as follows :

$$a_1 = \frac{1}{h}\{X^1[R]\,X[R]\} = \frac{1}{3}[1\,(1)\,3 + 0] = 1$$

$$a_2 = \frac{1}{h}\{X^2[R]\,X[R]\} = \frac{1}{3}[1\,(1)\,3 + 0] = 1$$

and

$$a_3 = \frac{1}{h}\{X^3[R]\,X[R]\} = \frac{1}{3}[1\,(1)\,3 + 0] = 1$$

Thus

$$\bar{D}[k] = \alpha^{*T}\,D[R]\,\alpha = \begin{bmatrix} D^1[R] & & 0 \\ & D^2[R] & \\ 0 & & D^3[R] \end{bmatrix}$$

In order to determine the similarity transformation, following matrix based upon the orthogonality theorem, is constructed,

$$G_i^j = \sum_R D^j[R]_{11}\,D[R] \tag{3.56}$$

where $D^j(R)_{ii}$ is the diagonal element of the irreducible representations $D^j[R]$ for $j = 1,2,3$ counting i once for every appearance of reducible representation $D[R]$. Then,

$$G_1^1 = \sum_R D^1[R]_{11}\,D[R]$$

$$= D^1[R_1]_{11}\,D[R_1] + D^1[R_2]_{11}\,D[R_2] + D^1]R_3]_{11}\,D[R_3]$$

$$= 1\begin{bmatrix} 1 & 0 & 0 \\ 0 & 1 & 0 \\ 0 & 0 & 1 \end{bmatrix} + 1\begin{bmatrix} 0 & 1 & 0 \\ 0 & 0 & 1 \\ 1 & 0 & 0 \end{bmatrix} + 1\begin{bmatrix} 0 & 0 & 1 \\ 1 & 0 & 0 \\ 0 & 1 & 0 \end{bmatrix} = \begin{bmatrix} 1 & 1 & 1 \\ 1 & 1 & 1 \\ 1 & 1 & 1 \end{bmatrix}$$

$$G_1^2 = \sum_R D^2[R]_{11}\,D[R_2]$$

$$= 1\begin{bmatrix} 1 & 0 & 0 \\ 0 & 1 & 0 \\ 0 & 0 & 1 \end{bmatrix} + a\begin{bmatrix} 0 & 1 & 0 \\ 0 & 0 & 1 \\ 1 & 0 & 0 \end{bmatrix} + a^2\begin{bmatrix} 0 & 0 & 1 \\ 1 & 0 & 0 \\ 0 & 1 & 0 \end{bmatrix} = \begin{bmatrix} 1 & a & a^2 \\ a^2 & 1 & a \\ a & a^2 & 1 \end{bmatrix}$$

and, $$C_1^3 = \sum_R D^3[R]_{11}\,D[R]$$

$$= 1\begin{bmatrix} 1 & 0 & 0 \\ 0 & 1 & 0 \\ 0 & 0 & 1 \end{bmatrix} + a^2\begin{bmatrix} 0 & 1 & 0 \\ 0 & 0 & 1 \\ 1 & 0 & 0 \end{bmatrix} + a\begin{bmatrix} 0 & 0 & 1 \\ 1 & 0 & 0 \\ 0 & 1 & 0 \end{bmatrix} = \begin{bmatrix} 1 & a^2 & a \\ a & 1 & a^2 \\ a^2 & a & 1 \end{bmatrix}$$

where $R_1 = E$, $R_2 = C_3^1$ and, $R_3 = C_3^2$.

The basis vector α_{mnp} denoting the pth linear independent vector corresponding to mth irreducible representation appearing for nth time, is constructed by scanning matrices G_1^j from the left hand side and picking first the linear independent columns of G_1^1 and then of G_1^2 and G_1^3. Hence,

$$\alpha_{111} = \begin{bmatrix} 1 \\ 1 \\ 1 \end{bmatrix}$$

which is normalized to unity on dividing by its norm $\sqrt{3}$

$$\alpha_{111} = \frac{1}{\sqrt{3}} \begin{bmatrix} 1 \\ 1 \\ 1 \end{bmatrix}$$

Similarly, α_{211} after normalization to unity is

$$\alpha_{211} = \frac{1}{\sqrt{3}} \begin{bmatrix} 1 \\ a^2 \\ a \end{bmatrix}$$

and α_{311} after normalization to unity is,

$$\alpha_{311} = \frac{1}{\sqrt{3}} \begin{bmatrix} 1 \\ a \\ a^2 \end{bmatrix}$$

Thus the matrix α comes out to be

$$\alpha = \begin{matrix} [\alpha_{111}\ \alpha_{211}\ \alpha_{311}] \\ D^1[R]\ D^2[R]\ D^3[R] \end{matrix} = \frac{1}{\sqrt{3}} \begin{bmatrix} 1 & 1 & 1 \\ 1 & a^2 & a \\ 1 & a & a^2 \end{bmatrix}$$

This matrix α is the familiar symmetrical component transformation matrix, and is the same as that derived through eigenvalues approach. Thus the symmetric 3-phase power systems networks displaying rotational symmetries and referred to as rotating elements, can be diagonalized by the unitary matrix α, known as symmetrical component transformation matrix.

3.5.4 Network Possessing both Rotational and Reflection Symmetries

The network elements having both rotational as well as reflection symmetries are referred as stationary elements typical example being that of transposed transmission lines. Let us take a 3-phase symmetric network possessing rotational as well as reflection symmetries. The network structure, and hence the network equation, remains invariant for permutations of port voltages which cause similar permutations in the port currents. These permutations among port variables correspond to the physical rotations of the network through 120, 240 and 360 electrical degrees, and reflections of the network about axes of symmetries. These axes of symmetries are assumed to coincide with the magnetic axis of phases a, b and c, respectively. In all, there are six symmetry elements, viz., C_3^1, C_3^2 and E representing rotations through 120°, 240° and 360° (elec), respectively, and δ_a, δ_b and δ_c representing reflections about phases a, b and c, respectively (Fig. 3.15). These symmetry operations can be represented by the following permutation matrices:

$$D[E]=\begin{bmatrix}1&0&0\\0&1&0\\0&0&1\end{bmatrix}\quad D[C_3^1]=\begin{bmatrix}0&1&0\\0&0&1\\1&0&0\end{bmatrix}$$

$$D[C_3^2]=\begin{bmatrix}0&0&1\\1&0&0\\0&1&0\end{bmatrix}\quad D[\delta_a]=\begin{bmatrix}1&0&0\\0&0&1\\0&1&0\end{bmatrix}$$

$$D[\delta_b]=\begin{bmatrix}0&0&1\\0&1&0\\1&0&0\end{bmatrix}\quad D[\delta_c]=\begin{bmatrix}0&1&0\\1&0&0\\0&0&1\end{bmatrix} \tag{3.57}$$

These operations constitute a group known as permutation group C_{3v}, consisting of three-fold rotations and three-fold reflections. Since the network equation remains invariant for permutations of port variables, from equation (3.33) (after substituting Z_{pq}^{abc} for A)

$$Z_{pq}^{abc}=D[R]\,[Z_{pq}^{abc}]\,D^{-1}[R]$$

for $R=E, C_3^1, C_3^2, \delta_a, \delta_b$ and δ_c.

With these substitutions, and comparing the coefficients of similar matrices, $D[R]\,[Z_{pq}^{abc}]\,D^{-1}[R]$ and Z_{pq}^{abc}, we have

$$Z_{pq}^{aa}=Z_{pq}^{bb}=Z_{pq}^{cc}=Z_{pq}^{S}$$

and,
$$Z_{pq}^{ab}=Z_{pq}^{bc}=Z_{pq}^{ac}=Z_{pq}^{cb}=Z_{pq}^{ba}=Z_{pq}^{m}$$

Thus, the coefficient matrix Z_{pq}^{abc} of a 3-phase element $p-q$ displaying both rotational as well as reflection symmetries will be of the form,

$$Z_{pq}^{abc}=\begin{bmatrix}Z_{pq}^{s}&Z_{pq}^{m}&Z_{pq}^{m}\\Z_{pq}^{m}&Z_{pq}^{s}&Z_{pq}^{m}\\Z_{pq}^{m}&Z_{pq}^{m}&Z_{pq}^{s}\end{bmatrix} \tag{3.58}$$

The similarity transformation matrix α can be developed based on the representation theory of finite group where permutation matrices are shown in equation (3.57). Since there are six distinct elements, the order of the group is six but there are only three classes,

$$D[E], \{D[C_3^1], D[C_3^2]\}, \text{ and } \{D[\delta_a], D[\delta_b], D[\delta_c]\}$$

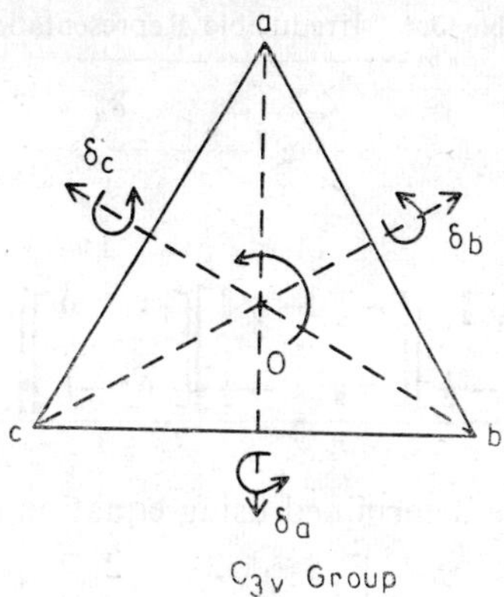

Fig. 3.15 Three-ϕ stationary elements

in the group. Therefore, the number of irreducible representations will be three. If l_1, l_2 and l_3 are the dimensions of these irreducible representations, then from the equation (3.53),

$$l_1^2 + l_2^2 + l_3^2 = 6$$

The solution of the above equation is,

$$l_1 = l_2 = 1, \text{ and } l_3 = 2$$

i.e., there are two irreducible representations, $D^1[R]$ and $D^2[R]$, of dimension 1 and one irreducible representation $D^3[R]$ of dimension 2. Since,

$$\sum_R [X^i(R)]^2 = h = 6$$

the characters of each of the six elements of the first two representations whose dimensions are 1, will be 1 or -1, but that of the third representation whose dimension is 2, will be 2, 1 or, -1, 0. The character table is given in Table 3.3.

Table 3.3 Character Table

C_{3v}	E	C_3^1	C_3^2	δ_a	δ_b	δ_c
$X^1[R]$	1	1	1	1	1	1
$X^2[R]$	1	1	1	-1	-1	-1
$X^3[R]$	2	-1	-1	0	0	0

It is evident that $D^1[R]$ and $D^2[R]$ whose dimensions are 1 are the same as that of their characters, but $D^3[R]$, whose dimension is 2 and characters 2, 1, -1 and rest zeros, can be determined using orthogonality theorem given by equation (3.52). The irreducible representation of the group is shown in Table 3.4.

Table 3.4 Irreducible Representation

C_{3v}	E	C_3^1	C_3^2	δ_a	δ_b	δ_c
$D^1[R]$	1	1	1	1	1	1
$D^2[R]$	1	1	1	-1	-1	-1
$D^3[R]$	$\begin{bmatrix} 1 & 0 \\ 0 & 1 \end{bmatrix}$	$\begin{bmatrix} -\frac{1}{2} & \sqrt{\frac{3}{2}} \\ -\sqrt{\frac{3}{2}} & -\frac{1}{2} \end{bmatrix}$	$\begin{bmatrix} -\frac{1}{2} & -\sqrt{\frac{3}{2}} \\ \sqrt{\frac{3}{2}} & -\frac{1}{2} \end{bmatrix}$	$\begin{bmatrix} 1 & 0 \\ 0 & -1 \end{bmatrix}$	$\begin{bmatrix} -\frac{1}{2} & -\sqrt{\frac{3}{2}} \\ -\sqrt{\frac{3}{2}} & \frac{1}{2} \end{bmatrix}$	$\begin{bmatrix} -\frac{1}{2} & \sqrt{\frac{3}{2}} \\ \sqrt{\frac{3}{2}} & \frac{1}{2} \end{bmatrix}$

Matrix G_i^j, can be determined using equation (3.56)

$$G_1^1 = \sum_R D^1[R]_{11} D[R] = \begin{bmatrix} 2 & 2 & 2 \\ 2 & 2 & 2 \\ 2 & 2 & 2 \end{bmatrix} \text{ for } R = E,$$

$$C_3^1, C_3^2, \delta_a, \delta_b, \delta_c, \; G_1^2 = \sum_R D^2[R]_{11} D[R] = \begin{bmatrix} 0 & 0 & 0 \\ 0 & 0 & 0 \\ 0 & 0 & 0 \end{bmatrix}$$

$$G_1^3 = \sum_R D^3[R]_{11} D[R] = \begin{bmatrix} 2 & -1 & -1 \\ -1 & \frac{1}{2} & \frac{1}{2} \\ -1 & \frac{1}{2} & \frac{1}{2} \end{bmatrix}$$

$$G_2^3 = \sum_R D^3[R]_{22} D[R] = \begin{bmatrix} 0 & 0 & 0 \\ 0 & \frac{3}{2} & -\frac{3}{2} \\ 0 & -\frac{3}{2} & \frac{3}{2} \end{bmatrix}$$

Following the same procedure, the basis vector α_{mnp}, after normalization to unity, can be obtained as,

$$\alpha_{111} = \frac{1}{\sqrt{3}} \begin{bmatrix} 1 \\ 1 \\ 1 \end{bmatrix}; \quad \alpha_{211} = \begin{bmatrix} 0 \\ 0 \\ 0 \end{bmatrix}$$

$$\alpha_{311} = \frac{1}{\sqrt{6}} \begin{bmatrix} 2 \\ -1 \\ -1 \end{bmatrix} = \frac{1}{\sqrt{3}} \begin{bmatrix} \sqrt{2} \\ -\sqrt{\frac{1}{2}} \\ -\sqrt{\frac{1}{2}} \end{bmatrix}$$

$$\alpha_{321} = \frac{1}{\sqrt{18/4}} \begin{bmatrix} 0 \\ \frac{3}{2} \\ -\frac{3}{2} \end{bmatrix} = \frac{1}{\sqrt{3}} \begin{bmatrix} 0 \\ \sqrt{\frac{3}{2}} \\ -\sqrt{\frac{3}{2}} \end{bmatrix}$$

Hence,

$$\alpha = \underset{D^1[R]\ \ D^3[R]\ \ D^3[R]}{\{\alpha_{111}\ \alpha_{311}\ \alpha_{321}\}} = \frac{1}{\sqrt{3}}\begin{bmatrix} 1 & \sqrt{2} & 0 \\ 1 & -\sqrt{\tfrac{1}{2}} & \sqrt{\tfrac{3}{2}} \\ 1 & -\sqrt{\tfrac{1}{2}} & -\sqrt{\tfrac{3}{2}} \end{bmatrix} \tag{3.59}$$

The number of times the irreducible representations $D^i(R)$ appear at the blocks of $\bar{D}(R)$ is confirmed from equation (3.55)

$$a_1 = \tfrac{1}{6}[1(1)3 + 1(1)0 + 1(1)0 + 1(1)1 + 1(1)1 + 1(1)1] = 1$$

$$a_2 = \tfrac{1}{6}[1(1)3 + 1(1)0 + 1(1)0 + 1(-1)1 + 1(-1)1 + 1(-1)1] = 0$$

$$a_3 = \tfrac{1}{6}[1(2)3 + 1(-1)0 + 1(-1)0 + 1(0)1 + 1(0)1 + 1(0)1] = 1$$

That is,

$$\bar{D}R = \alpha^{*T} DR\alpha = \begin{bmatrix} D^1[R] & 0 \\ 0 & D^3[R] \end{bmatrix} \tag{3.60}$$

Equation (3.59) is the familiar Clarke's components transformation matrix. However in case we choose complex basis, the irreducible representations of the group will be given by Table 3.5.

Table 3.5 Irreducible Representations with Complex Basis

C_{3v}	E	C_3^1	C_3^2	δ_a	δ_b	δ_c
$D^1[R]$	1	1	1	1	1	1
$D^2[R]$	1	1	1	-1	-1	-1
$D^3[R]$	$\begin{bmatrix}1 & 0\\0 & 1\end{bmatrix}$	$\begin{bmatrix}a & 0\\0 & a^2\end{bmatrix}$	$\begin{bmatrix}a^2 & 0\\0 & a\end{bmatrix}$	$\begin{bmatrix}0 & 1\\1 & 0\end{bmatrix}$	$\begin{bmatrix}0 & a^2\\a & 0\end{bmatrix}$	$\begin{bmatrix}0 & a\\a^2 & 0\end{bmatrix}$

The elements of the matrix G_i^i can be found in a similar fashion,

$$G_1^1 = \sum_R D^1[R]_{11} D[R] = \begin{bmatrix} 2 & 2 & 2 \\ 2 & 2 & 2 \\ 2 & 2 & 2 \end{bmatrix}$$

$$G_1^2 = \sum_R D^2[R]_{11} D[R] = \begin{bmatrix} 0 & 0 & 0 \\ 0 & 0 & 0 \\ 0 & 0 & 0 \end{bmatrix}$$

$$G_1^3 = \sum_R D^3[R]_{11} D[R] = \begin{bmatrix} 1 & a & a^2 \\ a^2 & 1 & a \\ a & a^2 & 1 \end{bmatrix}$$

and,

$$G_a^3 = \sum_R D^3[R]_{22} D[R] = \begin{bmatrix} 1 & a^2 & a \\ a & 1 & a^2 \\ a^2 & a & 1 \end{bmatrix}$$

Following the same procedure, the basis vector $^r\alpha_{mnp}$ can be found, and after normalization to unity,

$$\alpha_{111} = \frac{1}{\sqrt{12}} \begin{bmatrix} 2 \\ 2 \\ 2 \end{bmatrix} = \frac{1}{\sqrt{3}} \begin{bmatrix} 1 \\ 1 \\ 1 \end{bmatrix}$$

$$\alpha_{311} = \frac{1}{\sqrt{3}} \begin{bmatrix} 1 \\ a^2 \\ a \end{bmatrix}, \text{ and } \alpha_{321} = \frac{1}{\sqrt{3}} \begin{bmatrix} 1 \\ a \\ a^2 \end{bmatrix}$$

Hence the matrix is

$$\alpha = \underset{D^1[R]\; D^3[R]\; D^3[R]}{[\alpha_{111}\; \alpha_{311}\; \alpha_{321}]} = \frac{1}{\sqrt{3}} \begin{bmatrix} 1 & 1 & 1 \\ 1 & a^2 & a \\ 1 & a & a^2 \end{bmatrix} \tag{3.61}$$

Equation (3.61) is the familiar symmetrical components transformation matrix.

Proposition: The unitary matrix α which turns out to be Clarke's components by using real basis and which turns out to be symmetrical components by using complex basis, diagonalizes the coefficient matrix of 3-phase symmetric elements displaying rotational and reflection symmetries commonly known as stationary elements typical example being that of transposed transmission line.

3.5.5 *Transformation Matrices*

As discussed earlier, if both the 3-phase excitation (i. e. generation) and 3-phase network (i. e. transmission and distribution network) are balanced, the 3-phase system may be solved as a single phase system using positive sequence parameter since all the three voltages or currents per phase are equal in magnitude and displaced by an equal angle of 120 electrical degrees. This is exactly the situation when we conduct load flow or planning studies of a given power system.

However, during disturbances say during faults, the excitation becomes unbalanced even though the three phase network is balanced. In such situation it is advantageous to transform the phasor quantites into component quantities with the help of linear power invariant complex transformation matrix or system. Thus the linear complex transformation matrix or system transform the field of phasor quantities (viz. phasor voltages and currents) into the field of components quantities. This transformation has several advantages which will be discussed later.

The first attempt in this direction was made by a famous scientist Dr. C. L. Fortescue during the year 1918. According to his theory, a three phase unbalanced phasors of a 3-phase system can be resolved into three balanced system of phasors as shown below:

(1) Positive sequence components consisting of three phasors of equal magnitude but displaced by 120° electrical degrees and having the same phase sequence as the original unbalanced phasors.

(2) Negative sequence components consisting of three equal phasors but displaced by 120 electrical degrees and having the phase sequence opposite to the original 3-phase unbalanced phasors.

(3) Zero sequence components having three equal phasors with zero phase difference with each other.

The positive, negative and zero sequence components for the voltages (the same thing is ture for currents also) are shown below:

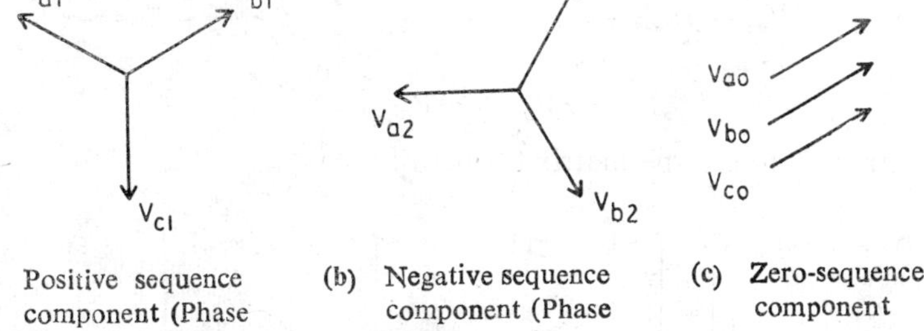

(a) Positive sequence component (Phase sequence abc)

(b) Negative sequence component (Phase sequence acb)

(c) Zero-sequence component

Fig. 3.16 Symmetrical components

If the original phasor is V^{abc}, they can be expressed in terms of component quantities as shown below:

$$V^a = V_{a1} + V_{a2} + V_{a0}$$

$$V^b = V_{b1} + V_{b2} + V_{b0}$$

$$V^c = V_{c1} + V_{c2} + V_{c0} \quad (3.62)$$

Now the positive sequence voltage,

$$|V_{a1}| = |V_{b1}| = |V_{c1}|$$

but if, $\quad V_{a1} = |V_{a1}| \angle 0$

then $\quad V_{b1} = |V_{a1}| \angle 240$

and $\quad V_{c1} = |V_{a1}| \angle 120 \quad (3.63)$

Now if we can define an operator "α" (i.e. complex number) with unit magnitude and angle 120° electrical degrees, then multiplying any complex number or phasor by this operator α, will revolve the original phasor through 120°, in counter-clockwise direction without changing the magnitude of the original phasor. Hence, by this definition,

$$\alpha = 1\angle 120 = e^{j2\pi/3}$$

$$= \cos 120^\circ + j \sin 120^\circ = -0.5 + j\,0.866 \qquad (3.64)$$

Similarly

$$\alpha^2 = \alpha \times \alpha = 1\ \angle 120\ 1\ \angle 120$$
$$= 1\ \angle 240$$
$$= -0.5 - j\,0.866$$
$$= \alpha^* \qquad (3.65)$$

$$\alpha^3 = 1 \qquad (3.66)$$

and

$$1 + \alpha + \alpha^2 = 0 \qquad (3.67)$$

With this choice of the operator "α" ($\alpha = 1\ \angle 120^\circ$) we can write the equation (3.62) for the original phasor $V^{a,b,c}$ in terms of component voltages as shown:

$$V_a = V_{a1} + V_{a2} + V_{a0}$$

$$V_b = V_{b1} + V_{b2} + V_{b0} = |\,V_{a1}\,|\ \angle 240 + |\,V_{a2}\,|\ \angle 120 + V_{a0}$$
$$= \alpha^2 V_{a1} + \alpha V_{a2} + V_{a0}$$

Similarly,

$$V_c = \alpha\, V_{a1} + \alpha^2 V_{a2} + V_{a0} \qquad (3.68)$$

i.e. equation (3.68) in the matrix form will be,

$$\begin{bmatrix} V_a \\ V_b \\ V_e \end{bmatrix} = \begin{bmatrix} 1 & 1 & 1 \\ 1 & \alpha^2 & \alpha \\ 1 & \alpha & \alpha^2 \end{bmatrix} \begin{bmatrix} V_{a0} \\ V_{a1} \\ V_{a2} \end{bmatrix} \qquad (3.69)$$

and in the symbolic form, we get

$$\overline{V}^{a,b,c}_{\text{phase}} = [A]\,\overline{V}^{0,1,2}_{\text{component}} \qquad (3.70)$$

where A is known as symmetrieal component matrix which transforms phasor voltages $V^{a,b,c}_{\text{phase}}$ into component valtages $V^{0,1,2}_{\text{component}}$.

A being a non-singular square matrix, its inverse exists and hence we have,

$$A^{-1} = \tfrac{1}{3}\begin{bmatrix} 1 & 1 & 1 \\ 1 & \alpha & \alpha^2 \\ 1 & \alpha^2 & \alpha \end{bmatrix} \qquad (3.71)$$

Thus

$$\begin{bmatrix} V_{a0} \\ {}_{1} \\ V_{a2} \end{bmatrix} = \tfrac{1}{3}\begin{bmatrix} 1 & 1 & 1 \\ 1 & \alpha & \alpha^2 \\ 1 & \alpha^2 & \alpha \end{bmatrix} \begin{bmatrix} V^a \\ V^b \\ V^c \end{bmatrix} \qquad (3.72)$$

The main demerit of this original symmetrical component matrix "A" is that it is power variant, i.e. it is not orthogonal (i.e. unitary) matrix.

3.5.6 *Symmetrical Components "T_S"*

We have developed earlier the unitary matrix α for 3-phase rotating elements solely based upon symmetry considerations. The unitary matrix α as derived in equation (3.51) is as shown below,

$$\alpha = T_S = \frac{1}{\sqrt{3}}\begin{bmatrix} 1 & 1 & 1 \\ 1 & \alpha^2 & \alpha \\ 1 & \alpha & \alpha^2 \end{bmatrix} \tag{3.73}$$

The matrix T_S is unitary i.e.

$$T_S^{*T} \cdot T_S = T_S \cdot T_S^{*T} = T_S \cdot T_S^{-1} = I = U$$

i.e.

$$T_S^{-1} = T_S^{*} = \frac{1}{\sqrt{3}}\begin{bmatrix} 1 & 1 & 1 \\ 1 & \alpha & \alpha^2 \\ 1 & \alpha^2 & \alpha \end{bmatrix} \tag{3.74}$$

As discussed in Section 3.5.3, the symmetrical components transformation matrix (equations 3.51 and 3.73) diagonalizes the coefficient matrix Z_{pq}^{abc} of 3-phase rotating elements i.e.

$$T_S^{*T}\, Z_{pp}^{abc}\, T_S \tag{3.75}$$

$$= \frac{1}{\sqrt{3}}\begin{bmatrix} 1 & 1 & 1 \\ 1 & \alpha & \alpha^2 \\ 1 & \alpha^2 & \alpha \end{bmatrix}\begin{bmatrix} Z_{pq}^{S} & Z_{pq}^{m1} & Z_{pq}^{m2} \\ Z_{pq}^{m2} & Z_{pq}^{S} & Z_{pq}^{m1} \\ Z_{pq}^{m1} & Z_{pq}^{m2} & Z_{pq}^{S} \end{bmatrix}\frac{1}{\sqrt{3}}\begin{bmatrix} 1 & 1 & 1 \\ 1 & \alpha^2 & \alpha \\ 1 & \alpha & \alpha^2 \end{bmatrix}$$

$$= \begin{bmatrix} Z_{pq}^{S} + Z_{pq}^{m1} + Z_{pq}^{m2} & & 0 \\ & Z_{pq}^{S} + \alpha^2 Z_{pq}^{m1} + \alpha Z_{pq}^{m2} & \\ 0 & & Z_{pq}^{S} + \alpha Z_{pq}^{m1} + \alpha^2 Z_{pq}^{m2} \end{bmatrix} \tag{3.76}$$

In equation (3.75), T_S and T_S^{*T} are substituted from equations (3.73) and (3.74) and Z_{pq}^{abc} which is the coefficient matrix for 3-phase rotating elements, is substituted from the equation (3.49). Here in the equation (3.76), $Z_{pq}^{S} + Z_{pq}^{m1} + Z_{pq}^{m2}$ is known as a zero sequence impedance and is denoted by Z_{pq}^{0}. Similarly $Z_{pq}^{S} + \alpha^2 Z_{pq}^{m1} + \alpha Z_{pq}^{m2}$ is known as a positive sequence impedance and is denoted by Z_{pp}^{+} or Z_{pq}^{1} and $Z_{pq}^{S} + \alpha Z_{pq}^{m1} + \alpha^2 Z_{pq}^{m2}$ is known as negative sequence impedance and is denoted by Z_{pq}^{-} or Z_{pq}^{2}.

Thus we conclude from equation (3.76) that the symmetrical component transformation matrix diagonalizes the coefficient matrix Z_{pq}^{abc} of the 3-phase rotating element. However, in this case $Z_{pq}^{0} \neq Z_{pq}^{1} \neq Z_{pq}^{2}$ i.e. sequence impedances are not equal.

Complex power : Symmetrical component transformations T_S is a linear power invariant transformation which transforms the field of phasors to the field of components, i.e.,

$$\bar{E}_{pq}^{abc} = [T_S]\,\bar{E}_{pq}^{0,1,2} \tag{3.77}$$

$$\bar{I}_{pq}^{abc} = [T_S]\,\bar{I}_{pq}^{0,1,2} \tag{3.78}$$

Here $\bar{E}_{pq}^{abc}$ and $\bar{I}_{pq}^{abc}$ are the column vectors of phasor voltages and current for phases *abc* of the element *p–q* and $\bar{E}_{pq}^{0,1,2}$ and $\bar{I}_{pq}^{0,1,2}$ are their transformed (i.e. zero, positive and negative sequence) components. Complex power in a 3-phase element

$$p-q = S_{pq} = P_{pq} - jQ_{pq} = \bar{E}_{pq}^{abc\,*T}\ \bar{I}_{pq}^{abc} \tag{3.79}$$

Substituting equations (3.77) and (3.78) in equation (3.79), we obtain

$$S_{p-q} = [T_S\,\bar{E}_{pq}^{0,1,2}]^{*T}\ T_S \bar{I}_{pq}^{0,1,2}$$

$$= \bar{E}_{pq}^{0,1,2\,*T}\ T_S^{*T}\ T_S \bar{I}_{pq}^{0,1,2}$$

$$= \bar{E}_{pq}^{0,1,2\,*T}\ \bar{I}_{pq}^{0,1,2} \quad (\text{as } T_S{}^{*}T_S = U = I)$$

$$= [E_{pq}^{0*}\ E_{pq}^{1}\ E_{pq}^{2}]^{*} \begin{bmatrix} I_{pq}^{0} \\ I_{pq}^{1} \\ I_{pq}^{2} \end{bmatrix}$$

$$= E_{pq}^{0*}\,I_{pq}^{0} + E_{pq}^{1*}\,I_{pq}^{1} + E_{pq}^{2*}\,I_{pq}^{2} \tag{3.80}$$

From equation (3.80), we conclude that symmetrical component matrix T_S is a linear, complex and power invariant transformations which transforms the field of phasors to the field of components.

3.5.7 *Clarke's Component "T_C"*

We have shown in Section 3.5.4 that a unitary matrix which turns out to be a Clarke's components by using real basis for representation and which turns out to be symmetrical components by using a complex basis of representation, diagonalizes the coefficient matrix Z_{pq}^{abc} of a 3-phase stationary elements. The Clarke's components transformation T_C as derived in the equation (3.59) is of the form,

$$\alpha = T_C = \frac{1}{\sqrt{3}} \begin{bmatrix} 1 & \sqrt{2} & 0 \\ 1 & -\sqrt{1/2} & \sqrt{3/2} \\ 1 & -\sqrt{1/2} & -\sqrt{3/2} \end{bmatrix} \tag{3.81}$$

This is on orthogonal matrix, i.e.

$$T_C{}^{T} \cdot T_C = U = I = T_C^{-1}\,T_C$$

and hence

$$T_C^{-1} = T_C^T = \frac{1}{\sqrt{3}} \begin{bmatrix} 1 & 1 & 1 \\ \sqrt{2} & -\sqrt{1/2} & -\sqrt{1/2} \\ 0 & +\sqrt{3/2} & -\sqrt{3/2} \end{bmatrix} \tag{3.82}$$

The Clarke's components transformation matrix T_C diagonalizes the coefficient matrix Z_{pq}^{abc} of a 3-phase stationary elements such as that of transposed transmission line. Using equation (3.33) and after substituting T_C for $D[R]$ and Z_{pq}^{abc} for A, we have

$$T_C^{-1} [Z_{pq}^{abc}] T_C = T_C^T [Z_{pq}^{abc}] T_C \tag{3.83}$$

$$= \frac{1}{\sqrt{3}} \begin{bmatrix} 1 & 1 & 1 \\ \sqrt{2} & -\sqrt{1/2} & -\sqrt{1/2} \\ 0 & \sqrt{3/2} & -\sqrt{3/2} \end{bmatrix} \begin{bmatrix} Z_{pq}^S & Z_{pq}^m & Z_{pq}^m \\ Z_{pq}^m & Z_{pq}^S & Z_{pq}^m \\ Z_{pq}^m & Z_{pq}^m & Z_{pq}^S \end{bmatrix}$$

$$\frac{1}{\sqrt{3}} \begin{bmatrix} 1 & \sqrt{2} & 0 \\ 1 & -\sqrt{1/2} & +\sqrt{3/2} \\ 1 & -\sqrt{1/2} & -\sqrt{3/2} \end{bmatrix} = \begin{bmatrix} Z_{pq}^S + 2Z_{pq}^m & & 0 \\ & Z_{pq}^S - Z_{pq}^m & \\ 0 & & Z_{pq}^S - Z_{pq}^m \end{bmatrix} \tag{3.84}$$

We have substituted T_C and T_C^T in equation (3.83) from equations (3.81) and (3.82) and Z_{pq}^{abc} for 3-phase stationary elements from equation (3.58).

In the equation (3.84) $Z_{pq}^S + 2Z_{pq}^m$ is known as zero sequence component, $Z_{pq}^S - Z_{pq}^m$ is known as α-component and $Z_p^S - Z_{pq}^m$ is known as β-component i.e. in this case α-component is equal to β-component.

Clarke's components transformation T_C is a linear real transformation which transforms the field of phasors to the field of 0, α, β components, i.e. for any element

$$\bar{E}_{pq}^{abc} = [T_C]\, \bar{E}_{pq}^{0,\alpha,\beta} \tag{3.85}$$

and

$$\bar{I}_{pq}^{abc} = [T_C]\, \bar{I}_{pq}^{0,\alpha,\beta} \tag{3.86}$$

Now,

complex power in any 3-phase element $p-q = S_{pq} = P_{pq} - jQ_{pq} = \bar{E}_{pq}^{abc*T} I_{pq}^{abc}$

$$= [E_{pq}^a \; E_{pq}^b \; E_{pq}^c]^* \begin{bmatrix} I_{pq}^a \\ I_{pq}^b \\ I_{pq}^c \end{bmatrix} \tag{3.87}$$

Substituting equations (3.85) and (3.86) in equation (3.87), we get

$$= S_{pq} = [T_C \bar{E}_{pq}^{0,\alpha,\beta}]^{*T} T_C \bar{I}_{pq}^{0,\alpha,\beta}$$

$$= \bar{E}_{pq}^{0,\alpha,\beta\,*T} T_C^{*T} T_C \bar{I}_{pq}^{0,\alpha,\beta}$$

$$= \bar{E}_{pq}^{0,\alpha,\beta *T} \bar{I}_{pq}^{0,\alpha,\beta}$$

$$= [E_{pq}^0 \; E_{pq}^\alpha \; E_{pq}^\beta]^* \begin{bmatrix} I_{pq}^0 \\ I_{pq}^\alpha \\ I_{pq}^\beta \end{bmatrix} \tag{3.88}$$

This is because T_C is an orthogonal matrix, i.e.

$$T_C^{*T} T_C = T_C^T T_C T_C = T_C^{-1} T_C = I = U$$

From equation (3.88) we conclude that the Clarke's components transformation is a linear power invariant real transformation which diagonalizes the cofficient matrix of 3-phase stationary elements.

As shown in Section 3.5.4, symmetrical component transformation T_S also diagonalizes the coefficient matrix of 3-phase stationary elements, i.e.,

$$T_S^{*T} [Z_{pq}^{abc}] T_S$$

$$= \frac{1}{\sqrt{3}} \begin{bmatrix} 1 & 1 & 1 \\ 1 & \alpha & \alpha^2 \\ 1 & \alpha^2 & \alpha \end{bmatrix} \begin{bmatrix} Z_{pq}^S & Z_{pq}^m & Z_{pq}^m \\ Z_{pq}^m & Z_{pq}^S & Z_{pq}^m \\ Z_{pq}^m & Z_{pq}^m & Z_{pq}^S \end{bmatrix} \frac{1}{\sqrt{3}} \begin{bmatrix} 1 & 1 & 1 \\ 1 & \alpha^2 & \alpha \\ 1 & \alpha & \alpha^2 \end{bmatrix}$$

$$= \begin{bmatrix} Z_{pq}^S + 2Z_{pq}^m & & 0 \\ & Z_{pq}^S - Z_{pq}^m & \\ 0 & & Z_{pq}^S - Z_{pq}^m \end{bmatrix} \tag{3.89}$$

In equation (3.89) $Z_{pq}^S + 2Z_{pq}^m$ is a zero sequence impedance and $Z_{pq}^S - Z_{pq}^m$ is a positive sequence as well as negative sequence impedance. That is for 3-phase stationary elements, positive sequence impedance is equal to the negative sequence impedance.

3.5.8 *Park's Component 'T_p'*

This matrix which transforms the phasor quantities to 0, *d* and *q* quantities, is defined as follows:

$$T_p = \begin{bmatrix} 1 & \cos\beta & \sin\beta \\ 1 & \cos(\beta-120) & \sin(\beta-120) \\ 1 & \cos(\beta+120) & \sin(\beta+120) \end{bmatrix} \tag{3.90}$$

This transformation matrix is useful in the synchronous machines modelling i.e. in transforming the machine equations in the phasor form into the field of 0, d and q quantities. The symmetrical component T_S and Clarke's component T_C are used in the analysis of power system networks.

However, before we go further, let us discuss the common features between these transformation matrices or systems. Following are the common features in all the three matrices;

(i) All of them have the entry "1" in column 1.

(ii) Sum of the column 2 and 3 are zero separately.

(iii) All the three columns are linearly independent and hence they form the basis vectors.

We shall see in the literature many types of transformations and all of them have the same common features as given above. In this chapter, we have developed solely based on symmetries, transformation matrices used in power systems only i.e. T_S and T_C and discussed their effects both on 3-phase stationary elements and also on rotating elements. We have taken up the case when the excitation is unbalanced but 3-phase network is balanced. And then we have seen that while in the phasor form, the different phasors are coupled but after the transformation to the field of component quantities, the different component quantities become uncoupled. In other words, while the earlier matrices describing the system (i.e. impedance or admittance matrix) were full, after transformation they become diagonal. We shall see in Chapter 9, that in addition to diagonalyzing inductance matrix in the case of machine dynamic equations, the Park's transformation also transform the differential equations with time dependent coefficients to the differential equations with constant coefficients.

3.6 Analysis of Multiphase Power System Networks

3. 6. 1 Introductory Remarks

Multiphase power system networks appear to have bright future over convensional 3-phase system. Amongst these multiphase systems, 4-phase and 6-phase systems are the most promising. Linear power invariant transformations with real and also with complex elements along with sequence-impedances and power are presented for these multiphase networks for simplifying their analysis. Multiphase power system networks, which are inherently symmetric, can be broadly classified into two categories; viz. (i) Rotating elements, and (ii) Stationary elements.

3.6.2 Rotating Elements

Multiphase power system networks which possess symmetry operations of proper rotations only, are known as rotating elements. Exploiting these symmetries and using representation theory of finite groups (Appendix 2), block diagonalizing transformation matrices can be constructed.

Since symmetry elements of such networks constitute a regular abelian (i.e. commulative) cyclic group, block diagonalizing transformation matrices can also be developed using eigenvalue approach.

Four-Phase Networks: The network equations under steady state conditions will be as shown,

$$\bar{v}_{pq}^{nbcd} = [Z_{pq}]^{abcd}\, \bar{i}_{pq}^{abcd} \tag{3.91}$$

where $\bar{v}_{pq}^{abcd}$ and $\bar{i}_{pq}^{abcd}$ are the column vectors of voltage drops and current through the 4-phase element $p-q$ for the phases a, b, c and d and Z_{pq}^{abcd} is a 4×4 coefficient matrix of element $p-q$. The coefficient matrix in this case will be cyclic i.e.

$$[Z_{pq}]^{abcd} = \begin{bmatrix} Z_{pq}^{aa} & Z_{pq}^{ab} & Z_{pq}^{ac} & Z_{pq}^{ad} \\ Z_{pq}^{ba} & Z_{pq}^{bb} & Z_{pq}^{bc} & Z_{pq}^{bd} \\ Z_{pq}^{ca} & Z_{pq}^{cb} & Z_{pq}^{cc} & Z_{pq}^{cd} \\ Z_{pq}^{da} & Z_{pq}^{db} & Z_{pq}^{dc} & Z_{pq}^{dd} \end{bmatrix} = \begin{bmatrix} Z_{pq}^{S} & Z_{pq}^{m1} & Z_{pq}^{m2} & Z_{pq}^{m3} \\ Z_{pq}^{m3} & Z_{pq}^{S} & Z_{pq}^{m1} & Z_{pq}^{m2} \\ Z_{pq}^{m2} & Z_{pq}^{m3} & Z_{pq}^{S} & Z_{pq}^{m1} \\ Z_{pq}^{m1} & Z_{pq}^{m2} & Z_{pq}^{m3} & Z_{pq}^{S} \end{bmatrix} \tag{3.92}$$

The transformation matrix A_c which will diagonalize the above cyclic matrix will be as shown,

$$A_c = \frac{1}{\sqrt{4}} \begin{bmatrix} 1 & 1 & 1 & 1 \\ 1 & i & -1 & -i \\ 1 & -1 & 1 & -1 \\ 1 & -i & -1 & i \end{bmatrix} \quad \text{where } i = \sqrt{-1} \tag{3.93}$$

Since $A_c^{*T} A_c = I = U = A_c^{-1} A_c$ i.e. the transformation matrix A_c is unitary hence $A_c^{-1} = A_c^{*T}$. We will have sequenece impedance after transformation of the form,

$$Z_{pq}^{0,1,2,3} = A_c^{*T} Z_{pq}^{abcd} A_c = \begin{bmatrix} Z_{pq}^{S} + Z_{pq}^{m1} + Z_{pq}^{m2} + Z_{pq}^{m3} & & & 0 \\ & Z_{pq}^{S} + iZ_{pq}^{m1} - Z_{pq}^{m2} - iZ_{pq}^{m3} & & \\ & & Z_{pq}^{S} - Z_{pq}^{m1} + Z_{pq}^{m2} - Z_{pq}^{m3} & \\ 0 & & & Z_{pq}^{S} - iZ_{pq}^{m1} - Z_{pq}^{m2} - iZ_{pq}^{m3} \end{bmatrix} \tag{3.94}$$

Six-Phase Networks: The network equation in this case will be,

$$\bar{v}_{pq}^{abcdef} = [Z_{pq}^{abcdef}]\, \bar{i}_{pq}^{abcdef} \tag{3.95}$$

The coefficient matrix Z_{pq}^{abcdef} will also be cyclic. The transformation matrix in this case will be of the form,

$$A_c = \frac{1}{\sqrt{6}} \begin{bmatrix} 1 & 1 & 1 & 1 & 1 & 1 \\ 1 & -1 & \alpha & \alpha^* & -\alpha & -\alpha^* \\ 1 & 1 & -\alpha^* & -\alpha & -\alpha^* & -\alpha \\ 1 & -1 & -1 & -1 & 1 & 1 \\ 1 & 1 & -\alpha & -\alpha^* & -\alpha & -\alpha^* \\ 1 & -1 & \alpha^* & \alpha & -\alpha^* & -\alpha \end{bmatrix} \quad \begin{aligned} &\text{where } \alpha = e^{j2\pi/6} \\ &\qquad = 0.5+j0.866 \\ &\text{and } \alpha^*=0.5-j0.866 \end{aligned} \tag{3.96}$$

Since $A_c^{*T} A_c = I = U = A_c^{-1} A_c$, the transformation matrix A_c is unitary. We also have

$$Z_{pq}^{0,1,2,3,4,5} = A_c^{*T} Z_{pq}^{abcdef} A_c \begin{bmatrix} Z_{pq}^0 & & 0 \\ & Z_{pq(1)} & \\ 0 & & \cdot \\ & & \cdot \end{bmatrix} \tag{3.97}$$

where $Z_{pq}^0 = Z_{pq}^S + Z_{pq}^{m1} + Z_{pq}^{m2} + Z_{pq}^{m3} + Z_{pq}^{m4} + Z_{pq}^{m5}$ is the zero sequence impedance and $Z_{pq(1)}$, $Z_{pq(2)}$ etc. are the first, second sequence impedances.

3.6.3 Stationary Elements

Multiphase power system networks which possess in addition to the symmetry operations of proper rotations, reflection symmetries also, are known as stationary elements; typical example being that of transposed transmission lines. These symmetry operations constitute a group and thus by using representation theory of finite groups, block diagonalizing transformation matrices can be constructed.

Four-Phase Stationary Elements (*i.e. 4-Phase Transposed Transmission Line*): Choosing complex basis for representation, we will obtain the complex transformation A_c (eqn. 3.93) but by choosing real basis for representation, we will obtain the following transformation matrix A_r with real elements, i.e.

$$A_r = \frac{1}{\sqrt{4}} \begin{bmatrix} 1 & 1 & \sqrt{2} & 0 \\ 1 & -1 & 0 & \sqrt{2} \\ 1 & 1 & -\sqrt{2} & 0 \\ 1 & -1 & 0 & -\sqrt{2} \end{bmatrix} \tag{3.98}$$

The transformation matrix A_r is orthogonal i.e.

$$A_r^T A_r = I = U = A_r^{-1} A_r \qquad \text{i.e. } A_r^{-1} = A_r^T$$

The coefficient matrix Z_{pq}^{abcd} of the 4-phase stationary elements $p-q$ will be symmetric i.e.

$$Z_{pq}^{abcd} = \begin{bmatrix} Z_{pq}^{S} & Z_{pq}^{m} & Z_{pq}^{m} & Z_{pq}^{m} \\ Z_{pq}^{m} & Z_{pq}^{S} & Z_{pq}^{m} & Z_{pq}^{m} \\ Z_{pq}^{m} & Z_{pq}^{m} & Z_{pq}^{S} & Z_{pq}^{m} \\ Z_{pq}^{m} & Z_{pq}^{m} & Z_{pq}^{m} & Z_{pq}^{S} \end{bmatrix} \tag{3.99}$$

and after transformation with A_r. Z_{pq}^{abcd} becomes

$$Z_{pq}^{0,1,2,3} = A_r^T Z_{pa}^{abcd} A_r = \begin{bmatrix} Z_{pq}^{S} + 3Z_{pq}^{m} & & 0 \\ & Z_{pq}^{S} - Z_{pq}^{m} & \\ 0 & - & \\ & - & \end{bmatrix} \tag{3.100}$$

Six-Phase Stationary Elements: Here also by choosing real basis, we will obtain the representation A_r as shown below

$$\text{i.e.} \quad A_r = \frac{1}{\sqrt{6}} \begin{bmatrix} 1 & 1 & -\sqrt{\frac{1}{2}} & \sqrt{\frac{3}{2}} & \sqrt{\frac{1}{2}} & -\sqrt{\frac{3}{2}} \\ 1 & -1 & \sqrt{\frac{1}{2}} & \sqrt{\frac{3}{2}} & \sqrt{\frac{1}{2}} & \sqrt{\frac{3}{2}} \\ 1 & 1 & \sqrt{2} & 0 & -\sqrt{2} & 0 \\ 1 & -1 & -\sqrt{2} & 0 & -\sqrt{2} & 0 \\ 1 & 1 & -\sqrt{\frac{1}{2}} & -\sqrt{\frac{3}{2}} & \sqrt{\frac{1}{2}} & \sqrt{\frac{3}{2}} \\ 1 & -1 & \sqrt{\frac{1}{2}} & -\sqrt{\frac{3}{2}} & \sqrt{\frac{1}{2}} & -\sqrt{\frac{3}{2}} \end{bmatrix} \tag{3.101}$$

The matrix A_r is orthogonal i.e. $A_r^T A_r = I = A_r^{-1} A_r$ i.e. $A_r^{-1} = A_r^T$. The coefficient matrix Z_{pq}^{abcdef} becomes after transformation $Z_{pq}^{0,1,2,3,4,5}$ where

$$Z_{pq}^{0,1,2,3,4,5} = A_r^T Z_{pq}^{abcdef} A_r = \begin{bmatrix} [Z_{pq}^{S} + 5Z_{pq}^{m}] & 0 & & \\ 0 & [Z_{pq}^{S} - Z_{pq}^{m}] & & \\ & & \cdot & \\ & & & \cdot \\ & & & & \cdot \end{bmatrix} \tag{3.102}$$

3.6.4 *Feasibility of Six-Phase Transmission System*

Basic feasibility studies of six-phase transmission and HPOT systems have been attempted by Stewart and Wilson for normal steady state operation, overvoltages, and insulation requirements. Out of various conductor array configurations (Fig. 3.17), array 'A' represents the best utilization of spacing for the phasor relationships of six-phase systems, while arrays *B* and *C* are more conventional configurations, with possible application for uprating existing lines. The important findings of these studies are repeated here for ready reference.

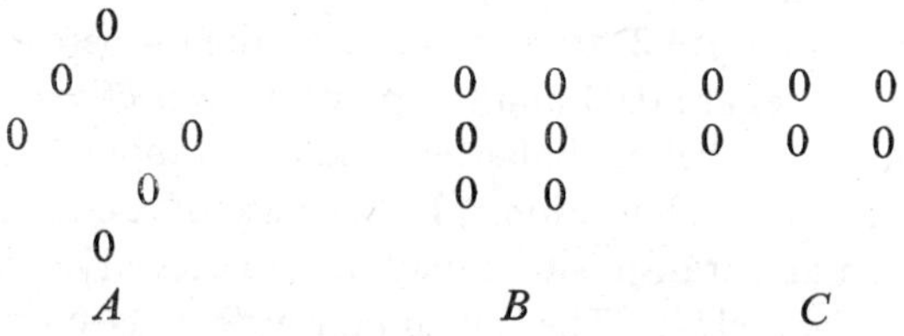

Fig. 3.17 Representative line conductor arrays

(i) Power transfer capability. If a three-phase double-circuit line is converted to six-phase line, the power capability can be increased by 73.2 percent.

(ii) Phase-to-ground and phase-to-phase (adjacent) voltages. The phase-to-phase voltage (for circular configuration *A*, Fig. 3.1A) decreases with the phase order. For higher phase orders, this becomes less than the phase-to-ground voltage. However, for a six-phase system both are equal.

(iii) Conductor clearance. The line conductor spacings can be reduced with increasing phase order since the phase-to-phase voltage decreases for constant phase-to-ground voltage. The degree of reduction is limited by motion of the individual conductors owing to ice, wind, and fault currents, etc.

(iv) Surge impedance loading (SIL). The surge impedance loading is approximately proportional to the phase-order increase and reaches saturation beyond six phases. Therefore, if SIL is the circuit rating criterion, the phase order beyond six becomes questionable.

(v) Thermal loading. The thermal loading follows a straight-line relationship with phase order. Thus, if the thermal loading is the criterion for circuit rating, the capacity increase is proportional to the number of phases.

(vi) Positive sequence surge impedance, inductive reactance, and X_0/X_1 ratio. This is somewhat higher for six-phase line than for three-phase line.

(vii) Transposition. While three-phase lines can be freely transposed, six-phase lines are difficult to transpose. The only transposition for a six-phase line is obtained by rotating the entire conductor array over the length of the line.

(viii) Current imbalance. For the circular array, the negative sequence generator current is negligible both for three- and six-phase lines. However, with reference to a sample system (ref. 15, Fig. 5), the line current imbalance is nearly 4 percent when energized as three-phase, and is reduced below 0.5 percent for six-phase. Shielding reduces this to under 0.02 percent. The asymmetric 2×3 conductor rectangular arrays for two three-phase circuits, depending on the relative phasing, produce maximum negative sequence currents in the generator of the order of 0.15 and 3.1 percent. The line current imbalance is around 4 percent for the alternate phasing relationship. For a six-phase untransposed configuration, the corresponding figures are 2 percent and less than 4 percent. With transposition, these figures are 0.03 and 0.1 percent, respectively. As is obvious, six-phase circuits are better balanced than two three-phase circuits with the same conductor configuration. In the case of a circular array, this is further improved and transposition may be unnecessary.

(ix) Electric fields: The maximum surface electric field decreases with phase order, whereas the maximum ground electric field increases with phase order. The addition of shield wires increases the surface field on the conductors and reduces the ground-level field. The electric field for one conductor open and unfaulted (single pole switching) shows small variation as compared with the same conductor open and grounded.

(x) Radio and audible noise. The performance of a six-phase system is better than two three-phase circuits having the same number of conductors.

(xi) Fault overvoltages. The fault overvoltages for a six-phase system are slightly higher than for a comparable three-phase system. For phase orders higher than six, the fault overvoltages are comparable to those of a three-phase system.

(xii) Switching surges. Phase-to-ground switching surges for three-phase and six-phase lines are approximately the same for the same conditions, with less than 4 percent difference. The phase-to-phase surges become increasingly important relative to phase-to-ground surges as the phase order increases. This may also set a practical upper limit to the phase order achievable.

(xiii) Rate of rise of recovery voltage (RRRV). RRRV across the breaker terminals during opening is found to be less on a six-phase system than for a three-phase system with comparable equivalent short-circuit MVA.

(xiv) Lightning performance. It has been estimated that, owing to to the reduced dimensions of six-phase line, the number of strokes to the line will be of the order of 20 percent less than for the double-circuit three-phase line. For the same phase-to-ground voltage, conventional three-phase double-circuit line and compact six-phase circuit may have very similar back-flashover performance. The probability of phase-to-phase flashover is much higher on a six-phase line. For well-shielded lines, this will be of no significance. The overall lightning performance of high phase order lines will be comparable to conventional three-phase lines assuming effective shielding.

(xv) Terminal insulation levels. The terminal insulation levels tend to be slightly higher for six-phase systems than for three-phase systems.

The above-mentioned findings indicate the general feasibility of HPOT systems, particularly the six-phase systems. Recently reported methods of compact transmission line design, employing new semiconductive glaze porcelain insulators, insulating cross-arms, limitation of internal overvoltages, and control of conductor galloping for reduction of tower head dimensions, further enhance the potentially more efficient use of six-phase and HPOT systems. However, extensive experiments are required to validate these analytical findings before they can be used for designing practical systems.

Problems

3.1 For the following sample power system network (as shown by one line diagram), draw its impedance diagram and also its reactance diagram. What assumptiont are made in drawing the reactance diagram and for what type of studies, this diagram is useful.

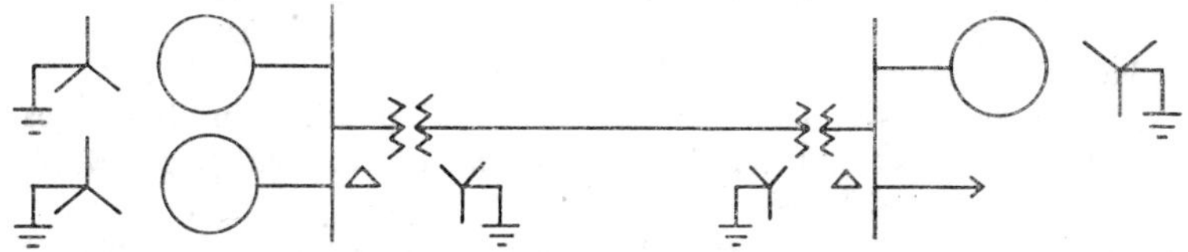

Fig. 1 One line diagram (Example 3.1)

3.2 The conductor of a 3-phase line is equilaterally spaced. Each conductor is a solid wire having a diameter of 0.25 cm. The spacing between conductors is 2.5 meters. Calculate the inductance per phase in h/m.

3.3 Find the capacitance and capacitive reactance per km of a three phase line operated at 50 cycles/second and arranged as shown in the figure below. Each conductor has a diameter of 0.45 cm.

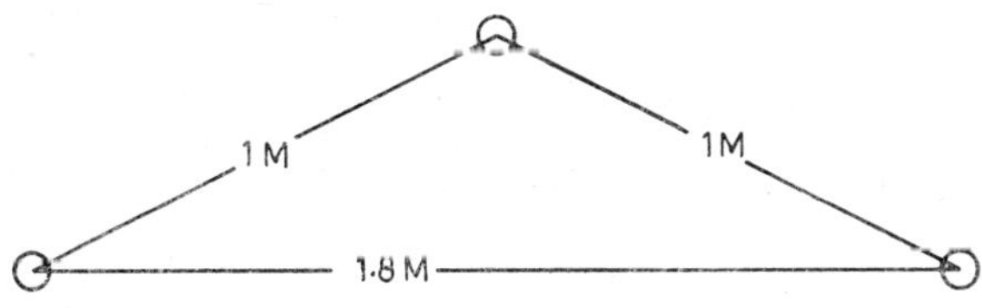

Fig. 2. Example 3.3

3.4 What are the different types of transformation you are familiar with. Write these transformations and also their features. Why symmetrical component transformation can diagonalize the coefficient matrix of both 3-phase rotating as well as stationary elements while the Clarke's component transformation can diagonalize only the coefficient matrix of 3-phase stationary elements. Discuss and prove this statement.

References

1. M. Hamermesh, *Group Theory and its Applications to Physical Problems*, Addison Wesley, 1962.
2. D.N. Kerns, "Analysis of symmetrical waveguide junctions", *Journal of Research of National Bureau of Standards*, Vol. 46, No. 4, April 1951, p. 267.
3. W.K. Kahn, "Scattering equivalent circuits for common symmetrical junctions", *IRE Trans. on Circuit Theory*, Vol. CT-3, June 1956, p. 121.
4. H. Rubin, "Symmetric Basis Functions for Networks with Arbitrary Geometrical Symmetries", *IEEE Trans. on Circuit Theory*, Vol. CT-18, No. 5. September 1971, p. 547.
5. H. Rubin and H.E. Meadows, "Controllability in linear time variable networks with arbitrary symmetry groups", *The Bell System Technical Journal*, February 1972, p. 507.
6. M.U. Siddiqui and V.P. Sinha, "Permutation invariant systems and their application in the filtering of finite discrete data", *IEEE International Conference on Accoustics*, Speach and Signal Processing, May 1977.
7. L. Mirsky, *An Introduction to Linear Algebra*, Oxford-Clarendon Press, 1961.
8. F.A. Cotton, *Chemical Applications of Group Theory*. 2nd ed., Wiley, New Delhi, 1976.
9. P.S. Alexandroff, *Introduction to the Group Theory*, (Translated by H. Perfect and G.M. Peterson), Blackie and Sons, London and Glasgow.
10. W.D. Stevension, *Elements of Power System Analysis*, McGraw-Hill, N.Y.
11. L.P. Singh and V.P. Sinha, "Analysis of multiphase power system networks", Proceeding IFAC Symposium in Large Scale Power Systems, pp. 160–167, held at New Delhi, Aug. 1979.
12. L.P. Singh and V.P. Sinha, "Group theoretic considerations in the steady state analysis of power system networks", *Journal of I.E.* (*India*), pp. 57–64, Vol. 60, Pt. EL1, Aug. 1979.
13. S.P. Nanda, S.N. Tiwari and L.P. Singh, 'Fault Analysis of Six-Phase Systems', Electric Power Systems Research, 4(1981), pp. 201–211.
14. S.N. Tiwari and L.P. Singh, 'Mathematical Modelling and Analysis of Multiphase Systems' presented at the *IEEE PES 1981 Trans. and Dist. Conference* held at Minneapolis, Sept. 20–25, 1981 and in PAS Trans. of IEEE June 1982.
15. J.R. Stewart and D.D. Wilson, "High phase order transmission—a feasibility analysis", Part I—Steady state considerations, *IEEE Trans. PAS-97*, (1978), pp. 2300–2307.
16. J.R. Stewart and D.D. Wilson, "High phase order transmission—a feasibility analysis", Part II—Over voltages and insulation, *IEEE Trans. PAS-97* (1978), pp. 2308–2317.

Chapter 4

Short Circuit Studies

4.1 Introductory Remarks

Short circuit studies and hence the fault analysis are very important for the power system studies since they provide data such as voltages and currents during and after the various types of faults which are necessary in designing the protective schemes of the power system. Current that flows in the power system components just after the occurrence of faults, that flows a few cycles later and the steady state value (i.e. sustained value of) fault currents differ very much from each other as discussed in the Chapter 3.

Protective scheme, basically consists of protective relays and switch gears such as circuit breakers etc. The protective relay which function as a sensing device is a very important component of the protective scheme as it is the relay which senses the fault, determines the location of the fault and then actuate the proper circuit breaker to operate and thus interrupt the circuit in which the fault has occurred. And hence three intervals of time actually lapse before the circuit breaker is called upon to operate. Therefore in designing the circuit breaker and hence protective scheme, all the three component currents are important such as the current that fixes the momentary rating of the circuit breaker, the current that flows a few cycles later when the circuit breaker actually operate and which fixes the interrupting rating of the circuit breaker and also the sustained i.e. steady state value of the fault current.

There are different types of faults in the power system which can broadly be divided into *symmetrical* and *unsymmetrical faults*. The currents and voltages resulting from various type of faults occurring at different locations through out the power system network must be calculated in order to provide sufficient data for designing the protective scheme i.e. both the protective relays and circuit breakers. Because of extensive calculations of voltages and currents due to different faults at different locations of the power system, it became necessary to use some sort of computer and today most of the short circuit studies and hence fault analysis are performed on the digital computer. However the early approach to the short circuit studies employed the bus frame of reference in the admittance form but this method could not become popular because it was time consuming. Today, most of the short circuit studies and hence fault calculations of the power system are formulated in the bus frame of reference using bus impedance matrix "Z_{BUS}" because this method is simpler and less time consuming. We shall discuss different types of fault both through fault impedance and also without any fault impedance, i.e. bolted fault and finally derive algorithms to determine the fault currents and voltages due to these faults.

4.2 Types of Faults

Different types of faults in the power system are:

(i) 3-phase direct short circuit or 3-phase fault through fault impedance (i.e. LLL or LLLG),

(ii) 1-phase to ground fault with or without fault impedance (LG)

(iii) Line to line direct short-circuit or short circuit with fault impedance (LL), and

(iv) Double line to ground fault with or without fault impedance (LLG).

Out of the four types of faults as listed above, the first one which is the least common but most severe, is a symmetrical fault i.e. after the fault, the system remains symmetrical, i.e. balanced. However, the last three faults are unsymmetrical, i.e. after the fault, the voltages and currents become unbalanced, thus it becomes a case of balanced network with unbalanced excitation. In addition to the above, there may be several other types of faults such as opening of one or two conductor etc. We shall derive and discuss the algorithms to calculate the fault currents and voltages etc. due to the above faults. We shall see that in the case of unsymmetrical faults, which is the case of unbalanced excitation in the 3-phase balanced network, it becomes simpler to work with component quantities since after transformation to component quantities, the different component quantities become uncoupled i.e. coefficient matrix describing the system, become diagonal after the transformation. Due to this, the different component quantities say zero, positive and negative sequence quantities are such that the 0-sequence current causes only 0-sequence voltage drops and similarly positive and negative sequence currents causes positive and negative sequence voltage drops only. Thus each sequence quantities may be assumed to flow in the like sequence impedance only and hence we have sequence networks. Now from the actual power system network, drawing different sequence network is very simple. For example only the positive sequence network will have a voltage source, since excitation voltage of the synchronous generator 'E_g' or prefault voltage 'V_F' are balanced, i.e. these are only positive sequence voltages. There is no voltage source in the negative or 0-sequence networks, Moreover the neutral of the system is the reference for positive and negative sequence network but ground is the reference for the 0-sequence network and hence only 0-sequence current flow, if the circuit from the neutral to the ground is complete, in the line between the neutral of the power system and the ground. For a fault at the terminals of an unloaded generator with excitation voltage 'E_g', the following will be the sequence voltage drops, (since we have seen in the Chapter 3 that the symmetrical component transformationdia gonalizes both the 3-phase stationary and rotating elements):

$$\begin{bmatrix} V_{a0} \\ V_{a1} \\ V_{a2} \end{bmatrix} = \begin{bmatrix} 0 \\ E_g \\ 0 \end{bmatrix} - \begin{bmatrix} Z_0 & & 0 \\ & Z_1 & \\ 0 & & Z_2 \end{bmatrix} \begin{bmatrix} I_{a0} \\ I_{a1} \\ I_{a2} \end{bmatrix} \tag{4.1}$$

where 0, 1 and 2 refers to zero, positive and negative sequence quantities. Similarly for a fault in the power system, we have the following relations:

$$\begin{bmatrix} V_{a0} \\ V_{a1} \\ V_{a2} \end{bmatrix} = \begin{bmatrix} 0 \\ V_F \\ 0 \end{bmatrix} - \begin{bmatrix} Z_0 & & 0 \\ & Z_1 & \\ 0 & & Z_2 \end{bmatrix} \begin{bmatrix} I_{a0} \\ I_{a1} \\ I_{a2} \end{bmatrix} \tag{4.2}$$

where V_F is the prefault voltage which is balanced.

We can easily determine how the different networks are connected when any of the above fault occurs. We will only give the connection diagrams for sequence networks to simulate different type of faults without giving the proof. (For proofs, the readers may refer to any book on power system analysis at the undergraduate level.)

Three-phase faults

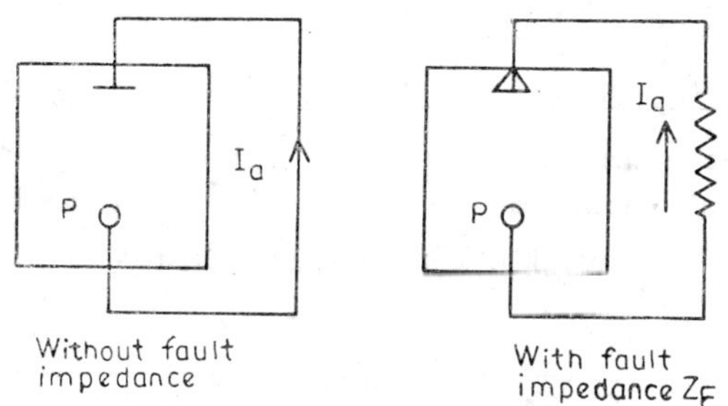

Fig. 4.1 Positive sequence network

Here P ○ is a point where fault occurs.

4.3 Short Circuit Studies of a Large Power System Networks

After the brief review, we shall discuss and derive the methods to simulate these faults on the digital computer.

Assumptions

(i) Representing each machine by a constant voltage source behind proper reactances which may be X'', X' or X.

(ii) Neglecting all the shunt connections such as static loads, line charging and transformer magnetizing circuits. With this assumption, the power system network become open circuited and thus the normal load currents (i.e. prefault currents) are automatically neglected and therefore, all the prefault bus voltages will have the same magnitude and

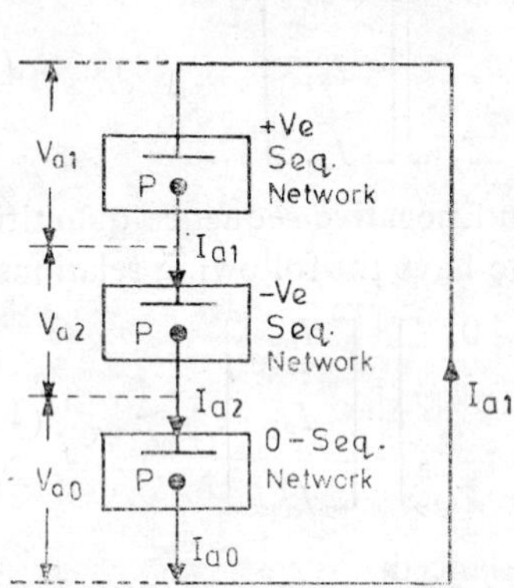

(a) 1-phase to ground fault without fault impedance.

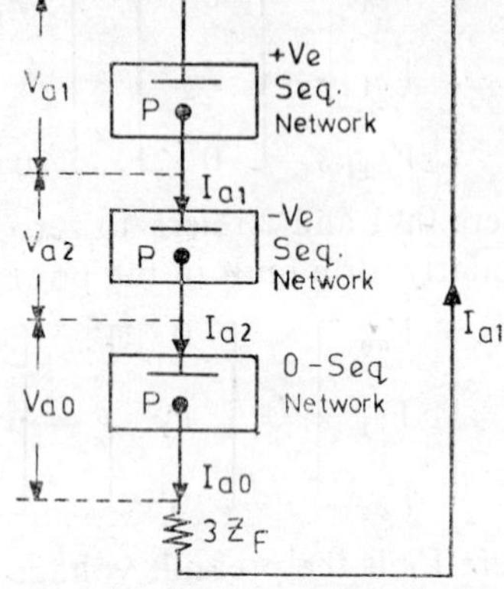

(b) 1-phase to ground fault with fault impedance.

Fig. 4.2 Single line to ground fault

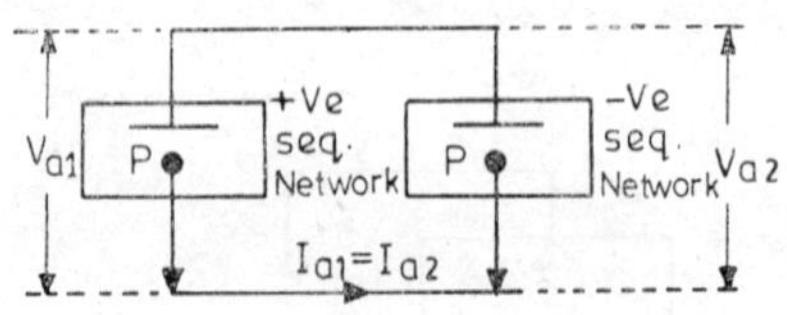

(a) Line to line fault without fault impedance

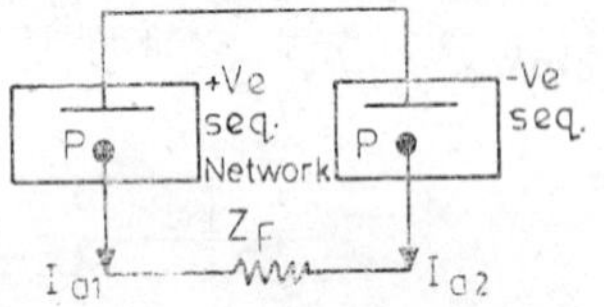

(b) Line to line fault with fault impedance.

Fig. 4.3 Line to line fault

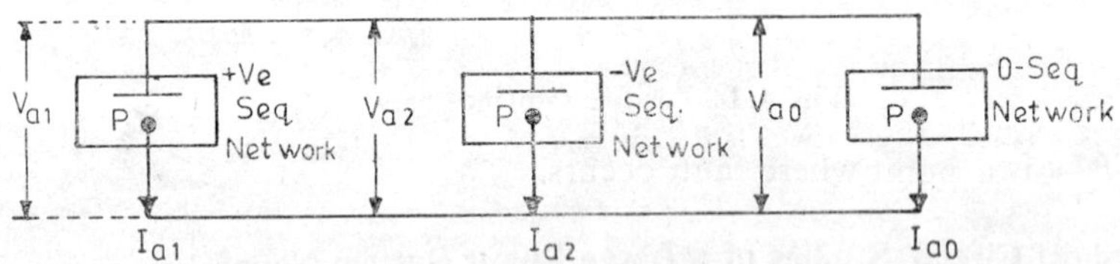

(a) Double line to ground fault without fault impedance.

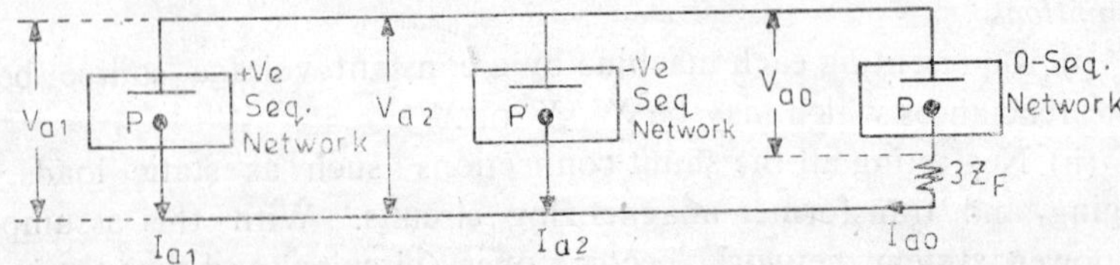

(b) Double line to ground fault with faultim pedance.

Fig. 4.4 Double line to ground fault

phase angle. Thus to work on per unit system, the prefault bus voltages are set equal to $1\angle 0$.

(iii) Setting all the transformers to nominal taps (i.e transformer tappings are neglected). Since we work in per unit system, with this representation, transformers will automatically be out of circuit.

(iv) Normally neglecting winding resistances and line resistances etc. With this assumption, the system will contain only reactances and hence the power system is represented by its most simplified reactance diagram. This assumption is usually made for short circuit studies on the D.C. network analyzer. However, for conducting studies on the digital computer, this assumption is usually not made.

(v) Equating the positive sequence impedance equal to negative sequence impedance also for 3-phase rotating elements even though these sequence impedances are equal only in the case of 3-phase stationary elements.

With these assumptions, the given power system network as shown in Fig. 4.5, will have the simplified representation as in Fig. 4.6.

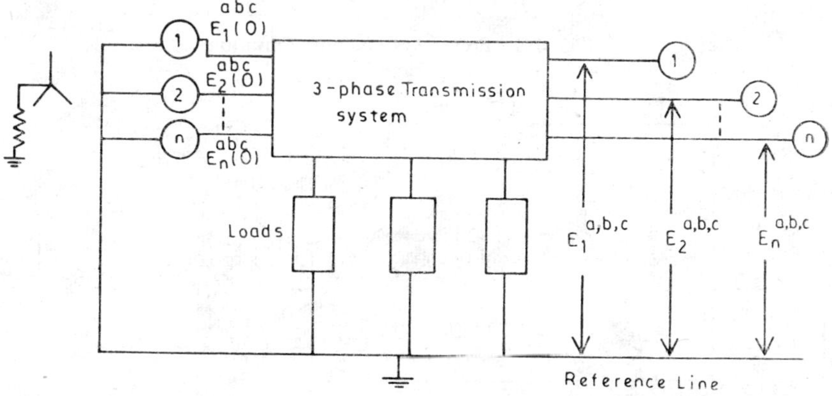

Fig. 4.5 Actual power system (3-phase network).

As discussed earlier, we shall formulate the short circuit problem in the bus frame of reference using Z_{BUS}. The Z_{BUS} matrix is formed including ground node as the reference node. Moreover, by taking into account all the assumptions, the power system network will become as shown in the Fig. 4.7 for a fault at any bus p. The Z_{BUS} matrix includes the parameters of both the rotating elements as well as stationary elements, i.e. parameters of machines, transformer and transmission systems. Actually this representation which is derived by the Thevenin's theorem, include Z_{BUS} matrix as defined earlier in series with the open circuit voltages (i.e. $E^{abc}_{1(0)}$, $E^{abc}_{2(0)}$ etc.) where open circuit voltages will be equal to the prefault bus voltages. This is true since all the shunt connections are neglected for the purpose of short circuit studies with the result the normal load current will automatically be neglected and thus the prefault voltages

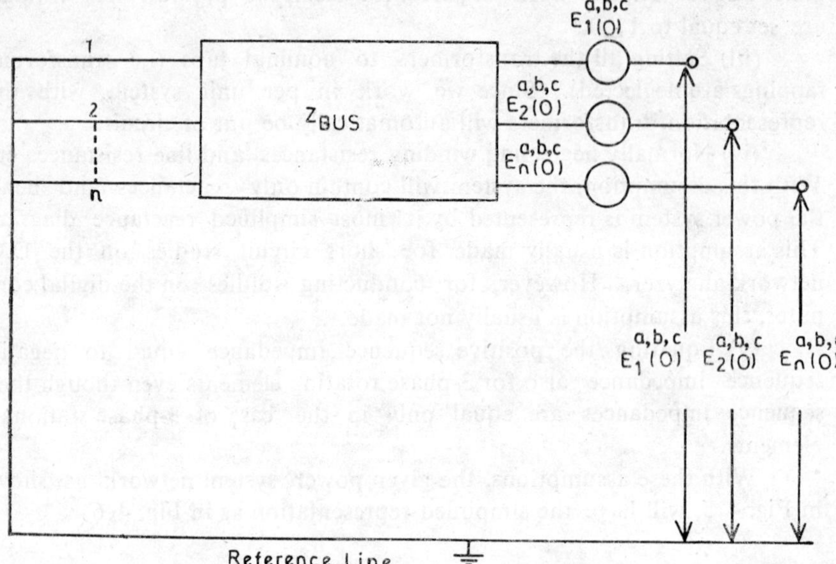

Fig. 4.6 Power system representation.

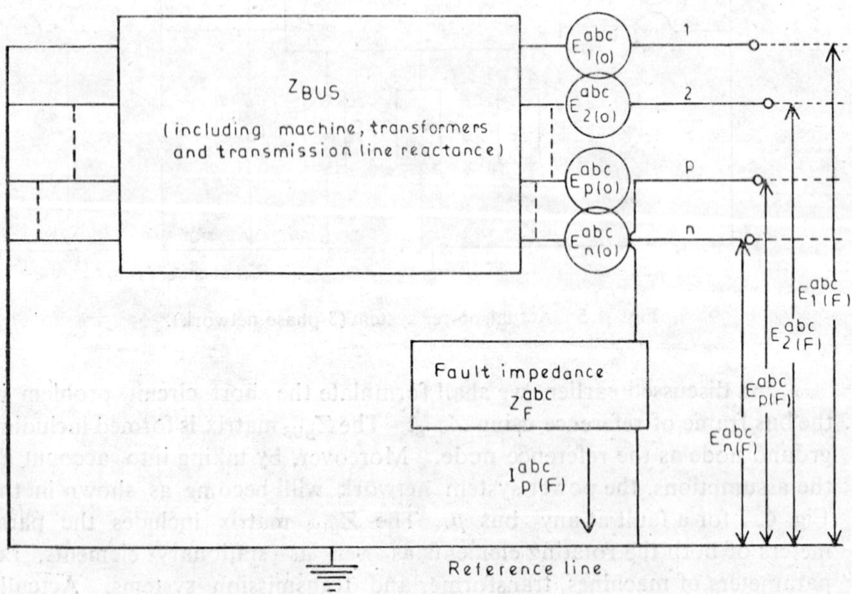

Fig. 4.7 Power system representation for a fault at bus *p*

of all the buses will be equal in magnitude and also will have the same phase angle. This prefault voltage will naturally be the open circuit voltage since shunt connections are removed, i.e. system is open circuited before the fault occurs.

The performance equation in the bus frame of reference using Z_{BUS}, of a power system when fault occurs at any bus p (Fig. 4.7) will be given by :

$$E^{abc}_{BUS\,(F)} = E^{abc}_{BUS\,(0)} - Z^{abc}_{BUS}\, I^{abc}_{BUS\,(F)} \tag{4.3}$$

where $E^{abc}_{BUS\,(F)}$ is the column vector of the unknown bus voltage after the fault

$$= \begin{bmatrix} E^{abc}_{1\,(F)} \\ E^{abc}_{2\,(F)} \\ \vdots \\ E^{abc}_{n\,(F)} \end{bmatrix} \tag{4.4}$$

and $E^{abc}_{BUS\,(0)}$ is the column vector of known prefault bus voltages, i.e. open circuit bus voltages

$$= \begin{bmatrix} E^{abc}_{1\,(0)} \\ E^{abc}_{2\,(0)} \\ \vdots \\ E^{abc}_{n\,(0)} \end{bmatrix} \tag{4.5}$$

and the unknown bus currents when fault occurs at any bus p will be

$$I^{abc}_{BUS\,(F)} = \begin{bmatrix} 0 \\ 0 \\ I^{abc}_{p\,(F)} \\ 0 \\ 0 \end{bmatrix} \tag{4.6}$$

Since prefault currents are neglected and therefore, for a fault at the bus p, only currents (i.e. fault current) at the bus p will flow.

The three phase bus impedance matrix which include ground as reference will be

$$Z^{abc}_{BUS} = \begin{bmatrix} Z^{abc}_{11} & Z^{abc}_{12} & \cdots & Z^{abc}_{1p} & Z^{abc}_{1n} \\ Z^{abc}_{21} & Z^{abc}_{22} & \cdots & Z^{abc}_{2p} & Z^{abc}_{2n} \\ \vdots & \vdots & & \vdots & \vdots \\ Z^{abc}_{p1} & \cdots & \cdots & \cdots & Z^{abc}_{pn} \\ Z^{abc}_{n1} & \cdots & \cdots & Z^{abc}_{np} & Z^{abc}_{nn} \end{bmatrix} \tag{4.7}$$

Thus equation (4.3) after substituting for eqns. (4.4), (4.5), (4.6) and (4.7) becomes,

$$E^{abc}_{BUS(F)} = E^{abc}_{BUS(0)} - Z^{abc}_{BUS}\, I^{abc}_{BUS(F)} \tag{4.8}$$

i.e.,

$$\begin{bmatrix} E^{abc}_{1(F)} \\ E^{abc}_{2(F)} \\ \vdots \\ E^{abc}_{P(F)} \\ \vdots \\ E^{abc}_{n(F)} \end{bmatrix} = \begin{bmatrix} E^{abc}_{1(0)} \\ E^{abc}_{2(0)} \\ \vdots \\ E^{abc}_{P(0)} \\ \vdots \\ E^{abc}_{n(0)} \end{bmatrix} - \begin{bmatrix} Z^{abc}_{11} & \cdots & Z^{abc}_{1p} & \cdots & Z^{abc}_{1n} \\ Z^{abc}_{22} & \cdots & Z^{abc}_{2p} & \cdots & Z^{abc}_{2n} \\ \vdots & & \vdots & & \vdots \\ Z^{abc}_{p1} & \cdots & \cdots & & Z^{abc}_{pn} \\ \vdots & & \vdots & & \vdots \\ Z^{abc}_{n1} & & \cdots & & Z^{abc}_{nn} \end{bmatrix} \begin{bmatrix} 0 \\ 0 \\ \vdots \\ I^{abc}_{P(F)} \\ \vdots \\ 0 \end{bmatrix} \tag{4.9}$$

From equation (4.9), we get

$$\begin{aligned} E^{abc}_{1(F)} &= E^{abc}_{1(0)} - Z^{abc}_{1P}\, I^{abc}_{P(F)} \\ E^{abc}_{2(F)} &= E^{abc}_{2(0)} - Z^{abc}_{2P}\, I^{abc}_{P(F)} \\ &\vdots \\ E^{abc}_{P(F)} &= E^{abc}_{P(0)} - Z^{abc}_{PP}\, I^{abc}_{P(F)} \\ &\vdots \\ E^{abc}_{n(F)} &= E^{abc}_{n(0)} - Z^{abc}_{nP}\, I^{abc}_{P(F)} \end{aligned} \tag{4.10}$$

Now for a fault at the bus p, the 3-phase voltage vector $E^{abc}_{P(F)}$ is given by

$$E^{abc}_{P(F)} = Z^{abc}_{F}\, I^{abc}_{P(F)} \tag{4.11}$$

where Z^{abc}_{F} is a 3×3 fault impedance matrix which depends upon the type of fault etc. But since from eqn. (4.10),

$$E^{abc}_{P(F)} = E^{abc}_{P(0)} - Z^{abc}_{PP}\, I^{abc}_{P(F)} \tag{4.12}$$

Comparing eqns. (4.11) and (4.12), we get

$$Z^{abc}_{F}\, I^{abc}_{P(F)} = E^{abc}_{P(0)} - Z^{abc}_{PP}\, I^{abc}_{P(F)}$$

i.e.

$$Z^{abc}_{F}\, I^{abc}_{P(F)} + Z^{abc}_{PP}\, I^{abc}_{P(F)} = E^{abc}_{P(0)} \tag{4.13}$$

or

$$[Z^{abc}_{F} + Z^{abc}_{pp}]\, I^{abc}_{P(F)} = E^{abc}_{p(0)} \tag{4.14}$$

From equation (4.14), we get

$$I^{abc}_{P(F)} = [Z^{abc}_{F} + Z^{abc}_{PP}]^{-1}\, E^{abc}_{P(0)} \tag{4.15}$$

and

$$E^{abc}_{P(F)} = Z^{abc}_{F}\, I^{abc}_{P(F)} = Z^{abc}_{F}\, [Z^{abc}_{F} + Z^{abc}_{PP}]^{-1}\, E^{abc}_{P(0)} \tag{4.16}$$

Similarly, the 3-phase bus voltages after the fault at the buses other than the faulted bus p, is:

$$E^{abc}_{i(F)} = E^{abc}_{i(0)} - Z^{abc}_{iP}\, I^{abc}_{P(F)} \qquad i = 1, 2, \ldots, n; \quad i \neq p \tag{4.17}$$

Now substituting for $I^{abc}_{P(F)}$ from eqn. (4.15) in eqn. (4.17), we get

$$E^{abc}_{i(F)} = E^{abc}_{i(0)} - Z^{abc}_{iP}\,[Z^{abc}_F + Z^{abc}_{PP}]^{-1}\,E^{abc}_{P(0)} \tag{4.18}$$

However, in case, we are given the fault admittance Y^{abc}_F (fault admittance at the faulted bus p), we have the relation at the faulted bus p,

$$I^{abc}_{P(F)} = Y^{abc}_F\,E^{abc}_{P(F)} \tag{4.19}$$

where Y^{abc}_F is a 3-phase fault admittance matrix.

Hence for a faulted bus P, we get from equation (4.10),

$$E^{abc}_{P(F)} = E^{abc}_{P(0)} - Z^{abc}_{PP}\,I^{abc}_{P(F)} \tag{4.20}$$

Substituting equation (4.19) in equation (4.20), we get

$$E^{abc}_{p(F)} = E^{abc}_{P(0)} - Z^{abc}_{PP}\,Y^{abc}_F\,E^{abc}_{P(F)} \tag{4.21}$$

or

$$E^{abc}_{P(F)} + Z^{abc}_{PP}\,Y^{abc}_F\,E^{abc}_{P(F)} = E^{abc}_{P(0)}$$

i.e.

$$[U + Z^{abc}_{PP}\,Y^{abc}_F]\,E^{abc}_{p(F)} = E^{abc}_{P(0)}$$

or

$$E^{abc}_{P(F)} = [U + Z^{abc}_{PP}\,Y^{abc}_F]^{-1}\,E^{abc}_{P(0)} \tag{4.22}$$

But

$$I^{abc}_{P(F)} = Y^{abc}_F\,E^{abc}_{P(F)}$$

Hence

$$I^{abc}_{P(F)} = Y^{abc}_F\,[U + Z^{abc}_{PP}\,Y^{abc}_F]^{-1}\,E^{abc}_{P(0)} \tag{4.23}$$

Similarly 3-phase bus voltages after the fault at the buses other than the faulted bus P, is given by

$$\begin{aligned} E^{abc}_{i(F)} &= E^{abc}_{i(0)} - Z^{abc}_{iP}\,I^{abc}_{P(F)} \\ &= E^{abc}_{i(0)} - Z^{abc}_{iP}\,Y^{abc}_F\,[U + Z^{abc}_{PP}\,Y^{abc}_F]^{-1}\,E^{abc}_{P(0)} \end{aligned} \tag{4.24}$$

$$i = 1, 2, \ldots, n$$
$$i \neq p$$

Hence after the fault occurs through the fault impedance/adm., we calculate first of all the bus voltages after the fault. After this, we calculate the fault current flowing in the elements of the power system (i.e. primitive) network. Thus the fault current (prefault current i.e. normal load current is neglected in the short circuit studies) in any element i–j of the power system network will be,

$i^{abc}_{ij(F)}$ = Fault current in the element i–j of the power system network

$$= y^{abc}_{ijp\sigma}\,[(E^{abc}_{\rho(F)}) - E^{abc}_{\sigma(F)}] \tag{4.25}$$

where $E^{abc}_{\rho(F)}$ and $E^{abc}_{\sigma(F)}$ are the bus voltages after the fault for any bus ρ and σ and $y^{abc}_{ij\rho\sigma}$ is the 3 phase element of the primitive network matrix,

i.e.

$$y^{abc}_{ij\rho\sigma} = \begin{bmatrix} y^{aa}_{ij\rho\sigma} & y^{ab}_{ij\rho\sigma} & y^{ac}_{ij\rho\sigma} \\ y^{ba}_{ij\rho\sigma} & y^{bb}_{ij\rho\sigma} & y^{bc}_{ij\rho\sigma} \\ y^{ca}_{ij\rho\sigma} & y^{cb}_{ij\rho\sigma} & y^{cc}_{ij\rho\sigma} \end{bmatrix} \tag{4.26}$$

Here $y^{ab}_{ij\rho\sigma}$ is the mutual admittance between the element $i-j$ of the phase a and element $\rho-\sigma$ of the phase b and so on. Moreover, $\rho-\sigma$ is the element $i-j$ and also the elements mutually coupled to $i-j$.

During these faults, the 3-phase excitation becomes unbalanced but the 3-phase network (as the 3-phase lines are transposed) is balanced. In such cases if we transform the phasor quantities to the component quantities by means of some linear, power invariant transformations say symmetrical components etc., the 3-phase coefficient matrix for both stationary as well as rotating elements become diagonalized and hence analysis become simpler.

4.3.1 Transformation to Symmetrical Components (i.e. positive, negative and zero sequence components)

We have already proved in the previous chapter that,

$$Z^{0,1,2}_{\text{comp}} = T^{*T}_{S}\, Z^{abc}_{\text{Phase}}\; T_{S}$$

where T_s is a symmetrical component matrix.

Using the above relations, we will obtain the following relations. *Primitive impedance matrix* Z^{abc}_{pq} after transformation becomes $Z^{0,1,2}_{pq}$ i.e.

$$Z^{0,1,2}_{pq} = T^{*T}_{s}\, Z^{abc}_{pq}\, T_{s} = \begin{bmatrix} Z^{0}_{pq} & 0 & 0 \\ 0 & Z^{1}_{pq} & 0 \\ 0 & 0 & Z^{2}_{pq} \end{bmatrix} \tag{4.27}$$

Now for 3-phase stationary element $Z^{1}_{pq} = Z^{2}_{pq}$ and we assume that for 3-phase rotating elements also,

$$Z^{1}_{pq} = Z^{2}_{pq}$$

Similarly any element $y^{abc}_{ij\rho\sigma}$ of the primitive network is transformed to $y^{0,1,2}_{ij\rho\sigma}$ which is also diagonal, i.e.

$$y^{0,1,2}_{ij\rho\sigma} = T^{*T}_{s}\, y^{abc}_{ij\rho\sigma}\, T_{s} = \begin{bmatrix} y^{0}_{ij\rho\sigma} & 0 & 0 \\ 0 & y^{1}_{ij\rho\sigma} & 0 \\ 0 & 0 & y^{2}_{ij\rho\sigma} \end{bmatrix} \tag{4.28}$$

Similarly each element Z^{abc}_{ij} of the bus impedance matrix can be diagonalized. The fault impedance matrix Z^{abc}_{F} can also be transformed to $Z^{0,1,2}_{F}$. However, $Z^{0,1,2}_{F}$ will only be diagonal if the fault is balanced. Same thing is true for Y^{abc}_{F} also.

Since we assume that all the bus voltages before the fault are equal in magnitude and have the same phase angle (this will be true as prefault i.e. normal load current is neglected) and also since we work in the per unit system, we set all the bus voltages equal to $1\angle 0$, Therefore, three phase bus voltage for any bus i before fault is $E^{abc}_{i(0)} = 1\angle 0$ (refer to Fig. 4.8).

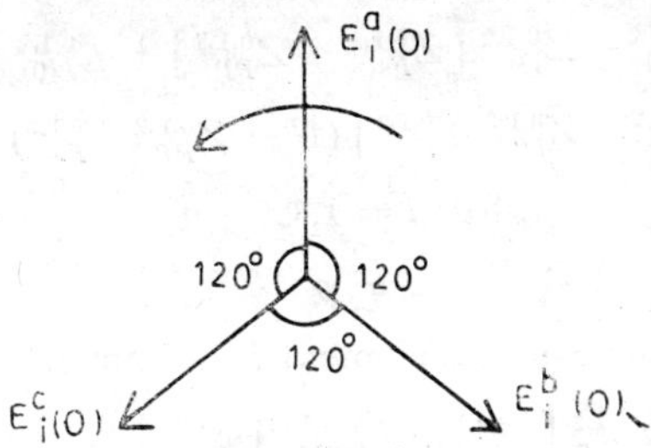

Fig. 4.8 Voltage phasor diagram.

i.e. $E^a_{i(0)} = E^b_{i(0)} = E^c_{i(0)} = 1$ p.u.

Let
$$\alpha = 1 \angle 120$$
then
$$\alpha^2 = 1 \angle 240$$

Therefore,
$$E^a_{i(0)} = |E^a_{i(0)}| \angle 0 = 1 \angle 0$$
$$E^b_{i(0)} = |E^a_i| \angle 240 = 1 \angle 240 = \alpha^2$$
and
$$E^c_{i(0)} = |E^a_i| \angle 120 = 1 \angle 120$$
$$= \alpha$$

Hence we get,
$$E^{abc}_{i(0)} = 1 \angle 0$$
$$= \begin{bmatrix} E^a_{i(0)} \\ E^b_{i(0)} \\ E^c_{i(0)} \end{bmatrix} = \begin{bmatrix} 1 \\ \alpha^2 \\ \alpha \end{bmatrix} \tag{4.29}$$

Transfering the above phasor voltages into the field of component quantities with the help of symmetrical component transformation matrix T_s, we get,
$$E^{0,1,2}_{i(0)} = (T^*_S)^T E^{abc}_{i(0)} = T_S^{-1} E^{abc}_{i(0)}$$
$$= T_S^{-1} \begin{bmatrix} 1 \\ \alpha^2 \\ \alpha \end{bmatrix} = \begin{bmatrix} 0 \\ \sqrt{3} \\ 0 \end{bmatrix} \tag{4.30}$$

for $i = 1, \ldots, p, \ldots, n$.

Thus we can write all the earlier equations into the component quantities. Hence for the faulted bus, p, we have from the equations (4.15), (4.23), (4.16) and (4.22)

$$I^{0,1,2}_{P(F)} = [Z^{0,1,2}_F + Z^{0,1,2}_{PP}]^{-1} E^{0,1,2}_{P(0)}$$
$$I^{0,1,2}_{P(F)} = Y^{0,1,2}_F [U + Z^{0,1,2}_{PP} Y^{0,1,2}_F]^{-1} E^{0,1,2}_{P(0)}$$
$$E^{0,1,2}_{P(F)} = Z^{0,1,2}_F [Z^{0,1,2}_F + Z^{0,1,2}_{PP}]^{-1} E^{0,1,2}_{P(0)}$$
$$E^{0,1,2}_{P(F)} = [U + Z^{0,1,2}_{PP} Y^{0,1,2}_{F(0)}]^{-1} E^{0,1,2}_{P(0)} \tag{4.31}$$

The bus voltages at the buses other than the faulted bus P are thus

$$E_{i(F)}^{0,1,2} = E_{i(0)}^{0,1,2} - Z_{iP}^{0,1,2}\,[Z_F^{0,1,2} + Z_{PP}^{0,1,2}]^{-1}\,E_{P(0)}^{0,1,2}$$

$$E_{i(F)}^{0,1,2} = E_{i(0)}^{0,1,2} - Z_{iP}^{0,1,2}\,Y_F^{0,1,2}\,[(U + Z_{PP}^{0,1,2}\,Y_F^{0,1,2})]^{-1}\,E_{P(0)}^{0,1,2}$$

$$\text{for } i = 1, \ldots, n$$

$$i \neq p \tag{4.32}$$

The fault current in the 3-phase element $i-j$ from equation (4.25), is

$$i_{ij(F)}^{0,1,2} = Y_{ij\rho\sigma}^{0,1,2}\,[\,E_{\rho(F)}^{0,1,2} - E_{\sigma(F)}^{0,1,2}\,] \tag{4.33}$$

4.4 Algorithms for Calculating System Conditions after the Occurrence of Faults

We have shown earlier that the different types of faults in a power system network are,

1. 3-phase short circuit
2. 3-phase to ground fault
3. Double line to ground fault
4. Line to line fault
5. Single line (i.e. phase) to ground fault

We shall take up all these cases of faults one by one and develop algorithms to calculate the bus voltages and fault currents. First of all we shall consider the cases of these faults through finite fault imp. i.e. Z_f (or fault admittance Y_f).

4.4.1 3-phase to ground fault through fault impedance Z_f per phase

Let for the faulted bus p (3-phase bus) the fault occurs through impedance Z_f and let there is also a finite ground impedance Z_g as shown in Fig. 4.9.

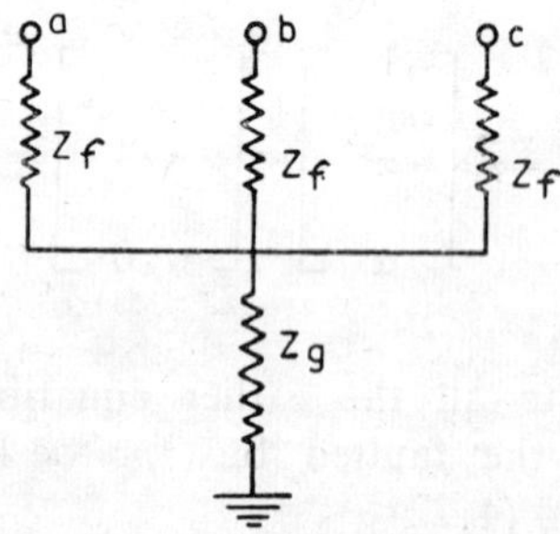

Fig. 4.9 3-phase to ground fault at any bus p.

For the above case, $Z_F^{a,b,c}$ (i.e. fault impedance matrix) in the bus frame of reference, is found by open circuit test as discussed in Chapter 2. Following is the procedure to obtain fault impedance matrix Z_F^{abc} in the bus frame of reference.

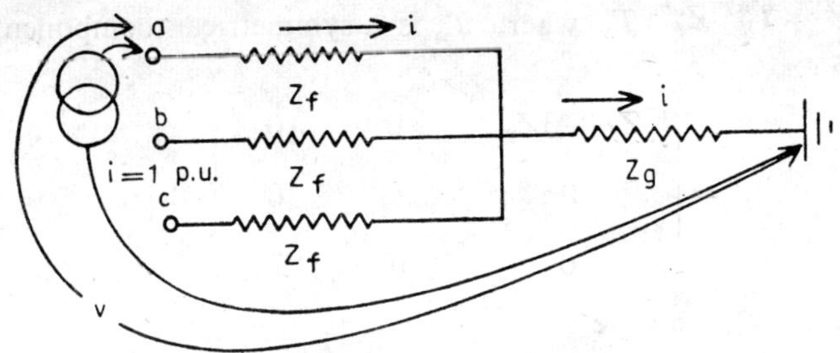

Fig. 4.10 For calculating diagonal elements of Z_F^{abc}.

Referring to Fig. 4.10, to calculate the diagonal element say Z_F^{aa}, we apply $i = 1$ p.u. at the phase 'a' and calculate the voltage at 'a' keeping phases b and c open circuited. Thus we have from Fig. 4.10,

$$v = [Z_f + Z_g]\, i \quad \text{as } i = 1 \text{ p.u.}$$

Therefore, we get,

$$Z_F^{aa} = Z_f + Z_g$$
$$= Z_F^{bb} = Z_F^{cc} \tag{4.34}$$

For off diagonal elements, we apply a current of 1 p.u. at the phase 'a' and calculate voltage at phase 'b' keeping other phases open circuited.

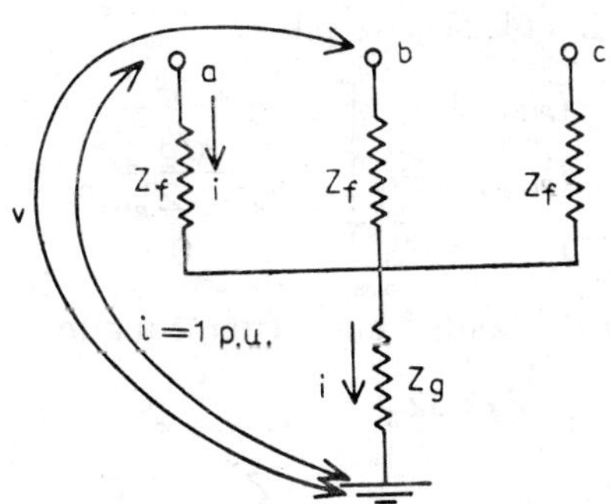

Fig. 4.11 For calculating off diagonal elements of Z_F^{abc}.

Thus referring to Fig. 4.11, we have

$$V = Z_g i$$

as $\quad i = 1$ p.u.

Hence the off diagonal element

$$Z_F^{ab} = Z_g = Z_F^{bc} = Z_F^{ca} \text{ etc.} \tag{4.35}$$

Thus the fault impedance matrix Z_F^{abc} in the bus frame of reference is as shown,

$$Z_F^{abc} = \begin{bmatrix} Z_f+Z_g & Z_g & Z_g \\ Z_g & Z_f+Z_g & Z_g \\ Z_g & Z_g & Z_f+Z_g \end{bmatrix} \tag{4.36}$$

Now, $Z_F^{0,1,2} = T_S^{*T} Z_F^{abc} T_S$ where T_S is a symmetrical component transformations

$$= \begin{bmatrix} Z_f + 3Z_g & 0 & 0 \\ 0 & Z_f & 0 \\ 0 & 0 & Z_f \end{bmatrix} \quad (4.37)$$

From eqn. (4.31), we have

$$I_{P(F)}^{0,1,2} = [Z_F^{0,1,2} + Z_{PP}^{0,1,2}]^{-1} E_{P(0)}^{0,1,2} \quad (4.38)$$

Substituting the $Z_F^{0,1,2}$ from eqn. (4.37) and $E_{P(0)}^{0,1,2}$ from eqn. (4.30) in eqn. (4.38), we get,

$$\begin{bmatrix} I_{P(F)}^{(0)} \\ I_{P(F)}^{(1)} \\ I_{P(F)}^{(2)} \end{bmatrix} = \begin{bmatrix} Z_f+3Z_g+Z_{PP}^0 & 0 & 0 \\ 0 & Z_f+Z_{PP}^1 & 0 \\ 0 & 0 & Z_f+Z_{PP}^2 \end{bmatrix}^{-1} \begin{bmatrix} 0 \\ \sqrt{3} \\ 0 \end{bmatrix} \quad (4.39)$$

But as per assumptions, $Z_{PP}^1 = Z_{PP}^2$ (i.e. positive sequence impedance = negative sequence impedance both for stationary and as well as rotating elements).

Solving equation (4.39), we get

$$\begin{bmatrix} I_{P(F)}^0 \\ I_{P(F)}^1 \\ I_{P(F)}^2 \end{bmatrix} = \begin{bmatrix} 0 \\ \dfrac{\sqrt{3}}{Z_f+Z_{PP}^1} \\ 0 \end{bmatrix} \quad (4.40)$$

And the voltages at the faulted bus p from equation (4.31),

$$\begin{bmatrix} E_{P(F)}^0 \\ E_{P(F)}^1 \\ E_{P(F)}^2 \end{bmatrix} = \begin{bmatrix} Z_f+3Z_g & 0 & 0 \\ 0 & Z_f & 0 \\ 0 & 0 & Z_f \end{bmatrix} \begin{bmatrix} 0 \\ \dfrac{\sqrt{3}}{Z_f+Z_{PP}^1} \\ 0 \end{bmatrix}$$

$$= \begin{bmatrix} 0 \\ \dfrac{\sqrt{3}\ Z_f}{Z_f+Z_{PP}^1} \\ 0 \end{bmatrix} \quad (4.41)$$

The voltages at the buses, other than the faulted bus p from equation (4.32),

$$\begin{bmatrix} E_{i(F)}^0 \\ E_{i(F)}^1 \\ E_{i(F)}^2 \end{bmatrix} = \begin{bmatrix} 0 \\ \sqrt{3} \\ 0 \end{bmatrix} - \begin{bmatrix} Z_{iP}^0 & 0 & 0 \\ 0 & Z_{iP}^1 & 0 \\ 0 & 0 & Z_{iP}^1 \end{bmatrix} \begin{bmatrix} 0 \\ \dfrac{\sqrt{3}}{Z_f+Z_{PP}^1} \\ 0 \end{bmatrix}$$

$$= \sqrt{3}\begin{bmatrix} 0 \\ 1 - \dfrac{Z^1_{iP}}{Z_f + Z^1_{PP}} \\ 0 \end{bmatrix} \tag{4.42}$$

Now we have to calculate the fault current in the elements of the power system network. However in this case, $Y^1_{ij\rho\sigma} = 0$ except for the element

$$\rho\sigma = ij$$

Thus the fault current in any element $i-j$, from eqn. (4.33),

$$\begin{bmatrix} i^0_{ij(F)} \\ i^1_{ij(F)} \\ i^2_{ij(F)} \end{bmatrix} = \begin{bmatrix} 0 \\ y^1_{ij,ij}\,[E^1_{i(F)} - E^1_{j(F)}] \\ 0 \end{bmatrix} \tag{4.43}$$

Here 1 refers to the positive sequence component. Thus in this case we have only positive sequence quantities i.e. positive sequence network consisting of positive sequence impedance (see Fig. 4.1).

4.4.2 Line to Line Fault Through Z_f

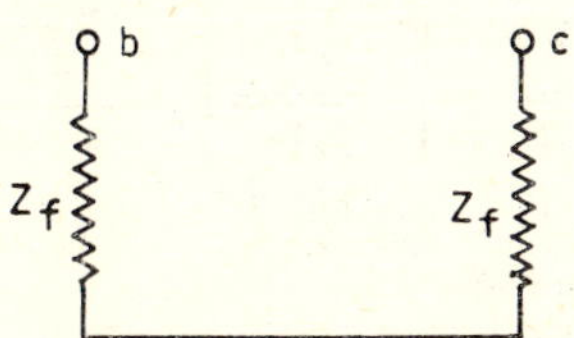

Fig. 4.12 Line to line fault.

Let us take a line to line fault at any bus p between any of the two phases b and c through a finite fault impedance Z_f per phase as shown in Fig. 4.12.

Clearly $Z_F^{a,b,c}$ is undefined. However, the elements of fault admittance matrix in the bus frame of reference, i.e. $Y_F^{a,b,c}$ is calculated by short circuit test as discussed in Chapter 2.

Referring to Fig. 4.13, for diagonal element, we apply a voltage

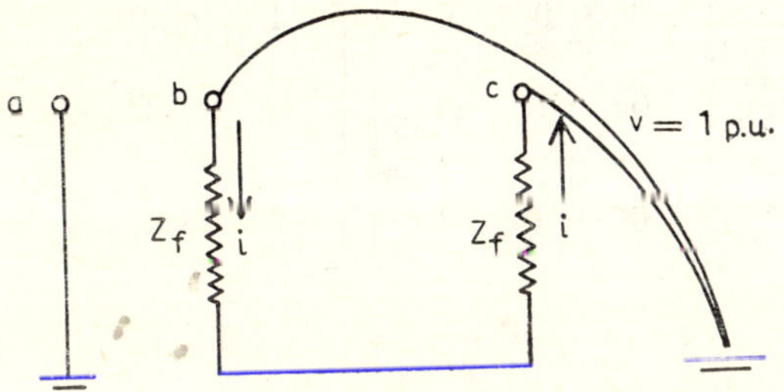

Fig. 4.13 Calculation of diagonal element Y_F^{bb}.

of 1 p.u. at the phase and calculate its current keeping other phases short circuited (i.e. connected to the ground).

Thus for Y_F^{bb} as shown in Fig. 4.13, we have

$$V = 2Z_f i$$

or $$i/V = 1/2Z_f = y_f/2 \quad \text{where} \quad y_f = 1/Z_f$$

and also, $$Y_F^{bb} = Y_F^{cc} = y_f/2 \tag{4.44}$$

In the similar fashion, it can be shown $Y_F^{aa} = 0$. However, to calculate the off diagonal element say, Y_F^{bc}, we apply the voltage 1 *p.u.* at the phase *b* and calculate the current at the phase *C* (which is the same as earlier but its direction is away from the bus), thus we have,

$$i/v = -y_f/2 = Y_F^{bc} = Y_F^{cb} \tag{4.45}$$

However, it can be shown that

$$Y_F^{ab} = Y_F^{ac} = 0 \tag{4.46}$$

Thus, the matrix Y_F^{abc} with the relations (4.44), (4.45) and (4.46) comes out to be,

$$Y_F^{a,b,c} = (y_f/2)\begin{bmatrix} 0 & 0 & 0 \\ 0 & 1 & -1 \\ 0 & -1 & 1 \end{bmatrix} \tag{4.47}$$

where $y_f = 1/Z_f$

and $Y_{F}^{0,1,2} = T_s^{*T}\, Y_F^{abc}\, T_s$

$$= \frac{y_f}{2}\begin{bmatrix} 0 & 0 & 0 \\ 0 & 1 & -1 \\ 0 & -1 & 1 \end{bmatrix} \tag{4.48}$$

From equation (4.31), we have

$$I_{p(F)}^{0,1,2} = Y_F^{0,1,2}\,[U + Z_{PP}^{0,1,2}\, Y_F^{0,1,2}]^{-1}\, E_{p(0)}^{0,1,2} \tag{4.49}$$

Substituting equation (4.48) in equation (4.49), we get,

$$\begin{bmatrix} I_{P(F)}^0 \\ I_{P(F)}^1 \\ I_{P(F)}^2 \end{bmatrix} = \frac{y_f}{2}\begin{bmatrix} 0 & 0 & 0 \\ 0 & 1 & -1 \\ 0 & -1 & 1 \end{bmatrix}\begin{bmatrix} 1 & 0 & 0 \\ 0 & 1+Z_{PP}^1\frac{y_f}{2} & -Z_{PP}^1\frac{y_f}{2} \\ 0 & -Z_{PP}^1\frac{y_f}{2} & 1+Z_{PP}^1\frac{y_f}{2} \end{bmatrix}^{-1} \times \begin{bmatrix} 0 \\ \sqrt{3} \\ 0 \end{bmatrix} \tag{4.50}$$

and from equation (4.31) and with proper substitution, we have,

$$\begin{bmatrix} E^0_{P(F)} \\ E^1_{P(F)} \\ E^2_{P(F)} \end{bmatrix} = \begin{bmatrix} 1 & 0 & 0 \\ 0 & 1+Z^1_{PP}\frac{y_f}{2} & -Z^1_{PP}\frac{y_f}{2} \\ 0 & -Z^1_{PP}\frac{y_f}{2} & 1+Z^1_{PP}\frac{y_f}{2} \end{bmatrix} \begin{bmatrix} 0 \\ \sqrt{3} \\ 0 \end{bmatrix} \tag{4.51}$$

For the buses, other than the faulted bus p, we have from equation (4.32),

$$\begin{bmatrix} E^0_{i(F)} \\ E^1_{i(F)} \\ E^2_{i(F)} \end{bmatrix} = \begin{bmatrix} 0 \\ \sqrt{3} \\ 0 \end{bmatrix} - \begin{bmatrix} Z^0_{iP} & 0 & 0 \\ 0 & Z^1_{iP} & 0 \\ 0 & 0 & Z^2_{iP} \end{bmatrix} \begin{bmatrix} I^0_{P(F)} \\ I^1_{P(F)} \\ I^2_{P(F)} \end{bmatrix} \tag{4.52}$$

After calculating the bus voltages after the fault, line current is calculated from equation (4.33). Above equations (i.e. equations (4.50), (4.51) and (4.52)) clearly indicate that only positive sequence and negative sequence quantities flow in this case. The sequence networks will be connected as shown in Fig. 4.3(b).

4.4.3 Double Line to Ground Fault

The double line to ground fault at the bus p between any of its phases b and c with the fault impedance Z_f per phase and ground impedance Z_g is shown in Fig. 4.14.

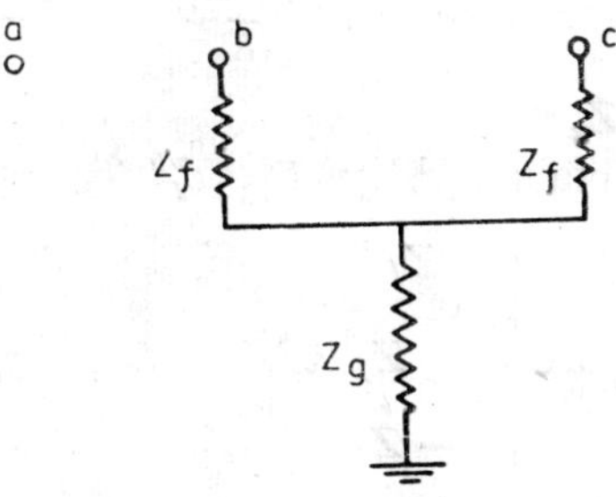

Fig. 4.14

The element $Z^{a,b,c}_F$ can be found by open circuit test, then clearly, diagonal element $Z^{aa}_F = \alpha$ but for Z^{bb}_F we apply a unit current source at the phase b and calculate its voltage keeping other phases open circuited as shown in Fig. 4.15. Then, we have

$$v = Z_f + Z_g = Z^{bb}_F = Z^{cc}_F$$

And for off diagonal element, we have $Z^{ab}_F = Z^{ac}_F = 0$ (as when we apply a current $i = 1$ p.u. at phase a, then voltages from phase b or c to ground

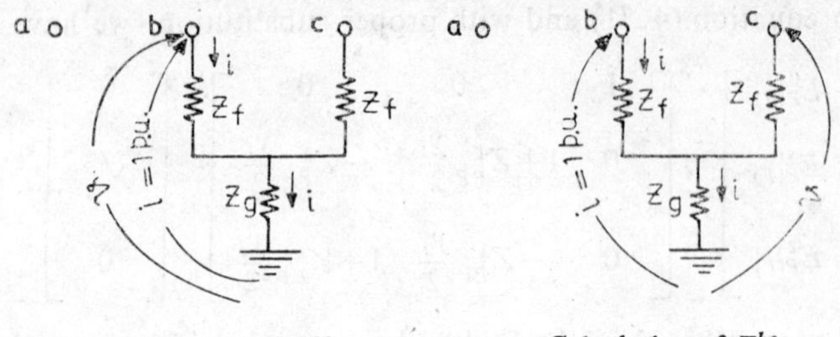

Calculation of Z_F^{bb} Calculation of Z_F^{bc}

Fig. 4.15

is zero). But for Z_F^{bc}, we apply $i = 1$ p.u. at the phase b and calculate voltage at the phase c keeping other phases open circuited, then we have $v = Z_{gi}$ but as $i = 1$ p.u. $v = Z_g = Z_F^{bc}$.

Thus we obtain,

$$Z_F^{a,b,c} = \begin{bmatrix} \alpha & 0 & 0 \\ 0 & Z_f + Z_g & Z_g \\ 0 & Z_g & Z_f + Z_g \end{bmatrix} \tag{4.53}$$

And,

$$Y_F^{abc} = Z_F^{abc^{-1}} = \begin{bmatrix} 0 & 0 & 0 \\ 0 & \dfrac{Z_f + Z_g}{K} & \dfrac{-Z_g}{K} \\ 0 & \dfrac{-Z_g}{K} & \dfrac{Z_f + Z_g}{K} \end{bmatrix} \tag{4.54}$$

where $K = Z_f^2 + 2Z_f Z_g$

From this, we obtain $Y_F^{0,1,2} = T_S^{*T} Y_F^{ab} T_S$ i.e.

$$Y_F^{0,1,2} = \frac{1}{3(Z_F^2 + 2Z_f Z_g)} \begin{bmatrix} 2Z_f & -Z_f & -Z_f \\ -Z_f & 2Z_f + 3Z_g & -(Z_f + 3Z_g) \\ -Z_f & -(Z_f + 3Z_g) & 2Z_f + 3Z_g \end{bmatrix} \tag{4.55}$$

Hence for the faulted bus p, we have from equation (4.31), after substituting $Y_F^{0,1,2}$ obtained in equation (4.55),

$$I_{P(F)}^{0,1,2} = Y_F^{0,1,2} [U + Z_{PP}^{0,1,2} \quad Y_F^{0,1,2}]^{-1} E_{P(0)}^{0,1,2} \tag{4.56}$$

and also,

$$E_{P(F)}^{0,1,2} = [U + Z_{PP}^{0,1,2} \quad Y_F^{0,1,2}]^{-1} E_{P(0)}^{0,1,2} \tag{4.57}$$

and for other buses,

$$E_{i(F)}^{0,1,2} = \begin{bmatrix} 0 \\ \sqrt{3} \\ 0 \end{bmatrix} - Z_{iP}^{0,1,2} I_{P(F)}^{0,1,2} \tag{4.58}$$

where

$$E_{P(0)}^{0,1,2} = \begin{bmatrix} 0 \\ \sqrt{3} \\ 0 \end{bmatrix}$$

In this case, we will have all the three sequence networks viz. positive, negative and zero sequence network connected in parallel as shown in Fig. 4.4(b).

4.4.4 1-phase (i.e. Line) to Ground Fault

Fault in phase '*a*' through Z_f is shown in Fig. 4.16. For calculating diagonal element Z_F^{aa} of the bus impedance matrix Z_F^{abc}, we apply $i = 1$ p.u. at phase a and calculate voltage at the phase a keeping other phases open circuited as shown in Fig. 4.16. Then we have,

$$v = Z_f i$$

$$v/i = Z_f \text{ but since } i = 1 \text{ p.u.}$$

$$v = Z_f = Z_F^{aa}$$

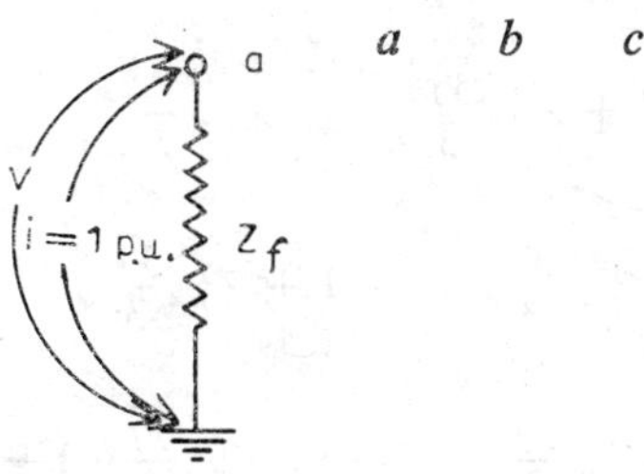

Fig. 4.16 Single line to ground fault at the phase *a*.

Following the same reasoning, for phase b and c, $Z_F^{bb} = Z_F^{cc} = \alpha$. All the off-diagonal elements calculated by open circuit test, will be,

$$Z_F^{ab} = Z_F^{bc} = Z_F^{ca} = 0$$

Thus the bus impedance matrix Z_F^{abc} comes out to be,

$$Z_F^{a,b,c} = \begin{bmatrix} Z_f & 0 & 0 \\ 0 & \alpha & 0 \\ 0 & 0 & \alpha \end{bmatrix} \tag{4.59}$$

and

$$Y_F^{abc} = Z_F^{abc^{-1}} = \begin{bmatrix} y_f & 0 & 0 \\ 0 & 0 & 0 \\ 0 & 0 & 0 \end{bmatrix} \tag{4.60}$$

and also $$Y_F^{0,1,2} = T_S^{*T} Y_F^{abc} T_S$$

$$= \frac{y_f}{3} \begin{bmatrix} 1 & 1 & 1 \\ 1 & 1 & 1 \\ 1 & 1 & 1 \end{bmatrix} \tag{4.61}$$

Using the same equation (4.31) for the faulted bus p, we have after the substitution of the equations (4.61),

$$\begin{bmatrix} I^0_{P(F)} \\ I^1_{P(F)} \\ I^2_{P(F)} \end{bmatrix} = \frac{y_f}{3} \begin{bmatrix} 1 & 1 & 1 \\ 1 & 1 & 1 \\ 1 & 1 & 1 \end{bmatrix} \begin{bmatrix} 1 + \frac{y_f}{3} Z^0_{PP} & \frac{y_f}{3} Z^0_{PP} & \frac{y_f}{3} Z^0_{PP} \\ \frac{y_f}{3} Z^1_{PP} & 1 + \frac{y_f}{3} Z^1_{PP} & \frac{y_f}{3} Z^1_{PP} \\ \frac{y_f}{3} Z^1_{PP} & \frac{y_f}{3} Z^1_{PP} & 1 + \frac{y_f}{3} Z^1_{PP} \end{bmatrix}^{-1} \times \begin{bmatrix} 0 \\ \sqrt{3} \\ 0 \end{bmatrix} \tag{4.62}$$

$$\begin{bmatrix} E^0_{P(F)} \\ E^1_{P(F)} \\ E^2_{P(F)} \end{bmatrix} = \begin{bmatrix} 1 + Z^0_{PP} \frac{y_f}{3} & Z^0_{PP} \frac{y_f}{3} & Z^0_{PP} \frac{y_f}{3} \\ Z^1_{PP} \frac{y_f}{3} & 1 + Z^1_{PP} \frac{y_f}{3} & Z^1_{PP} \frac{y_f}{3} \\ Z^2_{PP} \frac{y_f}{3} & Z^1_{PP} \frac{y_f}{3} & 1 + Z^1_{PP} \frac{y_f}{3} \end{bmatrix}^{-1} \begin{bmatrix} 0 \\ \sqrt{3} \\ 0 \end{bmatrix} \tag{4.63}$$

The voltages at the other buses are obtained from equation (4.32) i.e.

$$\begin{bmatrix} E^0_{i(F)} \\ E^1_{i(F)} \\ E^2_{i(F)} \end{bmatrix} = \begin{bmatrix} 0 \\ \sqrt{3} \\ 0 \end{bmatrix} - \begin{bmatrix} Z^0_{iP} & 0 & 0 \\ 0 & Z^1_{iP} & 0 \\ 0 & 0 & Z^1_{iP} \end{bmatrix} \begin{bmatrix} I^0_{P(F)} \\ I^1_{P(F)} \\ I^2_{P(F)} \end{bmatrix} \tag{4.64}$$

4.4.5 *Three-phase short circuit without ground with fault impedance Z_f per phase*

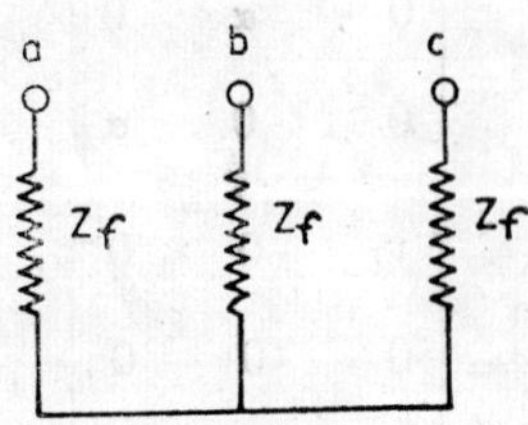

Fig. 4.17 3-phase fault.

The bus impedance matrix Z_F^{abc} is not defined.

Now to obtain $Y_F^{a,b,c}$ we do the short circuit test as shown in Fig. 4.18.

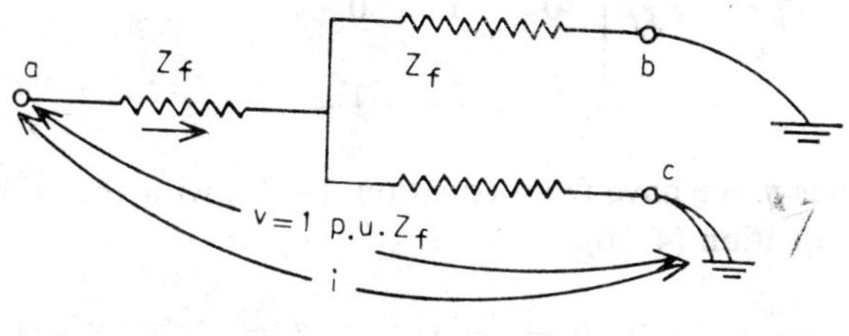

Fig. 4.18

Referring to Fig. 4.18, to obtain diagonal elements of the bus admittance Y_F^{abc} say Y_F^{aa}, we apply a unit voltage source at the phase a and calculate its current keeping other phases short circuited, then we have,

$$v = \left(Z_f + \frac{1}{\frac{1}{Z_f} + \frac{1}{Z_f}}\right) i = \frac{3}{2} Z_f i$$

i.e.
$$i/v = 2/3Z_f = \frac{2}{3} y_f; \quad \text{where} \quad y_f = \frac{1}{Z_f}$$

But since
$$v = 1 \text{ p.u.}; \quad Y_F^{aa} = \frac{2}{3} y_f$$

Thus all the diagonal elements

$$Y_F^{aa} = Y_F^{bb} = Y_F^{cc} = \frac{2}{3} y_f$$

But for the off diagonal elements say Y_F^{ab}, we apply a unit voltage of 1 p.u. at the phase a and calculate current at the phase b keeping other phases short circuited. In this case we have

$$i/v = -\tfrac{1}{3} y_f$$

But since
$$v = 1 \text{ p.u.}$$

$$Y_F^{ab} = -\tfrac{1}{3} y_f$$
$$= Y_F^{bc} = Y_F^{ac} \quad \text{etc.}$$

Thus, we have

$$Y_F^{a,b,c} = \frac{y_f}{3}\begin{bmatrix} 2 & -1 & -1 \\ -1 & 2 & -1 \\ 1 & -1 & 2 \end{bmatrix} \tag{4.65}$$

and with transformation,

$$Y_F^{0,1,2} = y_f \begin{bmatrix} 0 & 0 & 0 \\ 0 & 1 & 0 \\ 0 & 0 & 1 \end{bmatrix} \tag{4.66}$$

For faulted bus p, we have from equation (4.31) with the substitution of $Y_F^{0,1,2}$ from equation (4.66),

$$\begin{bmatrix} I^0_{P(F)} \\ I^1_{P(F)} \\ I^2_{P(F)} \end{bmatrix} = y_f \begin{bmatrix} 0 & 0 & 0 \\ 0 & 1 & 0 \\ 0 & 0 & 1 \end{bmatrix} \begin{bmatrix} 1 & 0 & 0 \\ 0 & 1+Z^1_{PP}y_f & 0 \\ 0 & 0 & 1+Z^1_{PP}y_f \end{bmatrix}^{-1} \begin{bmatrix} 0 \\ \sqrt{3} \\ 0 \end{bmatrix} \tag{4.67}$$

and also

$$\begin{bmatrix} E^0_{P(F)} \\ E^1_{P(F)} \\ E^2_{P(F)} \end{bmatrix} = \begin{bmatrix} 1 & 0 & 0 \\ 0 & 1+Z^1_{PP}y_f & 0 \\ 0 & 0 & 1+Z^1_{PP}y_f \end{bmatrix}^{-1} \begin{bmatrix} 0 \\ \sqrt{3} \\ 0 \end{bmatrix} \tag{4.68}$$

The bus voltages for the other buses from equation (4.32)

$$\begin{bmatrix} E^0_{i(F)} \\ E^1_{i(F)} \\ E^2_{i(F)} \end{bmatrix} = \begin{bmatrix} 0 \\ \sqrt{3} \\ 0 \end{bmatrix} - \begin{bmatrix} Z^0_{iP} & 0 & 0 \\ 0 & Z^1_{iP} & 0 \\ 0 & 0 & Z^1_{iP} \end{bmatrix} \begin{bmatrix} I^0_{P(F)} \\ I^1_{P(F)} \\ I^2_{P(F)} \end{bmatrix} \tag{4.69}$$

After calculating bus voltages after the fault, the fault currents in the element of the power system network is calculated by equation (4.33) i.e.

$$i^{abc}_{ij(F)} = y^{abc}_{ij\rho\sigma} \quad (E^{abc}_{\rho(F)} - E^{abc}_{\sigma(F)})$$

where $\rho\sigma$ is the element $i-j$ as well as those elements which are coupled to the element $i-j$ of the primitive network and $E^{abc}_{\rho(F)}$ etc. are the bus voltages after the fault.

Fault impedances/admittance matrices in the bus frame of reference which have been derived by open circuit or short circuit test are tabulated below.

Type of fault	Bus impedance matrix $Z_F^{0,1,2}$	Bus admittance matrix $Y_F^{0,1,2}$
3-phase fault to ground (eqn. 4.37)	$\begin{bmatrix} (Z_f+3Z_g) & 0 & 0 \\ 0 & Z_f & 0 \\ 0 & 0 & Z_f \end{bmatrix}$	
Line to line fault (eqn. 4.48)	...	$\frac{y_f}{2}\begin{bmatrix} 0 & 0 & 0 \\ 0 & 1 & -1 \\ 0 & -1 & 1 \end{bmatrix}$
Double line to ground fault (eqn. 4.55)	...	$\frac{1}{3(Z_f^2+2Z_fZ_g)}\begin{bmatrix} 2Z_f & -Z_f & -Z_f \\ -Z_f & 2Z_f+3Z_g & -(Z_f+3Z_g) \\ -Z_f & -(Z_f+3Z_g) & 2Z_f+3Z_g \end{bmatrix}$
Single line to ground fault (eqn. 4.61)	...	$y_f/3\begin{bmatrix} 1 & 1 & 1 \\ 1 & 1 & 1 \\ 1 & 1 & 1 \end{bmatrix}$
3-phase short circuit without ground (eqn. 4.66)	...	$y_f\begin{bmatrix} 0 & 0 & 0 \\ 0 & 1 & 0 \\ 0 & 0 & 1 \end{bmatrix}$

Now let us take in faults which are direct short circuit, i.e. when Z_f is zero. These are known as bolted fault.

4.5 Direct Short Circuit i.e. Bolted Fault

Here Z_f is zero i.e. Z_f and Y_f are undefined.

4.5.1 Single line to ground fault

The single line to ground fault in any phase *a* without any fault impedance is shown in Fig. 4.19.

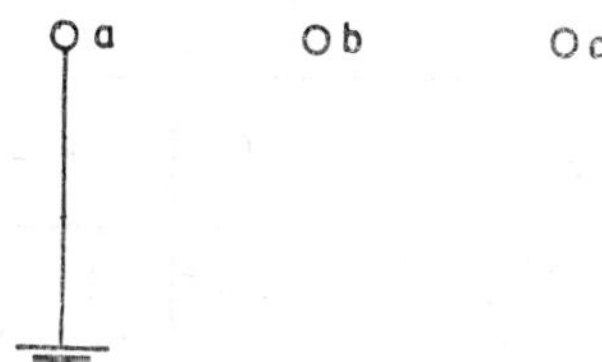

Fig. 4.19 Single line to ground fault in the phase *a*.

We have from equation (4.3),

$$E_{BUS\,(F)}^{abc} = E_{BUS\,(0)}^{abc} - Z_{BUS}^{abc}\, I_{BUS\,(F)}^{abc}$$

$$\begin{bmatrix} E_{P\,(F)}^{a} \\ E_{P\,(F)}^{b} \\ E_{P\,(F)}^{c} \end{bmatrix} = \begin{bmatrix} 1 \\ \alpha^2 \\ \alpha \end{bmatrix} - \begin{bmatrix} Z_{11} & Z_{12} & Z_{13} \\ Z_{21} & Z_{22} & Z_{23} \\ Z_{31} & Z_{32} & Z_{33} \end{bmatrix} \begin{bmatrix} I_{P\,(F)}^{a} \\ I_{P\,(F)}^{b} \\ I_{P\,(F)}^{c} \end{bmatrix} \tag{4.70}$$

From boundary conditions, we have

$$\begin{aligned} E_{P\,(F)}^{a} &= 0 \\ I_{P(F)}^{b} &= 0 \\ I_{P\,(F)}^{c} &= 0 \end{aligned} \tag{4.71}$$

Substituting the equation (4.71) in the equation (4.70), we get,

$$\begin{bmatrix} 0 \\ E_{P(F)}^{b} \\ E_{P\,(F)}^{c} \end{bmatrix} = \begin{bmatrix} 1 \\ \alpha^2 \\ \alpha \end{bmatrix} - \begin{bmatrix} Z_{11} & Z_{12} & Z_{13} \\ Z_{21} & Z_{22} & Z_{23} \\ Z_{31} & Z_{32} & Z_{33} \end{bmatrix} \begin{bmatrix} I_{P\,(F)}^{a} \\ 0 \\ 0 \end{bmatrix} \tag{4.72}$$

Thus we get from above, equation (i.e. equation (4.72))

$$\begin{aligned} 1 &= Z_{11}\, I_{P(F)}^{a} \\ E_{P(F)}^{b} &= \alpha^2 - Z_{21}\, I_{P(F)}^{a} \\ E_{P(F)}^{c} &= \alpha - Z_{31}\, I_{P(F)}^{a} \end{aligned} \tag{4.73}$$

Here

$$\begin{aligned} Z_{11} &= Z_{PP}^{aa} \\ Z_{12} &= Z_{PP}^{ab} = Z_{21} \\ Z_{13} &= Z_{PP}^{ac} = Z_{31} \quad \text{etc.} \end{aligned}$$

Similarly we obtain,

$$\bar{E}_{i(F)} = \bar{E}_{i(0)} - (Z_{iP})\, \bar{I}_{P(F)} \quad \text{for } i = 1, \ldots, n \quad i \neq p \tag{4.74}$$

4.5.2 *Line to Line Fault*

Let us take line to line fault between any two phases *b* and *c* without fault impedance as shown in Fig. 4.20.

Now from boundary conditions, we have

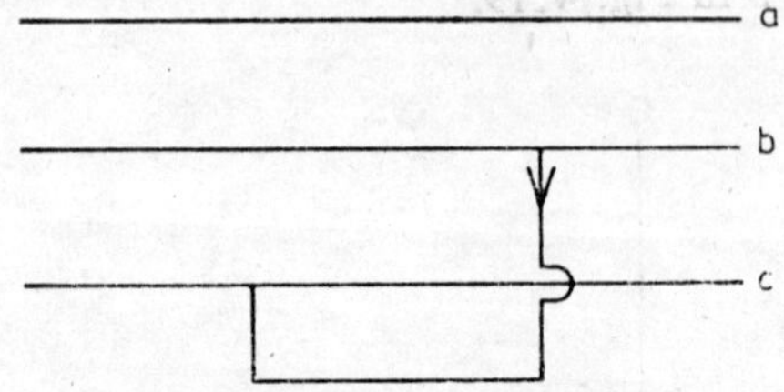

Fig. 4.20 Line to line fault.

$$E^b_{P(F)} = E^c_{P(F)}$$

$$I^b_{P(F)} = -I^c_{P(F)}$$

$$I^a_{P(F)} = 0 \tag{4.75}$$

We have from equation (4.3), for any bus p,

$$\bar{E}_{P(F)} = \bar{E}_{P(0)} - [Z_{PP}]\, \bar{I}_{P(F)} \tag{4.76}$$

Substituting the boundary conditions equation (4.75) in equation (4.76),

$$\begin{bmatrix} E^a_{P(F)} \\ E^c_{P(F)} \\ E^c_{P(F)} \end{bmatrix} = \begin{bmatrix} 1 \\ \alpha^2 \\ \alpha \end{bmatrix} - \begin{bmatrix} Z_{11} & Z_{12} & Z_{13} \\ Z_{21} & Z_{22} & Z_{23} \\ Z_{31} & Z_{32} & Z_{33} \end{bmatrix} \begin{bmatrix} 0 \\ -I^c_{P(F)} \\ I^c_{P(F)} \end{bmatrix} \tag{4.77}$$

where

$$Z_{11} = Z^{aa}_{PP}$$

$$Z_{12} = Z^{ab}_{PP}$$

$$Z_{13} = Z^{ac}_{PP} \text{ etc.}$$

4.5.3 *Double Line to Ground Fault*

The boundary conditions in this case are

$$I^a_{P(F)} = 0$$

$$E^b_{P(F)} = E^c_{P(F)} = 0 \tag{4.78}$$

Now the equation for the faulted bus p,

$$\bar{E}_{P(F)} = \bar{E}_{P(0)} - [Z_{PP}]\, \bar{I}_{P(F)}$$

Substituting the boundary conditions (eqn. 4.78), in the above equation, we have

$$\begin{bmatrix} E^a_{P(F)} \\ 0 \\ 0 \end{bmatrix} = \begin{bmatrix} 1 \\ \alpha^2 \\ \alpha \end{bmatrix} - \begin{bmatrix} Z_{11} & Z_{12} & Z_{13} \\ Z_{21} & Z_{22} & Z_{23} \\ Z_{31} & Z_{32} & Z_{33} \end{bmatrix} \begin{bmatrix} 0 \\ I^b_{P(F)} \\ I^c_{P(F)} \end{bmatrix} \tag{4.79}$$

Thus after the faults, we calculate the bus voltages and line currents. As discussed, these are useful in the design of the protective schemes of the power system.

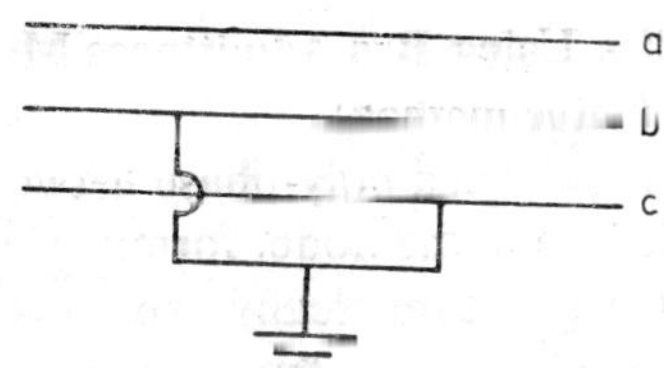

Fig. 4.21 Double line to ground fault.

4.6 Comparison between Symmetrical Components and Phase Coordinate Method of Short Circuit Studies

We have discussed in the previous sections the formulation of power system problem for the purpose of short circuit studies i.e. fault analysis using the technique of symmetrical components. As a matter of fact, most present day fault analysis of power system networks depend upon the use of symmetrical components theory. The method has been programmed for the analysis of a large number of common fault conditions and used successfully for many years. This is because the formulation in the field of components is quite convenient and useful in the cases where the network is symmetric i.e. balanced, only the excitation may become unbalanced after occurrence of unsymmetrical faults. In such situations as stated earlier, with the help of symmetrical components transformations, the field of phasors are transformed to the field of components because after transformations, we arrive at uncoupled set of equations. The method of symmetrical components is certainly straightforward and the solution in this case is obtained quite fast.

However, in cases where network itself is unbalanced prior to the fault, even after the transformation to the component quantities, equations remain coupled as they were before the transformations. Examples of such problems are the analysis of untransposed transmission lines, and analysis of other unbalanced system conditions such as analysis of unbalanced loads, single phase loading, single pole switching, bundled conductors, simultaneous faults such as cross country faults or breaker operation as each phase clears in turn and two or four phases supply. The use of method of symmetrical components in such cases is either unwieldy or cumbersome. However, by representing the condition of power system network in phase coordinates i.e. in phase voltage, phase current and in phase imp/adm. thereby preserving the physical identity of the system instead of transferring the phasor coordinates to the symmetrical components coordinates, these special problems as listed above can be represented along with the usual fault condition with equal facility. Using the system representation in the phase reference frame, a generalized analysis of power system network under fault conditions can be developed. The only drawback in the analysis of phasor coordinates is that it is time consuming i.e., solution in this case is not obtained straightforward like the method of symmetrical components, rather it takes several iterations to arrive at the solution point.

4.7 Short Circuit Studies Using Bus Admittance Matrix (i.e. by phase coordinates method)

Fault analysis of unbalanced poly-phase networks by the method of phase coordinates is based upon the nodel formulation and thus in this case the sparsity of Y_{BUS} matrix is automatically exploited. This method can be applied with ease both for complex faults as well as for simple faults regardless of the degree of unbalance in the system or the number or types of unbalanced faults.

In the phase coordinate method, polyphase network conditions are represented by phase voltages, currents, impedance/admittances, thereby preserving the physical identity of the system instead of transferring the phase coordinates to the symmetrical components coordinates.

4.7.1 *Generator Representation with Earthed Neutral*

Figure 4. 21a illustrates schematically a general 3-phase circuit element containing sources of e.m.f. in each phases, and self and mutual impedances in and between phases, which can be summarised by the matrix elements of the series impedance matrix Z_{abc}. This matrix may be computed directly from the basic element data and geometry, or by transformation of previously computed symmetrical component impedance quantities, e.g.

$$Z_{abc} = \tfrac{1}{3} T Z_{012} T^* \tag{4.80}$$

where

$$T = \begin{bmatrix} 1 & 1 & 1 \\ 1 & \alpha^2 & \alpha \\ 1 & \alpha & \alpha^2 \end{bmatrix} \tag{4.81}$$

and $\alpha = 1\angle 120°$.

For the current and voltage reference directions shown, the nodal equations for the currents injected into nodes 1, 2, ..., 6 are

$$\begin{bmatrix} I_1 \\ I_2 \\ I_3 \\ I_4 \\ I_5 \\ I_6 \end{bmatrix} = \left[\begin{array}{c|c} Y_{abc} + Y_{123}\ \text{shunt} & -Y_{abc} \\ \hline -Y_{abc} & Y_{abc} + Y_{456}\ \text{shunt} \end{array}\right] \begin{bmatrix} V_1 \\ V_2 \\ V_3 \\ V_4 \\ V_5 \\ V_6 \end{bmatrix} + \left[\begin{array}{c|c} -Y_{abc} & \\ \hline & Y_{abc} \end{array}\right] \begin{bmatrix} E_a \\ E_b \\ E_c \\ E_a \\ E_b \\ E_c \end{bmatrix} \tag{4.82}$$

If any of the nodes are short-circuited together, then the equations are modified accordingly by the appropriate row and column additions. If nodes 4, 5 and 6 are joined to form a neutral point N of three machine windings $N-1$, $N-2$, $N-3$ say, then the six equations from eqn. (4.82) are reduced to four equations by the summation of rows 4, 5, and 6 $(I_N = I_4 + I_5 + I_6)$ with a further summation of columns

$$(V_N = V_4 = V_5 = V_6).$$

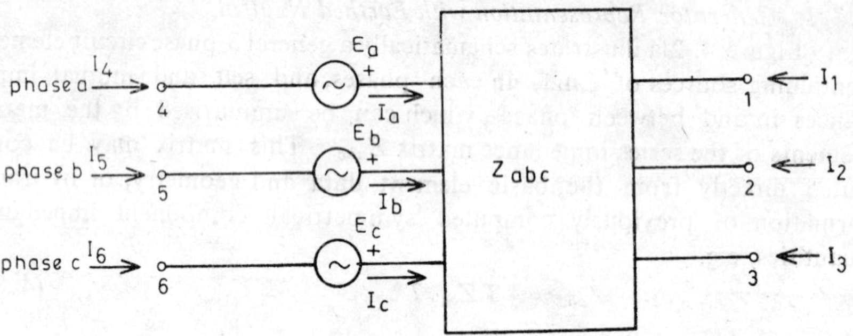

Fig. 4.21a General 3-phase system element.

For normal balanced machine designs where the machine may be adequately represented in the symmetrical-component analysis by the three uncoupled sequence impedance z_0, z_1 and z_2, or admittances y_0, y_1 and y_2, the phase admittance matrix Y_{abc} may be constructed from the sequence admittances from the relationships $Y_{abc} = \frac{1}{3}$.

$$\begin{bmatrix} y_0 + y_1 + y_2 & y_0 + \alpha y_1 + \alpha^2 y_2 & y_0 + \alpha^2 y_1 + \alpha y_2 \\ y_0 + \alpha^2 y_1 + \alpha y_2 & y_0 + y_1 + y_2 & y_0 + \alpha y_1 + \alpha^2 y_2 \\ y_0 + \alpha y_1 + \alpha^2 y_2 & y_0 + \alpha^2 y_1 + \alpha y_2 & y_0 + y_1 + y_2 \end{bmatrix}$$

$$= \begin{bmatrix} Y_{11} & Y_{12} & Y_{13} \\ Y_{21} & Y_{22} & Y_{33} \\ Y_{31} & Y_{32} & Y_{33} \end{bmatrix} \quad (4.83)$$

The equations of a star-connected machine with neutral N are thus, from eqns. (4.82) and (4.83)

$$\begin{bmatrix} I_1 \\ I_2 \\ I_3 \\ I_N \end{bmatrix} = \begin{bmatrix} Y_{11} & Y_{12} & Y_{13} & -y_0 \\ Y_{21} & Y_{22} & Y_{23} & -y_0 \\ Y_{31} & Y_{32} & Y_{33} & -y_0 \\ -y_0 & -y_0 & -y_0 & 3y_0 \end{bmatrix} \begin{bmatrix} V_1 \\ V_2 \\ V_3 \\ V_N \end{bmatrix}$$

$$+ \begin{bmatrix} -y_{11} & -y_{12} & -y_{13} & 0 \\ -Y_{21} & -Y_{22} & -Y_{23} & 0 \\ -Y_{31} & -Y_{32} & -Y_{33} & 0 \\ 0 & 0 & 0 & y_0 \end{bmatrix} \begin{bmatrix} E_1 \\ E_2 \\ E_3 \\ E_N \end{bmatrix} \quad (4.84)$$

where $E_1 = E_a$, $E_2 = E_b$, $E_3 = E_c$

and $E_N = E_1 + E_2 + E_3 = 0$ (4.85)

for balanced e.m.f.s per phase.

If the neutral is earthed solidly, the equation for the neutral node in eqn. (4.84) may be removed or included for solution subject to the constraint $V_N = 0$. If the neutral is earthed through an admittance Y_N, Y_N appears as a normal shunt admittance term in the element Y_{44} or Y_{NN}, which becomes $(3y_0 + Y_N)$.

The currents I_1, I_2, I_3 and I_N represent the net currents injected into the generator terminals; hence the currents injected into network, i.e. those leaving the generator terminals 1, 2, 3 and N, are $-I_1$, $-I_2$, $-I_3$ and $-I_N$, respectively. With an impedance earthed neutral, I_N is usually zero, the current to earth in the neutral connection being $Y_N V_N$; with a solidly earthed neutral, the current to earth is

$$I_N - y_0(V_1 + V_2 + V_3) \tag{4.86}$$

Finally, it is noted that if E_1, E_2 and E_3 are a balanced set of voltages, as is usually the case, then $E_1 = \alpha E_2 = \alpha^2 E_3$ and eqn. (4.84) reduce, with the sign change above noted, to the form for the currents injected into the network:

$$\begin{bmatrix} I_1 \\ I_2 \\ I_3 \\ I_N \end{bmatrix} = -\begin{bmatrix} Y_{11} & Y_{12} & Y_{13} & -y_0 \\ Y_{21} & Y_{22} & Y_{23} & -y_0 \\ Y_{31} & Y_{32} & Y_{33} & -y_0 \\ -y_0 & -y_0 & -y_0 & -y_{NN} \end{bmatrix} \begin{bmatrix} V_1 \\ V_2 \\ V_3 \\ V_N \end{bmatrix} + \begin{bmatrix} y_1 E_1 \\ y_1 E_2 \\ y_1 E_3 \\ -y_0 E_N \end{bmatrix} \tag{4.87}$$

where I_N is the current injected into the neutral node N, and is usually zero, E_N is the sum of the phase e.m.f.s, and is also usually zero, $(E_1 + E_2 + E_3 = 0)$, y_0 is the machine zero-sequence admittance and y_1 the positive-sequence admittance. The values of the voltage sources E_1, E_2 and E_3 are the appropriate phase-displaced transient or subtransient values according to the purpose of the study.

4.7.2 *Solution Technique*

The phase coordinate method consists of the following principal steps:

(a) system representation in phase frame of reference.
(b) assembly of the system nodal admittance matrix.
(c) modification of the system nodal admittance matrix to include, changes in the system configuration.
(d) formation and solution of nodal performance equation.

The general form of nodal admittance equation which corresponds to the system representation in phase frame of reference is as shown.

$$YV = \mathbf{I} \tag{4.88}$$

Equation (4.88) may be used to describe the 3-phase system where each bus bar in one line diagram of the balanced system is replaced by three equivalent separate-phase bus bars. Each voltage and current in equation (4.88) for the balanced system is replaced corresponding by three phase to earth voltages and three phase currents, with each element of the nodal admittance matrix being replaced by a 3-phase element represented by 3×3 nodal admittance submatrix.

The phase relationships at each bus-bar are, then say, for at any bus p, for example,

$$I_p = S_p^* / V_p \tag{4.89}$$

Equation (4.88) can be adjusted to the basic prefault form from which

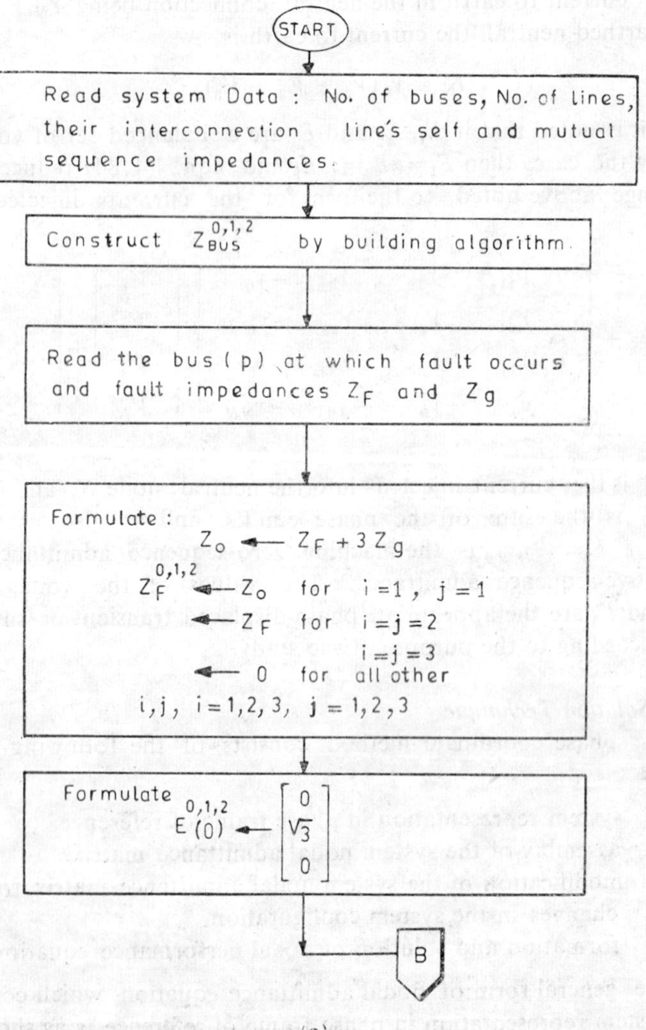

(a)

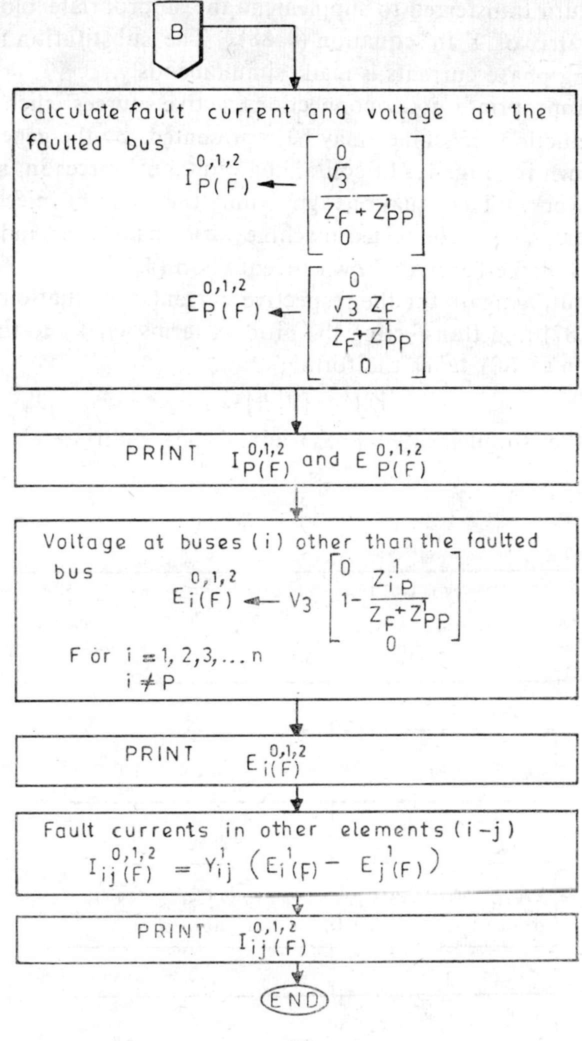

(b)

Fig. 4.22 Flow diagram for 3-phase-ground fault.

calculations for various faults commence by classifying the energy sources as active or passive, according to their behaviour during fault conditions. Some loads may be classified by passive admittances per phase, thus equation (4.89) for qth node becomes,

$$I_q = y'_{qo} V_q \tag{4.90}$$

Equation (4.88) may accordingly be modified by substituting load currents using equation (4.90) and then transferring the admittances across to supplement the diagonal elements of Y. If the load has unequal positive, negative and zero sequence admittances, the admittance Y'_{qo} in equation (4.90) may be replaced by an equivalent 3×3 phase admittance matrix which is in turn transferred to supplement the appropriate block diagonal 3×3 submatrix of Y in equation (4.88). The substitution for the three corresponding phase currents is made simultaneously.

With appropriate node connections, active sources such as synchronous and induction machines may be represented by the general network element shown in Fig. 4.21a containing voltage sources in series with a passive network. The equations governing the current injected into the network from a star connected machine with neutral grounded through impedance is of the form as shown in equation (4.87).

Substituting again for the respective currents in equation (4.88) from equation (4.87) and transferring the product terms $y_{ij} V_j$ to the left-hand-side, equation (4.88) takes the form

$$Y'V = [y_g E] = I' \tag{4.91}$$

where Y' is the supplemented phase admittance matrix and $[y_g E]$, i.e. I'

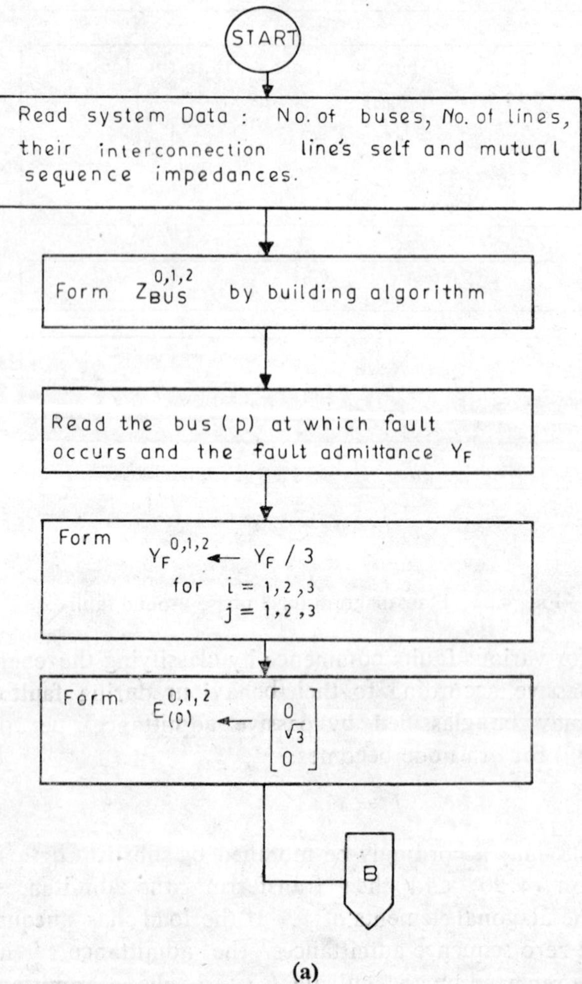

(a)

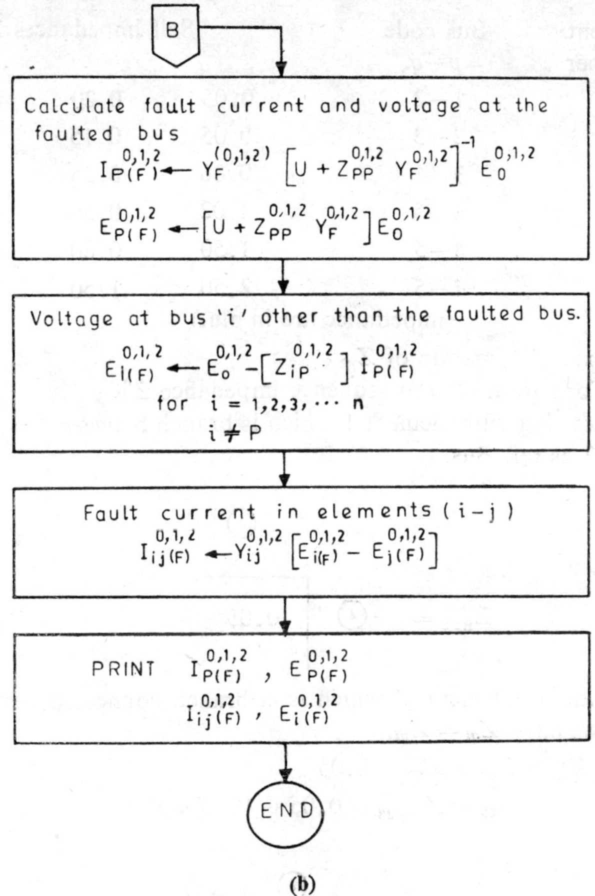

(b)

Fig. 4.23 Flow diagram for 1-phase-ground fault.

is a column matrix whose elements are of the form $y_g E$ or are zero. After this the usual fault analysis proceeds.

4.8 Numerical Example

For the sample network shown below, perform the short circuit studies and determine the bus voltage after the fault, line flow and fault level for (a) 3-phase to ground fault and (b) 1-phase to ground fault at bus (Assume direct short circuit) (5).

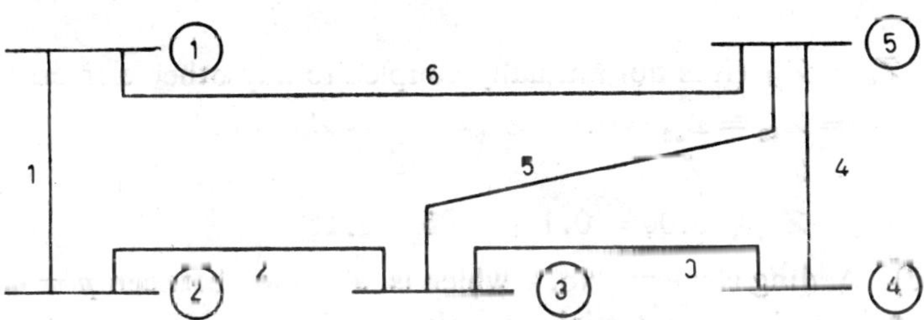

Element Number	Bus code $p-q$	Self impedances $Z_{pq}^{0,1,2}$		
1	1−2	0.05	0.20	0.20
2	2−3	0.05	0.15	0.15
3	3−4	0.06	0.25	0.25
4	4−5	1.02	0.50	0.50
5	3−5	1.50	0.80	0.80
6	1−5	2.50	1.50	1.50

(impedance are in p.u.)

Solution: Formation of Z_{BUS}

(1) Formation of zero sequence impedance Z_{BUS}^0

(a) Starting with element 1 which is branch between $p = 1$ to $q = 2$. Taking bus 1 as ref. Bus.

$$Z_{BUS} = \begin{array}{c|c} & ② \\ \hline ② & 0.05 \end{array}$$

(b) Adding element 2 which is a branch connected between $p = 2$ and $q = 3$, we have $Z_{qi} = Z_{pi}$.

$$Z_{23} = Z_{32} = Z_{22} = 0.05$$

$$Z_{33} = Z_{23} + Z_{23-23} = 0.05 + 0.05 = 0.10$$

$$Z_{BUS} = \begin{array}{c|cc} & ② & ③ \\ \hline ② & 0.05 & 0.05 \\ ③ & 0.05 & 0.10 \end{array}$$

(c) Adding element No. 3 which is a branch connected between $p =$ ③ and $q =$ ④

$Z_{qi} = Z_{pi}$ (it is not mutually coupled to any other element)

$$Z_{24} = Z_{42} = Z_{32} = 0.05, \; Z_{34} = Z_{43} = Z_{33} = 0.10$$

$$Z_{qq} = Z_{pq} + Z_{pq\,pq}$$

$$Z_{44} = Z_{34} + 0.06 = 0.1 + 0.06 = 0.16$$

(d) Adding element No. 4 which is a branch between $p = 4$, $q = 5$ and is uncoupled to the partial network.

$$Z_{52} = Z_{42} = 0.05;\ Z_{53} = Z_{43} = 0.10;\ Z_{54} = Z_{44} = 0.16$$

$$Z_{55} = Z_{45} + 1.02 = 0.16 + 1.02 = 1.18.$$

$Z_{BUS} =$

	②	③	④	⑤
②	0.05	0.05	0.05	0.05
③	0.05	0.10	0.10	0.10
④	0.05	0.10	0.16	0.16
⑤	0.05	0.10	0.16	0.18

(e) Adding element 5 which is a link connected between buses $p =$ ③ and $q =$ ⑤ and it is uncoupled.

$$Z_{1i} = Z_{pi} - Z_{qi};\ Z_{12} = Z_{32} - Z_{52} = 0.05 - 0.05 = 0;$$

$$Z_{13} = Z_{33} - Z_{53} = 0.10 - 0.10 = 0;$$

$$Z_{14} = Z_{34} - Z_{54} = 0.10 - 0.16 = -.06;$$

$$Z_{15} = Z_{35} - Z_{55} = 0.10 - 1.18 = -1.08;\ Z_{11} = Z_{p1} - Z_{q1} + Z_{pq\,pq};$$

$$Z_{11} = Z_{31} - Z_{51} + 1.50 = 2.58.$$

Therefore,

$$Z'_{22} = 0.05 - \frac{0 \times 0}{2.58} = 0.05:\ Z'_{23} = 0.05;\ Z'_{24} = 0.05;\ Z'_{25} = 0.05;$$

$$Z'_{33} = 0.10;\ Z'_{34} = 0.1;\ Z'_{35} = 0.1;\ Z'_{44} = 0.16 - \frac{0.06 \times 0.06}{2.58} = 0.1586$$

$$Z'_{45} = 0.16 - \frac{0.06 \times 1.08}{2.58} - 0.1348;\ Z'_{55} - 1.18\ \frac{1.08 \times 1.08}{2.58} - 0.7279$$

$$Z_{BUS} = \begin{array}{c|cccc} & (2) & (3) & (4) & (5) \\ \hline (2) & 0.05 & 0.05 & 0.05 & 0.05 \\ (3) & 0.05 & 0.10 & 0.10 & 0.10 \\ (4) & 0.05 & 0.10 & 0.1586 & 0.1348 \\ (5) & 0.05 & 0.10 & 0.1348 & 0.7279 \end{array}$$

(f) Adding element 6 connected between,

$p = 1$ to $q = 5$.

$Z_{1i} = -Z_{qi}$; $Z_{12} = -Z_{52} = -0.05$; $Z_{13} = -Z_{53} = -0.1$;

$Z_{14} = -Z_{54} = -0.1348$; $Z_{15} = -Z_{55} = -0.7277$

$Z_{11} = -Z_{q1} + Z_{pq\,pq} = -Z_{51} + 2.5 = 0.7279 + 2.5 = 3.2279$

$$Z'_{22} = 0.05 - \frac{0.05 \times 0.05}{3.2279} = 0.04922$$

$$Z'_{23} = 0.05 - \frac{0.05 \times 0.10}{3.2279} = 0.04845$$

$$Z'_{24} = 0.05 - \frac{0.05 \times 0.1348}{3.2279} = 0.04791$$

$$Z'_{25} = 0.05 - \frac{0.05 \times 0.7279}{3.2279} = 0.03872$$

$$Z'_{33} = 0.1 - \frac{0.1 \times 0.1}{3.2279} = 0.0969$$

$$Z'_{34} = 0.1 - \frac{0.1 \times 0.1348}{3.2279} = 0.09582$$

$$Z'_{35} = 0.1 - \frac{0.1 \times 7279}{3.2279} = 0.07745$$

$$Z'_{44} = 0.1586 - \frac{0.1348 \times 0.1348}{3.2279} = 0.15297$$

$$Z'_{45} = 0.1348 - \frac{0.1348 \times 0.7279}{3.2279} = 0.1044$$

$$Z'_{55} = 0.7279 - \frac{0.7279 \times 0.7279}{3.2279} = 0.5637$$

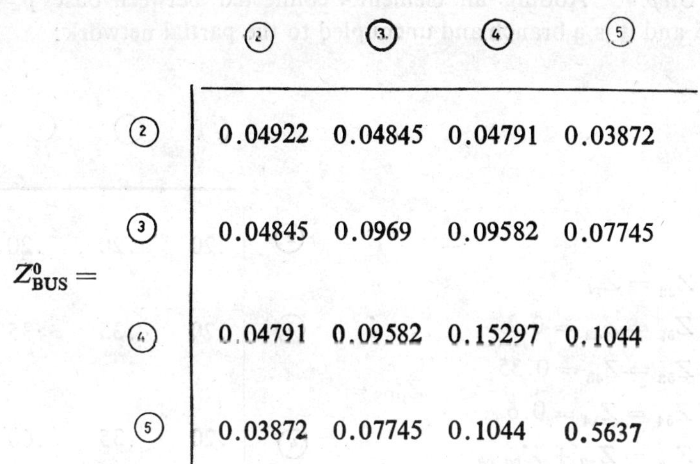

$Z^0_{BUS} =$

	②	③	④	⑤
②	0.04922	0.04845	0.04791	0.03872
③	0.04845	0.0969	0.09582	0.07745
④	0.04791	0.09582	0.15297	0.1044
⑤	0.03872	0.07745	0.1044	0.5637

(2) Formation of positive sequence impedance Z^1_{BUS}

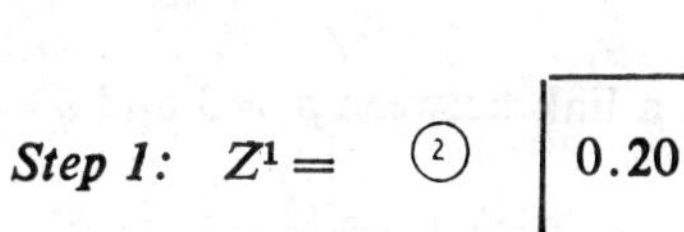

Step 1: $Z^1 =$

	②
②	0.20

Step 2: Adding an element 2 connected between $p = 2$, $q = 3$

$$Z_{32} = Z_{22} = 0.2$$
$$Z_{qq} = Z_{pq} + Z_{pq\,pq}$$
$$Z_{33} = Z_{23} + 0.15 = 0.2 + 0.15 = 0.35$$

Step 3: Adding an element 3 connected between buses $p = 3$ and $q = 4$

$$Z_{42} = Z_{32} = 0.2$$
$$Z_{43} = Z_{33} = 0.35$$
$$Z_{44} = Z_{34} + 0.25 = 0.35 + 0.25 = 0.6$$

$Z_{BUS} =$

	②	③	④
②	0.2	0.2	0.2
③	0.2	0.35	0.35
④	0.2	0.35	0.60

Step 4: Adding an element 4 connected between buses $p = 4$ and $q = 5$ and it is a branch and uncoupled to the partial network:

$Z_{22} = Z_{pl}$

$Z_{52} = Z_{42} = 0.2$

$Z_{53} = Z_{43} = 0.35$

$Z_{54} = Z_{44} = 0.6$

$Z_{qq} = Z_{pq} + Z_{pq\,pq}$

$Z_{55} = .6 + .5 = 1.1$

$Z_{BUS} =$

	②	③	④	⑤
②	.20	.20	.20	.20
③	.20	.35	.35	.35
④	.20	.35	.60	.60
⑤	.20	.35	.60	1.1

Step 5: Adding an element 5 which is a link between $p = 3$ and $q = 5$

$Z_{1i} = Z_{pl} - Z_{ql} = Z_{3i} - Z_{5i}$

$Z_{12} = Z_{32} - Z_{52} = 0.20 - 0.20 = 0$

$Z_{13} = Z_{33} - Z_{53} = 0.35 - 0.35 = 0$

$Z_{14} = Z_{34} - Z_{54} = 0.35 - 0.60 = -0.25$

$Z_{15} = Z_{35} - Z_{55} = 0.35 - 1.1 = -0.075$

$Z_{11} = Z_{p1} - Z_{q1} + Z_{pq\,pq}$

$= Z_{31} - Z_{51} + .8 = 0 + 0.75 + 0.8 = 1.55$

$Z'_{22} = 0.2,\ Z'_{33} = 0.35$

$Z'_{23} = 0.2.\ Z'_{34} = 0.35$

$Z'_{24} = 0.2,\ Z'_{35} = 0.35$

$Z'_{25} = 0.20$

$$Z'_{44} = 0.6 - \frac{0.25 \times 0.25}{1.55} = 0.5596$$

$$Z'_{45} = 0.6 - \frac{0.25 \times 0.75}{1.55} = 0.479$$

$$Z'_{55} = 1.1 - \frac{0.75 \times 0.75}{1.53} = 0.737$$

$$Z_{BUS}^{1'} = Z_{BUS}^{2} = \begin{array}{c|cccc} & (2) & (3) & (4) & (5) \\ \hline (2) & 0.2 & 0.2 & 0.2 & 0.2 \\ (3) & 0.2 & 0.35 & 0.35 & 0.35 \\ (4) & 0.2 & 0.35 & 0.559 & 0.479 \\ (5) & 0.20 & 0.35 & 0.429 & 0.737 \end{array}$$

Step 6: Adding element 6 which is a link connected between $p = 1$ and $q = 5$

$$Z_{1l} = -Z_{ql} = -Z_{5l}$$

$$Z_{12} = -Z_{52} = -0.20$$

$$Z_{13} = -Z_{53} = -0.35$$

$$Z_{14} = -Z_{54} = -0.479$$

$$Z_{15} = -Z_{55} = -0.737$$

$$Z_{11} = -Z_{q1} + Z_{pq\ pq} = -Z_{51} + 1.5 = 0.737 + 1.5 = 2{\cdot}237$$

$$Z'_{22} = 0.2 - \frac{0.2\times0.2}{2.237} = 0.1821$$

$$Z'_{23} = 0.2 - \frac{0.2\times0.35}{2.237} = 0.1687$$

$$Z'_{24} = 0.2 - \frac{0.2\times0.479}{2.237} = 0.1571$$

$$Z'_{25} = 0.2 - \frac{0.2\times0.737}{2.237} = 0.1341$$

$$Z'_{33} = 0.35 - \frac{0.35\times0.35}{2.237} = 0.2952$$

$$Z'_{34} = 0.35 - \frac{0.35\times0.479}{2.237} = 0.275$$

$$Z'_{35} = 0.35 - \frac{0.35\times0.737}{2.237} = 0.2346$$

$$Z' = 0.5596 - \frac{0.479\times0.479}{.237} = 0.4570$$

$$Z'_{45} = 0.479 - \frac{.479 \times .737}{2.237} = 0.3211$$

$$Z'_{55} = .737 - \frac{.737 \times .737}{2.237} = 0.4942$$

$Z^1_{BUS} = Z^2_{BUS} =$

	②	③	④	⑤
②	0.1821	0.1687	0.1571	0.1341
③	0.1687	0.1952	0.2750	0.2346
④	0.1571	0.2750	0.4570	0.3211
⑤	0.1341	0.2346	0.3211	0.4942

Fault Evaluation

(a) *3φ to GND Fault* (LLLG)
Assuming the fault impedance $Z_F = 0$

Total fault current at ⑤

$$I^{0,1,2}_{5(F)} = \begin{bmatrix} 0 \\ \frac{\sqrt{3}}{Z_{55}} \\ 0 \end{bmatrix} = \begin{bmatrix} 0 \\ \frac{\sqrt{3}}{0.4942} \\ 0 \end{bmatrix} = \begin{bmatrix} 0 \\ 0.023\sqrt{3} \\ 0 \end{bmatrix}$$

Bus voltages after fault,

$$E^{0,1,2}_{5(F)} = \begin{bmatrix} 0 \\ 0 \\ 0 \end{bmatrix}$$ since there is LLLG fault at ⑤ with $Z_F = 0$

$$E^{0,1,2}_{2(F)} = \begin{bmatrix} 0 \\ \sqrt{3} - \frac{Z^*_{35}\sqrt{3}}{0.4942} \\ 0 \end{bmatrix} = \begin{bmatrix} 0 \\ \sqrt{3} - \frac{0.1341}{0.4942}\sqrt{3} \\ 0 \end{bmatrix} = \begin{bmatrix} 0 \\ 0.7286\sqrt{3} \\ 0 \end{bmatrix}$$

$$E^{0,1,2}_{3(F)} = \begin{bmatrix} 0 \\ \sqrt{3} - \frac{0.2346}{0.4942}\sqrt{3} \\ 0 \end{bmatrix} = \begin{bmatrix} 0 \\ 0.5256\sqrt{3} \\ 0 \end{bmatrix}$$

$$E^{0,1,2}_{4(F)} = \begin{bmatrix} 0 \\ \sqrt{3} - \frac{0.3211}{0.4942}\sqrt{3} \\ 0 \end{bmatrix} = \begin{bmatrix} 0 \\ 0.3502\sqrt{3} \\ 0 \end{bmatrix}$$

Line Flows

$$i^{0,1,2}_{45(F)} = \begin{bmatrix} 0 \\ Y_{45,45}\,[E_{4(F)} - E^{1}_{5(F)}] \\ 0 \end{bmatrix} = \begin{bmatrix} 0 \\ 2(.3502\sqrt{3} - 0) \\ 0 \end{bmatrix}$$

$$= \begin{bmatrix} 0 \\ 0.7004\sqrt{3} \\ 0 \end{bmatrix}$$

$$i^{0,1,2}_{53(F)} = \begin{bmatrix} 0 \\ 1.25(0 - 0.5256\sqrt{3}) \\ 0 \end{bmatrix} = \begin{bmatrix} 0 \\ -0.657\sqrt{3} \\ 0 \end{bmatrix}$$

$$i^{0,1,2}_{15(F)} = \begin{bmatrix} 0 \\ 0 \\ 0 \end{bmatrix}$$

$$i^{0,1,2}_{12(F)} = \begin{bmatrix} 0 \\ 5(0 - 0.7286\sqrt{3}) \\ 0 \end{bmatrix} = \begin{bmatrix} 0 \\ -3.643\sqrt{3} \\ 0 \end{bmatrix}$$

$$i^{0,1,2}_{23(F)} = \begin{bmatrix} 0 \\ 6.66(0.7286\sqrt{3} - 0.5256\sqrt{3}) \\ 0 \end{bmatrix} = \begin{bmatrix} 0 \\ 1.352\sqrt{3} \\ 0 \end{bmatrix}$$

$$i^{0,1,2}_{43(F)} = \begin{bmatrix} 0 \\ 4.(0.3502\sqrt{3} - 0.5256\sqrt{3}) \\ 0 \end{bmatrix} = \begin{bmatrix} 0 \\ -0.7016\sqrt{3} \\ 0 \end{bmatrix}$$

(b) *Single line to ground fault* (*LG*)

Here also $Z_F = 0$

$$I_{5(F)}^{0,1,2} = \frac{\sqrt{3}}{Z_{55}^0 + 2Z_{55}^1}\begin{bmatrix}1\\1\\1\end{bmatrix}$$

$$= \frac{\sqrt{3}}{0.5637 + 2\times 0.4942}\begin{bmatrix}1\\1\\1\end{bmatrix} = \begin{bmatrix}0.6442\sqrt{3}\\0.6442\sqrt{3}\\0.6442\sqrt{3}\end{bmatrix}$$

$$E_{5(F)}^{0,1,2} = \frac{\sqrt{3}}{Z_{55}^{(0)} + 2Z_{55}^{(1)}}\begin{bmatrix}-Z_{55}^{(0)}\\Z_{55}^{0} + Z_{55}^{1}\\-Z_{55}^{2}\end{bmatrix}$$

$$= 0.6442\sqrt{3}\begin{bmatrix}-0.5637\\1.0579\\-0.4942\end{bmatrix} = \begin{bmatrix}-0.3631\sqrt{3}\\0.6815\sqrt{3}\\-0.31836\sqrt{3}\end{bmatrix}$$

$$E_{2(F)}^{0,1,2} = \begin{bmatrix}0\\\sqrt{3}\\0\end{bmatrix} - \frac{\sqrt{3}}{Z_{55}^0 + 2Z_{55}^1}\begin{bmatrix}Z_{52}^0\\Z_{52}^1\\Z_{52}^2\end{bmatrix}$$

$$= \begin{bmatrix}0\\\sqrt{3}\\0\end{bmatrix} - \begin{bmatrix}0.03872\\0.1341\\0.1341\end{bmatrix} 0.6442\sqrt{3}$$

$$= \begin{bmatrix}-0.0249\sqrt{3}\\0.9136\sqrt{3}\\-0.0863\sqrt{3}\end{bmatrix}$$

$$E_{3(F)}^{0,1,2} = \begin{bmatrix}0\\\sqrt{3}\\0\end{bmatrix} - 0.6442\sqrt{3}\begin{bmatrix}0.07745\\0.2346\\0.2346\end{bmatrix} = \begin{bmatrix}-0.04987\\0.849\\0.154\end{bmatrix}\sqrt{3}$$

$$E_{4(F)}^{0,1,2} = \begin{bmatrix}0\\\sqrt{3}\\0\end{bmatrix} - 0.6442\sqrt{3}\begin{bmatrix}0.1044\\0.3211\\0.3211\end{bmatrix} = \begin{bmatrix}-0.0672\sqrt{3}\\0.7933\sqrt{3}\\-0.2067\sqrt{3}\end{bmatrix}$$

Short circuit currents in the lines connected to the faulted bus

$$i_{53}^{0,1,2} = \begin{bmatrix} y^0_{53\,53}[E^0_{5(E)} - E^0_{3(E)}] \\ y^1_{53\,53}[E^1_{5(E)} - E^1_{3(E)}] \\ y^2_{53\,53}[E^2_{5(E)} - E^2_{3(F)}] \end{bmatrix}$$

$$= \begin{bmatrix} 0.6667[-0.3631\sqrt{3}+0.04982\sqrt{3}] \\ 1.25\,[0.6815\sqrt{3}-0.849\sqrt{3}] \\ 1.25\,[-0.31836\sqrt{3}+0.153\sqrt{3}] \end{bmatrix} = \begin{bmatrix} -0.2088\sqrt{3} \\ -0.2093\sqrt{3} \\ -0.2054\sqrt{3} \end{bmatrix}$$

$$i_{54(F)}^{0,1,2} = \begin{bmatrix} 0.98[-0.3631\sqrt{3}+0.672\sqrt{3}] \\ 2\,[0.6815\sqrt{3}-0.7933\sqrt{3}] \\ 2\,[-0.31836\sqrt{3}+0.2067\sqrt{3}] \end{bmatrix} = \begin{bmatrix} -0.2899\sqrt{3} \\ -0.2236\sqrt{3} \\ 0.2233\sqrt{3} \end{bmatrix}$$

$$i_{51(F)}^{0,1,2} = \begin{bmatrix} 0 \\ 0 \\ 0 \end{bmatrix}$$

$$i_{21(F)}^{0,1,2} = \begin{bmatrix} 20\,[-0.249\sqrt{3}] \\ 5\,[0.9136\sqrt{3}] \\ 5\,[-0.0863\sqrt{3}] \end{bmatrix} = \begin{bmatrix} -0.498\sqrt{3} \\ 4.368\sqrt{3} \\ -0.4315\sqrt{3} \end{bmatrix}$$

$$i_{54(F)}^{0,1,2} = \begin{bmatrix} 0.98\,[-0.3631\sqrt{3}+0.672\sqrt{3}] \\ 2\,[0.6815\sqrt{3}-0.7933\sqrt{3}] \\ 2\,[-0.31836\sqrt{3}+0.2067\sqrt{3}] \end{bmatrix} = \begin{bmatrix} -0.2899\sqrt{3} \\ -0.2236\sqrt{3} \\ -0.2233\sqrt{3} \end{bmatrix}$$

$$i_{51\,(F)}^{0,1,2} = \begin{bmatrix} 0 \\ 0 \\ 0 \end{bmatrix}$$

$$i_{21\,(F)}^{0,1,2} = \begin{bmatrix} 20[-0.249\sqrt{3}] \\ 5[0.9136\sqrt{3}] \\ 5[-0.0863\sqrt{3}] \end{bmatrix} = \begin{bmatrix} -0.498\sqrt{3} \\ 4.368\sqrt{3} \\ -0.4315\sqrt{3} \end{bmatrix}$$

$$i_{32\,(F)}^{0,1,2} = \begin{bmatrix} 20[-0.4987\sqrt{3}+0.249\sqrt{3}] \\ 6.667\sqrt{3}[0.849-0.9136] \\ 6.667\sqrt{3}[-0.154+0.0863] \end{bmatrix} = \begin{bmatrix} -0.4994\sqrt{3} \\ -0.4306\sqrt{3} \\ -0.4513\sqrt{3} \end{bmatrix}$$

$$i_{43\,(F)}^{0,1,2} = \begin{bmatrix} 16.67\sqrt{3}[-0.0672-0.04987] \\ 4\sqrt{3}[0.7933-0.849] \\ 4\sqrt{3}[-0.2067+0.154] \end{bmatrix} = \sqrt{3}\begin{bmatrix} -0.2888 \\ -0.2228 \\ -0.2108 \end{bmatrix}$$

Fault level

The maximum currents to be interrupted by the circuit Breaker at bus 5

$$= I_{53}^{0,1,2} + I_{54}^{0,1,2} + I_{51}^{0,1,2}$$

$$= \sqrt{3}\begin{bmatrix} -0.2088 \\ -0.2093 \\ -0.2054 \end{bmatrix} + \sqrt{3}\begin{bmatrix} -0.2899 \\ -0.2236 \\ -0.2233 \end{bmatrix} = \sqrt{3}\begin{bmatrix} -0.4987 \\ -0.4329 \\ -0.4287 \end{bmatrix}$$

Examples

4.1 For the sample network shown in Fig. 1, (a) form the Z_{BUS} matrix by building algorithm, (b) modify the Z_{BUS} matrix obtained in part (a) to include the addition of an element from bus 3 to bus 5 with an impedance 0.2 and coupled to the element 4 with a mutual impedance of 0.1 and (c) modify the Z_{BUS} matrix obtained in part (b) to remove the new element from bus 3 to 5.

Element No.	Bus Code $p-q$	Self imp. $Z_{pq}^{0,1,2}$		
1	1—2	0.04	0.15	0.15
2	2—3	0.05	0.20	0.20
3	3—4	0.10	0.50	0.50
4	4—5	0.50	1.50	1.50
5	2—5	2.00	1.20	1.20
6	1—5	0.10	0.10	0.10

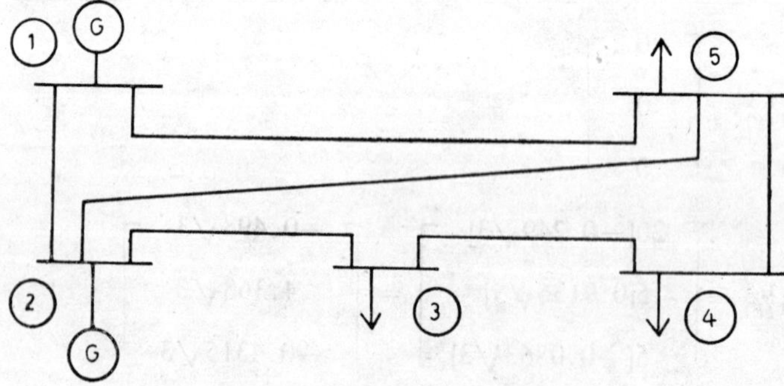

Fig. 1

4.2 For the sample network shown in Fig. 1, perform the short circuit studies and determine the total fault current, bus voltage after the fault and line flows for the (a) 3-phase fault at the bus 3, (b) line to line fault at bus 3 and (c) line to ground fault at the bus 3.

4.3 For the sample network shown in Fig. 2, perform the short circuit studies and determine bus voltages after the fault and line flows for (a) 3-phase to ground fault at the bus 5, and (b) Double line to ground fault at bus 5.

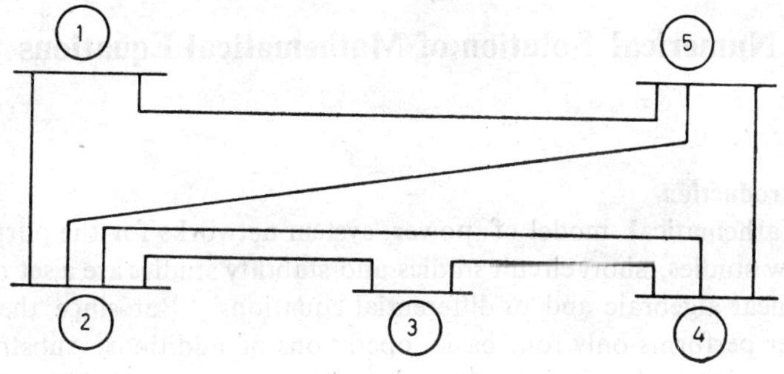

Fig. 2

Element No.	Bus Code $p-q$	Self imp. $Z_{pq}^{0,1,2}$		
1	1–2	0.05	0.15	0.15
2	2–3	0.06	0.25	0.25
3	3–4	0.15	0.60	0.60
4	4–5	0.60	1.50	1.50
5	2–5	2.00	1.25	1.25
6	2–4	0.20	0.15	0.15
7	1–5	0.25	0.15	0.15

References

1. Ahmad, H. El-Abiad, "Digital calculation of line to ground short circuits by matrix method", *Trans. AIEE*, Vol. 79, Part III, pp. 323-332, 1960.
2. Ahmed H. El-Abiad etc, "Calculation of short circuit using a high-speed digital computer", *Trans. AIEE*, Vol. 80, Part III, pp. 702-707, 1961.
3. H.E. Brown etc., "Digital calculation of 3-phase short circuits by matrix method", *Trans. AIEE*, Vol. 79, Part III, pp. 1277-1281, 1960.
4. R.T. Byerly etc., "Digital calculation of power system networks under faulted conditions", *Trans. AIEE*, Vol. 79, Part III, pp. 1297-1307, 1958.
5. G.W. Stagg etc., *Computer Methods in Power System analysis*, McGraw-Hill, NY.
6. Laughton, M.A., "Analysis of unbalanced polyphase networks by the method of phase coordinates part 2 fault analysis", *Proc. IEE*, Vol. 116, No. 5, pp. 857-865, May 1969.
7. Roy, L., "Generalized polyphase fault analysis program: Calculation of cross country fault", *Proc. IEE*, Vol. 126, No. 10, pp. 995-1001, Oct. 1979.
8. Laughton, M.A. and Saleh, A.O.M., "United phase-coordinate load flow and fault anylysis of polyphase network", *Electrical Power and Energy Systems*, Vol. 2, No. 4, pp 181-192, Oct. 1980.

Chapter 5

Numerical Solution of Mathematical Equations

5.1 Introduction

Mathematical model of power system networks for the purposes of load flow studies, short circuit studies and stability studies are a set of linear or nonlinear algebraic and or differential equations. But since the digital computer performs only four basic operations of additions, substractions, multiplications and divisions, therefore, in order to solve these mathematical equations on the digital computer, it is necessary to transform these linear or nonlinear algebraic and differential equation to a set of four basic operations of additions, substractions, multiplications and divisions with the help of numerical methods. We shall discuss only important numerical techniques to solve both algebraic and differential equations and illustrate them with the help of solved examples.

5.2 Solution of Algebraic Equations

There are different numerical techniques for the solution of algebraic equations. These algebraic equations can be expressed in the following form:

$$\begin{aligned} f_1(x_1, x_2, \ldots, x_n) &= y_1 \\ f_2(x_1, x_2, \ldots, x_n) &= y_2 \\ \vdots \quad & \quad \vdots \\ f_n(x_1, x_2, \ldots, x_n) &= y_n \end{aligned} \tag{5.1}$$

Here f_i are the functions relating the unknown variables x_i, with the known constants y_i. If any one of the f_i are nonlinear, the above is a set of nonlinear algebraic equations (equations involving power or product of the variables x_i). However, if all the f_i are linear, then above is a set of linear algebraic equations. The linear algebraic equations can be expressed as follows:

$$[A] \quad \overline{X} = \overline{Y} \tag{5.2}$$

where $[A]$ is the coefficient matrix of the physical system, $\overline{X}$ is a column vector of unknowns and $\overline{Y}$ is the column vector of known constants. The above (eqn. 5.2) can be expressed as,

$$\begin{bmatrix} a_{11} & a_{12} & \cdots & a_{1n} \\ a_{21} & a_{22} & \cdots & a_{2n} \\ & \vdots & & \\ a_{n1} & a_{n2} & \cdots & a_{nn} \end{bmatrix} \begin{bmatrix} x_1 \\ x_2 \\ \vdots \\ x_n \end{bmatrix} = \begin{bmatrix} y_1 \\ y_2 \\ \vdots \\ y_n \end{bmatrix} \tag{5.3}$$

Equation (5.2) is a nonhomogeneous equation. The homogeneous equations will be of the following form,

$$[A]\,\overline{X}=0 \tag{5.4}$$

From the coefficient matrix A, we can obtain the augmented matrix $\hat{A}$ by augmenting the coefficient matrix A in its $n+$ 1st column by known constants $\overline{Y}$, *i.e.*

$$\hat{A}=\begin{bmatrix} a_{11} & a_{12} & \cdots & a_{1n} & y_1 \\ a_{21} & a_{22} & \cdots & a_{2n} & y_2 \\ a_{n1} & a_{n2} & \cdots & a_{nn} & y_n \end{bmatrix} \tag{5.5}$$

The *necessary* and *sufficient condition* for a set of simultaneous equations (eqn. 5.2) to be *consistent* (*i.e.* to have solution) is that the rank of coefficient matrix $[A]$ must be equal to the rank of augmented matrix $[\hat{A}]$. However, in order to have a *unique solution,* the coefficient matrix A must be square and non singular *i.e.* its rank should be equal to the number of columns (*i.e.* rank be equal to the number of variables). The *unique solution* is *nontrivial* for a set of non-homogeneous equations and *trivial* (*i.e.* zero) for homogeneous equations. If the rank of the coefficient matrix A is less than the number of rows, some equations are *reduntant* and do not place any constraint on the solution. However, if the determinant of the coefficient matrix A is zero *i.e.* if the rank A is less than the number of columns (*i.e.* variables), there will be *infinite number* of solutions.

In order to choose a *numerical method* to solve a set of algebraic equations on the digital computer, following points must be considered.

(i) No. of steps needed to obtain the solution, *i.e.* speed by which solution is obtained.
(ii) Resultant accuracy, and
(iii) Computer memory limitations.

We shall first of all take a set of linear algebraic equations and discuss the numerical technique to solve these equations keeping in mind the above three factors.

5.3 Linear Algebraic Equations

The numerical techniques to solve a set of linear algebraic equation can broadly be classified into two main headings viz. (i) *Direct method* or *exact method*, and (ii) *Iterative technique.* In the case of direct method. the solution can be obtained in distinct no. of steps However, the number of steps required to obtain the solution depend upon the problem size, i.e. the order of the coefficient matrix A and the numerical method used. Hence it is possible to compare the direct methods (i.e. numerical methods) before use since the number of steps required to obtain solution is known in advance. The only error is the loss of significant digits which is not bounded. The solution will be far from nominal solution because the *round of error* goes on getting accumulated after each step. Round-off error results due to the substraction or division by two numbers which are nearly equal. If the round-off error is bounded, the solution obtained will be nearly exact. This is why this method is also known as *exact method.*

In the case of *iterative techniques,* the solution is obtained in an orderly fashion starting from its initial approximate solution i.e. initial guess. Here we start from the initial approximate solution and in an orderly fashion, the approximate solution converges to the nominal solution. Thus the rate of convergence i.e. number of steps needed to obtain solution depends upon the initial guess, problem size (i.e. no. of equations) and the iterative technique used. Thus depending upon these factors, the approximate solution may converge to the nominal solution, diverge or oscillate about the nominal solution. The round-off error in this case goes on getting corrected in each step i.e. in each iteration.

5.3.1 Direct Method

Let us take a set of linear algebraic equations of the form as shown below,

$$[A]\, X = \bar{b} \tag{5.6}$$

Equation (5.6) for the specific case of three equations in three variables will be as shown below,

$$\begin{aligned} a_{11}\, x_1 + a_{12}\, x_2 + a_{13}\, x_3 &= y_1 \\ a_{21}\, x_1 + a_{22}\, x_2 + a_{23}\, x_3 &= y_2 \\ a_{31}\, x_1 + a_{32}\, x_2 + a_{33}\, x_3 &= y_3 \end{aligned} \tag{5.7}$$

In the matrix form equation (5.7) can be expressed as

$$\begin{bmatrix} a_{11} & a_{12} & a_{13} \\ a_{21} & a_{22} & a_{23} \\ a_{31} & a_{32} & a_{33} \end{bmatrix} \begin{bmatrix} x_1 \\ x_2 \\ x_3 \end{bmatrix} = \begin{bmatrix} y_1 \\ y_2 \\ y_3 \end{bmatrix} \tag{5.8}$$

The solution of equation (5.8) can be obtained by a Direct method known as Cramer's rule as shown below.

$$x_1 = \frac{|A_1|}{|A|} \quad \text{where} \quad |A| \neq 0$$

$$x_2 = \frac{|A_2|}{|A|},$$

and

$$x_3 = \frac{|A_3|}{|A|} \tag{5.9}$$

Here,

$$|A_1| = \begin{vmatrix} y_1 & a_{12} & a_{13} \\ y_2 & a_{22} & a_{23} \\ y_3 & a_{32} & a_{33} \end{vmatrix} \quad |A_2| = \begin{vmatrix} a_{11} & y_1 & a_{13} \\ a_{21} & y_2 & a_{23} \\ a_{31} & y_3 & a_{33} \end{vmatrix}$$

and

$$|A_3| = \begin{vmatrix} a_{11} & a_{12} & y_1 \\ a_{21} & a_{22} & y_2 \\ a_{31} & a_{32} & y_3 \end{vmatrix}$$

Crammer's rule requires more computational efforts and at the same time magnitude of error is also large, with the result this method is not very popular. More popular direct methods are, the successive eliminations of variables where we obtain finally one equation in one variable whose solution is straight-forward.

The original method which is duc to Gauss is known as *Gauss Elimination method.* Later on, many more methods have been developed but all of them are derived from Gaussian Elimination methods and attempt have been made in these methods to minimise computational efforts, round off error and also the computer memory. We shall illustrate some of these methods with the help of examples.

5.3.2 *Gaussian Elimination Methods*

To illustrate this method, we shall again take a set of 3 linear algebraic equations in 3 variables as shown below,

$$a_{11}\, x_1 + a_{12}\, x_2 + a_{13}\, x_3 = y_1 \tag{5.10a}$$
$$a_{21}\, x_1 + a_{22}\, x_2 + a_{23}\, x_3 = y_2 \tag{5.10b}$$
$$a_{31}\, x_1 + a_{32}\, x_2 + a_{33}\, x_3 = y_3 \tag{5.10c}$$

Solution Procedure

Step 1: Divide the equation (5.10a) by its leading coefficient a_{11} ($a_{11} \neq 0$), hence we get

$$x_1 + \frac{a_{12}}{a_{11}}\, x_2 + \frac{a_{13}}{a_{11}}\, x_3 = \frac{y_1}{a_{11}} \tag{5.11a}$$

Substituting

$$a'_{1j} = a_{1j}/a_{11} \quad \text{for} \quad j = 2, 3$$

and $\quad y_1' = y_1/a_{11}$

in equation (5.11a), we obtain,

$$x_1 + a'_{12}\, x_2 + a'_{13} x_3 = y_1' \tag{5.11b}$$

Step 2 : Eliminate the variable x_1 from the following equations (i.e equations 2 and 3) by combining these equations with equation (1). This is done by multiplying the derived equation (5.11b) by the leading coefficient a_{21} (where $a_{21} \neq 0$) of equation (5.10b) and subtracting the resultant from equation (5.10b). Hence we obtain

$(a_{21} - a_{21})\, x_1 + (a_{22} - a_{21}\, a'_{12})\, x_2 + (a_{23} - a_{21}\, a'_{13})\, x_3 = y_2 - a_{21} y_1'$ (5.12a)

Again substituting,

$$a_{2j} - a_{21}\, a'_{1j} = a'_{2j} \quad \text{for } j = 2, 3$$

and

$$y_2' = y_2 - a_{21}\, y_1'$$

in equation (5.12a), we obtain,

$$a'_{22}\, x_2 + a'_{23}\, x_3 = y_2' \tag{5.12b}$$

Similarly x_1 is eliminated from equation (5.10c) by multiplying the derived equation (5.11b) by the leading coefficient a_{31} (where $a_{31} \neq 0$) of equation (5.10c) and subtracting the result from equation (5.10c). Thus we get,

$(a_{31} - a_{31})\, x_1 + (a_{32} - a_{31}\, a'_{12})\, x_2 + (a_{33} - a_{31}\, a'_{13})\, x_3 = y_3 - a_{31}\, y_1'$ (5.13a)

Again substituting

$$a'_{3j} = a_{3j} - a_{31}\, a'_{1j} \text{ for } j = 2, 3$$

and

$y_3' = y_3 - a_{31}\, y_1'$ in the equation (5.13a),

we obtain,

$$a'_{32}\, x_2 + a'_{33}\, x_3 = y_3' \tag{5.13b}$$

Thus with the elimination of x_2 from equations (5.10b) and (5.10c), the set of equations (5.10) becomes,

$$\begin{aligned} x_1 + a'_{12}\, x_2 + a'_{13}\, x_3 &= y_1' && \text{(a)} \\ a'_{22}\, x_2 + a'_{23}\, x_3 &= y_2' && \text{(b)} \\ a'_{32}\, x_2 + a'_{33}\, x_3 &= y_3' && \text{(c)} \end{aligned} \tag{5.14}$$

Step 3: Divide equation (b) of the set of equation (5.14) by its leading coefficient a'_{22} (where $a'_{22} \neq 0$). The result will be

$$x_2 + \frac{a'_{23}}{a'_{22}}\, x_3 = \frac{y'_2}{a'_{22}} \tag{5.15a}$$

Substituting

$$a'_{2j}/a'_{22} = a''_{2j} \text{ for } j = 3$$

and

$y'_2/a'_{22} = y''_2$ in equation (5.15a),

we get,

$$x_2 + a''_{23}\, x_3 = y''_2 \tag{5.15b}$$

Step 4: Eliminate x_2 from the following equations, i.e. from equation (c) of the set of equation (5.14) by combining the equation (c) of the set of equation (5.14) with the derived equation (5.15b). Thus we obtain,

$$(a'_{32} - a'_{32})\, x_2 + (a'_{33} - a'_{32}\, a''_{23})\, x_3 = y'_3 - a'_{32}\, y''_2 \tag{5.16a}$$

Substituting

$$a'_{3j} - a'_{32}\, a''_{2j} = a''_{3j} \quad \text{for } j = 3$$

and

$y'_3 - a'_{32} y''_2 = y''_3$ in equation (5.16a);

we get,

$$a''_{33}\, x_3 = y''_3 \tag{5.16b}$$

Step 5: Divide the equation (5.16b) by a''_{33}, to obtain

$$x_3 = y''_3 / a''_{33} \tag{5.17a}$$

Substituting

$$y'''_3 = y''_3 / a''_{33} \text{ in the equation (5.17a);}$$

we get,

$$x_3 = y'''_3, \tag{5.17b}$$

Hence final transformed equation becomes (i.e. by collecting equations (5.11b), (5.15b) and (5.17b))

$$\begin{aligned} x_1 + a'_{12}\, x_2 + a'_{13}\, x_3 &= y'_1 \quad &\text{(a)} \\ x_2 + a''_{23}\, x_3 &= y''_2 \quad &\text{(b)} \\ x_3 &= y'''_3 \quad &\text{(c)} \end{aligned} \tag{5.18}$$

Equation (5.18) is triangular set of equations and its solution will be the solution of equation (5.10). The process of obtaining this triangular set of equations are known as forward process. The solution is obtained by back substitution because in equation (5.18c), x_3 is already known (being one equation in one variable). Now substituting x_3 in the equation (5.18b), x_2 can be known and finally substituting both x_2 and x_3 in equation (5.18a), x_1 can be calculated. This process of obtaining solution is known as back substitution.

5.3.3 *Gauss Jordan Elimination Method*

This method is quite similar to that of the Gauss elimination method except that, in this case, we obtain a diagonal set of equations whereas in the case of Gauss elimination method, we obtain a triangular set of equations. Thus finally, in this case, the equations obtained are such, that there is no need of the back substitution because here we obtain one equation in one variable and hence solution is apparent. To illustrate this method, we will again take the same equations (i.e. equation (5.10)),

$$\begin{bmatrix} a_{11} & a_{12} & a_{13} \\ a_{21} & a_{22} & a_{23} \\ a_{31} & a_{32} & a_{33} \end{bmatrix} \begin{bmatrix} x_1 \\ x_2 \\ x_3 \end{bmatrix} = \begin{bmatrix} y_1 \\ y_2 \\ y_3 \end{bmatrix} \tag{5.19}$$

The first two steps here are the same as that of Gauss Elimination method viz. divide row 1 by a_{11} to obtain 1 in the place of a_{11} and eliminate the variable x_1 from the 2nd and 3rd rows by combining rows (2) and (3) with the row (1). Thus now we obtain

$$\begin{bmatrix} 1 & a'_{12} & a'_{13} \\ 0 & a'_{22} & a'_{23} \\ 0 & a'_{32} & a'_{33} \end{bmatrix} \begin{bmatrix} x_1 \\ x_2 \\ x_3 \end{bmatrix} = \begin{bmatrix} y_1 \\ y'_2 \\ y_3 \end{bmatrix} \tag{5.20}$$

In the next step, there is a change in the Gauss-Jordan method i.e. after eliminating x_2 from the following equation, i.e. from equation (3) onwards (just like Gauss elimination method), x_2 is also eliminated from the previous equations. Thus first of all we divide the row 2 by a'_{22} and then combine row (3) with the row (2) and also row (1) with the row (2) to eliminate x_2 both from the following equations (i.e. row 3) and previous equation (i.e. row 1). Hence after this operation, we get,

$$\begin{bmatrix} 1 & 0 & a_{13} \\ 0 & 1 & a''_{23} \\ 0 & 0 & a'_{33} \end{bmatrix} \begin{bmatrix} x_1 \\ x_2 \\ x_3 \end{bmatrix} = \begin{bmatrix} y_1 \\ y \\ y'_3 \end{bmatrix} \tag{5.21}$$

In the third step after eliminating x_3 from the following equations i.e. rows which are not in the present case, x_3 is also eliminated from the previous equations i.e. rows (rows 1 and 2) by combining these equations with the equation (3). Hence, we obtain,

$$\begin{bmatrix} 1 & 0 & 0 \\ 0 & 1 & 0 \\ 0 & 0 & 1 \end{bmatrix} \begin{bmatrix} x_1 \\ x_2 \\ x_3 \end{bmatrix} = \begin{bmatrix} y'''_1 \\ y'''_2 \\ y'''_3 \end{bmatrix} \tag{5.22}$$

Thus we obtain finally a diagonal set of equations i.e. one equation in one variable and there is no need of back substitution as in the case of Gauss Elimination method. However, arithmetic operations in this case are much more than Gauss elimination method because as it is clear from the example, every row in this case is operated three times for this three order system while in the case of Gauss elimination method, the equation one is operated only once, the equation second two times and only third equation three times.

5.3.4 *Method of Elimination using Pivotal Condensation*

In the methods of elimination, the main error is the *round off error* which is due to the *loss of significant digits* and this error goes on getting accumulated in each step. Hence in order to decrease or eliminate round off error which are normally due to dividing two numbers which are nearly equal, the entire coefficient matrix A (eqn. 5.8) is scanned and the largest element say a_{ij} is selected. The ith row is divided by this largest element a_{ij} to obtain 1 in the i-jth position. Now in the case of Gauss Jordan elimination method, by combining rows following the i-th row as well as the rows preceding ith rows, the variable x_j is eliminated from these rows. Then the reduced coefficient matrix A (i.e. leaving ith row and jth column) is scanned and again the same procedure is adopted. However, in the case of Gauss elimination method, if the element a_{ij} is the largest, it is brought to i-ith position by elimentary transformations and then in the similar manner, the equations are solved. This method which is known as *pivotal condensation* or the *method of maximum pivot strategy* is illustrated by the solved example 5.2.

We now give some solved examples to illustrate the method of eliminations.

Example 5.1: Use Gaussian elimination method to solve the following linear equations:

$$x_1 + \frac{1}{2}x_2 + \frac{1}{3}x_3 = 1$$

$$\frac{1}{2}x_1 + \frac{1}{3}x_2 + \frac{1}{4}x_3 = 0$$

$$\frac{1}{3}x_1 + \frac{1}{4}x_2 + \frac{1}{5}x_3 = 0$$

Solution: The above equations can be written in the matrix form as shown,

$$\begin{bmatrix} 1 & 1/2 & 1/3 \\ 1/2 & 1/3 & 1/4 \\ 1/3 & 1/4 & 1/5 \end{bmatrix} \begin{bmatrix} x_1 \\ x_2 \\ x_3 \end{bmatrix} = \begin{bmatrix} 1 \\ 0 \\ 0 \end{bmatrix}$$

i.e $\quad [A]\,\overline{X} = \overline{b}$

$$\text{The augmented matrix } [A \mid b] = \begin{bmatrix} 1 & 1/2 & 1/3 & 1 \\ 1/2 & 1/3 & 1/4 & 0 \\ 1/3 & 1/4 & 1/5 & 0 \end{bmatrix}$$

Eliminating x_1 from the equations (2) and (3), we get

$$\begin{bmatrix} 1 & 1/2 & 1/3 & 1 \\ 0 & 1/12 & 1/12 & -1/2 \\ 0 & 1/12 & 4/45 & -1/3 \end{bmatrix}$$

Dividing row 2 by 1/12 and eliminating x_2 from eqn. (3), we have

$$\begin{bmatrix} 1 & 1/2 & 1/3 & 1 \\ 0 & 1 & 1 & -6 \\ 0 & 0 & 1/180 & 1/6 \end{bmatrix}$$

Dividing row 3 by 1/180, we get

$$\begin{bmatrix} 1 & 1/2 & 1/3 & 1 \\ 0 & 1 & 1 & 6 \\ 0 & 0 & 1 & 30 \end{bmatrix}$$

Thus from the above (by the method of back substitution), we obtain

$$x_3 = 30$$

Substituting this in equation (2), we obtain

$$x_2 + x_3 = -6$$

$$x_2 = -6 - 30 = -36$$

Substituting x_2 and x_3 in equation (1) and solving, we obtain

$$x_1 + \frac{1}{2}x_2 + \frac{1}{3}x_3 = 1$$

$$x_1 - \frac{36}{2} + \frac{30}{3} = 1$$

or $\quad x_1 = 18 - 10 + 1 = 9 \quad$ Answer

Example 5.2: Solve the following linear algebraic equations by Gauss Jordan elimination method using maximum pivot strategy (i.e. pivotal condensation)

$$x_1 + 2x_2 + 8x_3 = -3$$

$$2x_1 - 4x_2 - 6x_3 = 0$$

$$-2x_1 + 4x_2 + 2x_3 = 4$$

Solution:

The coefficient matrix $A = \begin{bmatrix} 1 & 2 & 8 \\ 2 & -4 & -6 \\ -2 & 4 & 2 \end{bmatrix}$

Column vector of known coefficients $b = \begin{bmatrix} -3 \\ 0 \\ 4 \end{bmatrix}$

Thus the augmented matrix $[A : b] = \begin{bmatrix} 1 & 2 & 8 & -3 \\ 2 & -4 & -6 & 0 \\ -2 & 4 & 2 & 4 \end{bmatrix}$

The largest element *i.e.* pivot is $a_{13} = 8$ and hence dividing the first row by 8 and eliminating x_3 from the second and third row, we get

$$\begin{bmatrix} 1/8 & 1/4 & 1 & -3/8 \\ 11/4 & -5/2 & 0 & -9/4 \\ -9/4 & 7/2 & 0 & 19/4 \end{bmatrix}$$

Now scanning the reduced matrix (i.e. leaving row 1 and column 3), the largest element obtained is $a_{32} = 7/2$. Thus dividing the third row by 7/2 and eliminating x_2 from the first and the second row, we obtain

$$\begin{bmatrix} -4/112 & 0 & 1 & -80/112 \\ 61/44 & 0 & 0 & 8/7 \\ 9/14 & 1 & 0 & 19/14 \end{bmatrix}$$

Finally dividing the second row by 61/44 and eliminating x_1 from the first and the third row, we get

$$\begin{bmatrix} 0 & 0 & 1 & 2047/2989 \\ 1 & 0 & 0 & 352/477 \\ 0 & 1 & 0 & 4945/5978 \end{bmatrix}$$

Hence from the above, solution will be

$$x_1 = 352/477$$

$$x_2 = 4945/5978$$

and $$x_3 = 2047/2989$$

5.4 Iterative Techniques

In this technique, we have the initial guess and then by successive approximation, we obtain the nominal solution within specified tolerance. In a situation where the diagonal elements of the coefficient matrix are the largest (i.e. coefficient matrix with strong diagonal), this method converges normally to the nominal solution within reasonable time, i.e. iterations. However, in some cases approximate solution may diverge or oscillate about nominal solution. To illustrate the method *i.e.* technique, let us take a set of three equations in three variables as shown,

$$\begin{aligned} a_{11} x_1 + a_{12} x_2 + a_{13} x_3 &= y_1 \\ a_{21} x_1 + a_{22} x_2 + a_{23} x_3 &= y_2 \\ a_{31} x_1 + a_{32} x_2 + a_{33} x_3 &= y_3 \end{aligned} \tag{5.23}$$

The above set of equations can be written as

$$\begin{aligned} a_{11} x_1 &= y_1 - a_{12} x_2 - a_{13} x_3 \\ a_{22} x_2 &= y_2 - a_{21} x_1 - a_{23} x_3 \\ a_{33} x_3 &= y_3 - a_{31} x_1 - a_{32} x_2 \end{aligned}$$

i.e. $$x_1 = \frac{1}{a_{11}} (y_1 - a_{12} x_2 - a_{13} x_3)$$

$$x_2 = \frac{1}{a_{22}} (y_2 - a_{21} x_1 - a_{23} x_3)$$

$$x_3 = \frac{1}{a_{33}} (y_3 - a_{31} x_1 - a_{32} x_2) \tag{5.24}$$

The process start by selecting initial set of values for the unknowns *i.e.* by assuming x_1^0, x_2^0 and x_3^0. These values are substituted on the right hand side of equations (5.24) to obtain the values of unknowns after first iteration *i.e.* x_1^1, x_2^1 and x_3^1. The calculated values x_1^1, x_2^1 and x_3^1 become estimate for the second iteration i.e. x_1^1, x_2^1 and x_3^1 are substituted on the right hand side of equation (5.24) so as to obtain x_1^2, x_2^2 and x_3^2. In this way the process becomes automatic and continues till $|x_i^{k+1} - x_i^k| < \varepsilon$ where k is an iteration count and ε is a very small number whose values depend upon the accuracy desired. There are two common types of iterative techniques viz. (*i*) Gauss Iterative technique and (*ii*) Gauss Seidel iterative technique. We shall discuss them with the help of the same example (i.e. eqn. (5.24)).

5.4.1 *Gauss Iterative Technique*

Let us take equation (5.24)

$$x_1 = \frac{1}{a_{11}}(y_1 - a_{12}\,x_2 - a_{13}\,x_3)$$

$$x_2 = \frac{1}{a_{22}}(y_2 - a_{21}\,x_1 - a_{23}\,x_3)$$

$$x_3 = \frac{1}{a_{33}}(y_3 - a_{31}\,x_1 - a_{32}\,x_2)$$

Let x_1^0, x_2^0 and x_3^0 are initial solution (i.e. initial estimate). These are substituted on the right hand side to obtain x_1^1, x_2^1 and x_3^1 after the first iteration *i.e.*

$$x_1^1 = \frac{1}{a_{11}}(y_1 - a_{12}\,x_2^0 - a_{13}\,x_3^0)$$

$$x_2^1 = \frac{1}{a_{22}}(y_2 - a_{21}\,x_1^0 - a_{23}\,x_3^0)$$

$$x_3^1 = \frac{1}{a_{33}}(y_3 - a_{31}\,x_1^0 - a_{32}\,x_2^0) \tag{5.25}$$

x_1^1, x_2^1 and x_3^1 obtained after the first iteration are substituted on the right hand side of equation (5.25) to obtain x_1^2, x_2^2 and x_3^2 after the second iteration. The process is continued till

$$|x_i^{k+1} - x_i^k| < \varepsilon \qquad \text{for } i=1, 2, 3$$

where k is an iteration count. ε is a very small number which depends upon the accuracy desired.

However in this method the entire iteration is completed with the same values i.e. estimates. In other words, new values for unknowns are substituted only after the end of the iteration.

5.4.2. *Gauss Seidel Iterative Technique*

In the case of Gauss Seidel iterative technique, the substitution is made immediately in the same iteration, i.e. latest values of unknowns replace immediately the existing values without waiting for the iteration to complete which is unlike Gauss iterative technique where substitution is made only at the end of an iteration, with the result Gauss seidel method converges faster than Gauss iterative technique. To illustrate the method, let us take the same equation i.e.

$$x_1 = \frac{1}{a_{11}} (y_1 - a_{12}x_2 - a_{13}x_3)$$

$$x_2 = \frac{1}{a_{22}} (y_2 - a_{21}x_1 - a_{23}x_3)$$

$$x_3 = \frac{1}{a_{33}} (y_3 - a_{31}x_1 - a_{32}x_2)$$

Let x_1^0, x_2^0 and x_3^0 are the initial guess, then we have

$$x_1^1 = \frac{1}{a_{11}} (y_1 - a_{12}x_2^0 - a_{13}x_3^0)$$

$$x_2^1 = \frac{1}{a_{22}} (y_2 - a_{21}x_1^1 - x_{23}x_3^0)$$

$$x_3^1 = \frac{1}{a_{22}} (y_3 - a_{31}x_1^1 - a_{23}x_2^1) \tag{5.26}$$

For the first equation, x_2^0 and x_3^0 are used to calculate x_1^1 but in the second equation x_1^1 and x_3^0 are used while for the third equation x_1^1 and x_2^1 are substituted. In this way latest values of unknowns immediately replace existing values in the same iteration.

5.5 Solution of Nonlinear Algebraic Equations

Let us take the following equations

$$\begin{aligned} f_1(x_1, x_2, x_3, \ldots, x_n) &= y_1 \\ f_2(x_1, x_2, x_3, \ldots, x_n) &= y_2 \\ \vdots \qquad\qquad & \quad \vdots \\ f_n(x_1, x_2, x_3, \ldots, x_n) &= y_n \end{aligned} \tag{5.27}$$

Here f_s are the functions relating unknowns x_s with the known constants y_s. If any of the f_s are nonlinear, the above is a set of nonlinear algebraic equations (i.e. equations containing powers of unknowns and or their products). Such nonlinear algebraic equations cannot be solved by the Direct method i.e. by the method of elimination. This is because the method of elimination, where we must obtain finally a set of equations with one equation in one variable, if applied to the set of nonlinear algebraic equations, it is not always possible to eliminate the variables so as to reduce the equations with one equation with one variable.

Iterative techniques, both Gauss and Gauss Seidel can be used to find solution in the case of nonlinear algebraic equations also. For this, the nonlinear algebraic equations (5.27) can be written as shown below,

$$\begin{aligned} x_1 &= y_1 - \phi_1 (x_2, x_3, \ldots, x_n) \\ x_2 &= y_2 - \phi_2 (x_1, x_3, \ldots, x_n) \\ &\vdots \\ x_n &= y_n - \phi_n (x_1, x_2, \ldots, x_{n-1}) \end{aligned} \tag{5.28}$$

Here we choose initial estimates $x_1^0, x_2^0, \ldots, x_n^0$ and substitute these on the right hand side of equation (5.28) to obtain $x_1^1, x_2^1, \ldots, x_n^1$. The values obtained after the first iteration i.e. $x_1^1, x_2^1, \ldots, x_n^1$ become the estimate for the next iteration and in this way, the technique becomes automatic. The process is continued till

$$|x_i^{k+1} - x_i^k| < \varepsilon \tag{5.29}$$

5.5.1 *Newton Raphson's Method*

This method is used for the solution of the nonlinear algebraic equations. With the help of this method, a set of nonlinear algebraic equations are transformed into the set of linear algebraic equations. Let us take a set of nonlinear algebraic equations as shown below:

$$\begin{aligned} f_1 (x_1, x_2, x_3, \ldots, x_n) &= y_1 \\ f_2 (x_1, x_2, x_3, \ldots, x_n) &= y_2 \\ &\vdots \\ f_n (x_1, x_2, x_3, \ldots, x_n) &= y_n \end{aligned} \tag{5.30}$$

Let $x_1^0, x_2^0, x_3^0, \ldots, x_n^0$ is the initial estimate and let $\Delta x_1, \Delta x_2, \Delta x_3, \ldots, \Delta x_n$ are the corrections with the initial estimate $x_1^0, x_2^0, x_3^0, \ldots, x_n^0$ so that the equations (5.30) are satisfied i.e.

$$\begin{aligned} f_1 (x_1^0 + \Delta x_1, x_2^0 + \Delta x_2, \ldots, x_n^0 + \Delta x_n) &= y_1 \\ f_2 (x_1^0 + \Delta x_1, x_2^0 + \Delta x_2, \ldots, x_n^0 + \Delta x_n) &= y_2 \\ f_n (x_1^0 + \Delta x_1, x_2^0 + \Delta x_2, \ldots, x_n^0 + \Delta x_n) &= y_n \end{aligned} \tag{5.31}$$

Expanding the eqn. (5.31) by Taylor's theorem about the initial solution x^0 and neglecting terms which are powers of Δx and higher order derivatives (this is only correct provided the initial estimate x^0 is close to solution point otherwise there will be large truncation error), we obtain,

$$\begin{aligned} &f_1(x_1^0 + \Delta x_1, x_2^0 + \Delta x_2, \ldots, x_n^0 + \Delta x_n) \\ &= f_1(x_1^0, x_2^0, \ldots, x_n^0) + \Delta x_1 \frac{\partial f_1}{\partial x_1}\bigg|_0 + \Delta x_2 \frac{\partial f_1}{\partial x_2}\bigg|_0 + \ldots + \Delta x_n \frac{\partial f_1}{\partial x_n}\bigg|_0 \\ &= y_1 \end{aligned}$$

i.e. $$y_1 - f_1(x_1^0, x_2^0, \ldots, x_n^0) = \Delta x_1 \frac{\partial f_1}{\partial x_1} + \Delta x_2 \frac{\partial f_1}{\partial x_2} + \ldots + \Delta x_n \frac{\partial f_1}{\partial x_n} \tag{5.32}$$

Similarly,

$$f_2(x_1^0 + \Delta x_1, x_2^0 + \Delta x_2, \ldots, x_n^0 + \Delta x_n)$$

$$= f_2(x_1^0, x_2^0, \ldots, x_n^0) + \Delta x_1 \left.\frac{\partial f_2}{\partial x_1}\right|_0 + \Delta x_2 \left.\frac{\partial f_2}{\partial x_2}\right|_0 + \ldots + \Delta x_n \left.\frac{\partial f_2}{\partial x_n}\right|_0$$

$$= y_2$$

i.e. $$y_2 - f_2(x_1^0, x_2^0, \ldots, x_n^0) = \Delta x_1 \frac{\partial f_2}{\partial x_1} + \Delta x_2 \frac{\partial f_2}{\partial x_2} + \ldots + \Delta x_n \frac{\partial f_2}{\partial x_n} \quad (5.33)$$

and also

$$y_n - f_n(x_1^0, x_2^0, \ldots, x_n^0) = \Delta x_1 \frac{\partial f_n}{\partial x_1} + \Delta x_2 \frac{\partial f_n}{\partial x_2} + \ldots + \Delta x_n \frac{\partial f_n}{\partial x_n} \quad (5.34)$$

The above equations (eqns. (5.32), (5.33) and (5.34)) can be written in the matrix form as shown below,

$$\begin{bmatrix} y_1 - f_1(x_1^0, x_2^0, \ldots, x_n^0) \\ y_2 - f_2(x_1^0, x_2^0, \ldots, x_n^0) \\ \vdots \\ y_n - f_n(x_1^0, x_2^0, \ldots, x_n^0) \end{bmatrix}$$

$$= \begin{bmatrix} \partial f_1/\partial x_1 & \partial f_1/\partial x_2 & \ldots & \partial f_1/\partial x_n \\ \partial f_2/\partial x_1 & \partial f_2/\partial x_2 & \ldots & \partial f_2/\partial x_n \\ & \vdots & & \\ \partial f_n/\partial x_1 & \partial f_n/\partial x_2 & \ldots & \partial f_n/\partial x_n \end{bmatrix} \begin{bmatrix} \Delta x_1 \\ \Delta x_2 \\ \\ \Delta x_n \end{bmatrix} \quad (5.35)$$

which can be written symbolically as

$$\overline{D} = [J]\,\overline{C} \quad (5.36)$$

where $[J]$ is called jacobian matrix. Here the process starts by assuming initial estimate for the unknowns i.e. $x_1^0, x_2^0, x_3^0, \ldots, x_n^0$. With these initial estimates, the elements of matrix $\overline{D}$ and also the elements of Jacobian matrix $[J]$ are calculated. The elements of Jacobian are calculated by substituting $x_1^0, x_2^0, \ldots, x_n^0$ in the equation and then taking its partial derivative. The above equation (5.35) are a set of linear equations which can be solved either by iterative techniques or by the method of elimination (note that the method of elimination cannot be used in the case of nonlinear algebraic equations) for the correction vector $\overline{C}$ (i.e. Δx_i). The correction thus obtained, is used to update the initial estimate i.e.

$$x_i^1 = x_i^0 + \Delta x_i \text{ for } i = 1,2, \ldots n \quad (5.37)$$

This values of x_i^1 is again used to compute the elements of $\overline{D}$ and also of the Jacobian $[J]$. That is, in every iteration, Jacobian is built with the latest values of unknowns. The process is continued till solution converges. However, if the initial estimate is not close to solution point, there will be large *truncation error*.

5.6 Numerical Solution of Differential Equations

Linear (ordinary) differential equation can be expressed as follows:

$$F\left(x, y, \frac{dy}{dx}, \frac{d^2y}{dx^2}, \dots, \frac{d^my}{dx^m}\right) = 0 \tag{5.38}$$

where x is the *independent variable* and y is the *dependent variable.* The order of this differential equation is m because m is the highest order derivative. In case, conditions are satisfied at a single point, the above is said to be *initial value problem.* In case, conditions are satisfied at more than one point, the above is known as *boundary value problem.* Initial value problem of first order can be expressed as,

$$dy/dx = f(x, y),$$

$$y(x_0) = y_0 \tag{5.39}$$

Note: Differential equations of the order higher than 1 with given initial conditions, can be reduced to a system of first order differential equation (initial value problem) as shown by the following example.

$$x^2 \frac{d^2y}{dx^2} + x\frac{dy}{dx} + (x^2 - p^2)y = 0 \tag{5.40}$$

with initial conditions

$$y(x_0) = y_0 = A$$

and

$$\frac{dy}{dx}(x_0) = y'(x_0) = B \tag{5.41}$$

Let $\quad dy/dx = Z$

Hence substituting eqn. (5.41) in eqn. (5.40), we get,

$$x^2 \frac{dZ}{dx} + xZ + (x^2 - p^2)y = 0$$

i.e. $\quad dZ/dx = -\frac{1}{x^2}(xZ + (x^2 - p^2)y)$

and $\quad dy/dx = Z$

In some cases, even the boundary value problem can be reduced to a system of first order differential equation.

5.6.1 *Error*

Certain types of error come into picture in the numerical solution of differential equations. These are,

(*i*) *Truncation error* (i.e. Descritization error) which is due to the particular numerical technique used, and

(*ii*) *Round-off-error* which is due to the loss of significant digits in the solution process as we can retain only fixed number of digits after each operation.

5.6.2 *Methods of Numerical Solution*

(*i*) *Single Step Method* : In the case of single step method for determining solution at x_{i+1}, we consider dependence at only one earlier point say x_i. In this case algorithm will be of the following form,

$$y_{i+1} = y_i + h\underbrace{\phi\,(x_i, y_i, h)}_{\text{increment function}} \qquad (5.42)$$

In the case of initial value problem, we know dependence at the earlier point from initial condition and hence no difficulty arises in finding solution at x_i for the first order differential equation as shown,

$$dy/dx = y' = f(x, y)$$

with initial condition

$$y\,(x_0) = y_0$$

There is one more advantage of the single step method and it is that the step size can be changed at any stage.

(*ii*) *Multistep Method :* In case, while discussing solution at x_{i+1}, dependence at k earlier points is considered, it is said to be k-step method. In the case of multistep method, information at k earlier points is not available totally, it is available only at x_0 from the initial conditions. Thus in this case, in the beginning we have to use single step method to provide information at k steps and then we use multistep method. In addition, it is difficult to change step size in the multistep method.

We shall now discuss both the single step method and also the multistep method along with examples.

5.6.3 *Single Step Method*

Following are the most common single step methods.

(*i*) *Euler's methods :* Let us take an initial value problem as shown,

$$dy/dx = y' = f(x, y)$$

with initial condition

$$y\,(x_0) = y_0 \qquad (5.43)$$

Algorithm is

$$y_{i+1} = y_i + h\,f\,(x_i, y_i) \qquad (5.44)$$

Geometrical interpretation of the algorithm (5.44) is clear from Fig. 5.1.

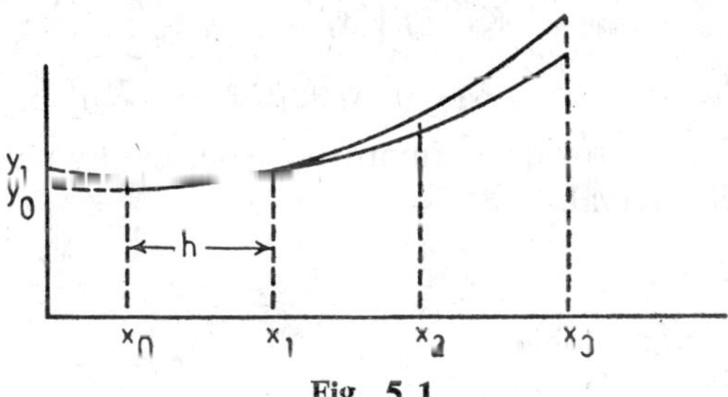

Fig. 5.1

Here the part of the curve in the interval h is being approximated by segment of the straight line whose slope is the same as that at the beginning of the subinterval. The method is only of theoretical importance.

(*ii*) *Modified Euler's Method:* This belongs to the method known as predictor-corrector method. Here the slope is approximated at the middle of the interval. The algorithm for predictor formula is

$$y_{i+1} = y_i + h\,f\,(x_i,\, y_i) \tag{5.45}$$

and the algorithm for the corrector formula is,

$$y_{i+1} = y_i + \frac{h}{2}\,[f\,(x_i,\, y_i) + f\,(x_{i+1},\, y_{i+1})] \tag{5.46}$$

It is clear from the above, that, the predictor formula is the same as the Euler's method where we obtain y_{i+1} at x_{i+1} with the initial value (or initial condition) of y_i at x_i. Then we correct the value y_{i+1} thus obtained with the help of corrector formula. This method is the same as *Runga Kutta method of order 2.*

(*iii*) *Runga Kutta Method of order 3 :* The algorithm to calculate y_{i+1} at x_{i+1} for an initial value problem using single step method is

$$y_{i+1} = y_i + h\,\phi\,(x_i,\, y_i,\, h) \tag{5.47}$$

Based upon this algorithm and using the indirect use of Taylor's theorem we obtain the following by comparing the coefficients.

$$y_{i+1} = y_i + \frac{h}{6}(K_1 + 4K_2 + K_3)$$

where

$$K_1 = f(x_i,\, y_i)$$

$$K_2 = f\left(x_i + \frac{h}{2},\, y_i + \frac{h}{2}\,K_1\right)$$

and

$$K_3 = f(x_i + h,\, y_i + 2h\,K_2 - h\,K_1). \tag{5.48}$$

(*iv*) *Classical Runga Kutta Method of Order 4:* Again expanding the function (eqn. 5.47) and matching the terms, we obtain

$$y_{i+1} = y_i + \frac{h}{6}\,(K_1 + 2K_2 + 2K_3 + K_4)$$

where

$$K_1 = f(x_i,\, y_i)$$

$$K_2 = f\,(x_i + \frac{h}{2},\, y_i + \frac{h}{2}\,K_1)$$

$$K_3 = f\left(x_i + \frac{h}{2},\, y_i + \frac{h}{2}\,K_2\right)$$

$$K_4 = f(x_i + h,\, y_i + h\,K_3) \tag{5.49}$$

All these algorithms provide an improved solution by taking average of the slope at half interval.

5.7 Stability

(i) Inherent unstable

(ii) Partial stable

(iii) Relative stable
(iv) Weak stable
(v) Strong stable
(vi) Absolute stable

A numerical method is said to be unstable, if error introduced at any other stage of computation due to round-off error or truncation error, or erroneous initial conditions, or due to any other reason, increase without bounds in subsequent calculations.

Stable Method: Growth of error remains bounded.

5.8 Multistep Method

In this method, to obtain y_{i+1}, information at K earlier points must be known. It is then referred as K-step method. This method is also known as Predictor-Corrector method based upon Predictor-Corrector formula.

In the case of *predictor formula*, information at x_{i+1}, f_{i+1} is not made use of; the result will be *explicit formula.* The predictor formula is actually an *open integration formula.* While in the case of *corrector formula,* information at x_{i+1}, f_{i+1} is also included. Hence corrector formula, which is an *implicit formula,* is similar to the *closed integration formula.* Actually corrector formula is obtained by making use of closed integration formula while predictor formula is obtained by making use of open integration formula. Combined together, they are known as *predictor-corrector formula* which actually gives rise to *multistep method.*

Multistep methods are *not self starting* since some of the values are missing initially. Actually we know only one value from the initial conditions and hence, earlier values are obtained by single step method. Algorithms for the predictor-corrector formula is obtained by following integration formula,

$$y_{i+1} = y_{i-K} + \int_{X_{i-K}}^{X_{i+1}} \phi(x)\,dx \tag{5.50}$$

Following are the important predictor-corrector formulae (i.e. multistep methods).

(*i*) *Fourth order Milne method*

Predictor formula:

$$y_{i+1} = y_{i-3} + \frac{4h}{3}(2f_i - f_{i-1} + 2f_{i-2}).$$

Corrector formula:

$$y_{i+1} = y_{i-1} + \frac{h}{3}(f_{i+1} + 4f_i + f_{i-1}) \tag{5.51}$$

(*ii*) *Sixth order Milne method*

Predictor formula:

$$y_{i+1} = y_{i-5} + \frac{3h}{10}(11f_i - 14f_{i-1} + 26f_{i-2} - 14f_{i-3} + 11f_{i-4})$$

Corrector formula:

$$y_{i+1} = y_{i-3} + \frac{2h}{45}(7f_{i+1} + 32f_i + 12f_{i-1} + 32f_{i-2} + 7f_{i-3}). \quad (5.52)$$

Modified Adams or Adams Multon method

Predictor formula:

$$y_{i+1} = y_i + \frac{h}{24}(55f_i - 59f_{i-1} + 37f_{i-2} - 9f_{i-3})$$

Corrector formula:

$$y_{i+1} = y_i + \frac{h}{24}(9f_{i-1} + 19f_i - 5f_{i-1} + f_{i-2}) \quad (5.53)$$

In all this predictor-corrector method, we use predictor formula only once and this is followed by the iterative use of corrector formula. We shall now illustrate these methods with the help of following examples.

Example 5.3: Solve the following initial value problem by modified Euler's method.

$$dy/dx = \dot{y} = xy$$

and i.e., $y(1) = 1$ i.e. at $x = 1, y = 1$

Find y at $x = 1.1, 1.2$.

Solution: Predictor $y_{i+1} = y_i + hf(x_i, y_i)$

Corrector $y_{i+1} = y_i + \frac{h}{2}(f(x_i, y_i) + f(x_{i+1}, y_{i+1}))$

At $x = 1$ i.e. at $x_i = 1$, $y_i = 1.0$

$$dy/dx = f(x_i, y_i) = x \cdot y = 1$$

(a) At $x = 1.1$, i.e. when $x_{i+1} = 1.1$

Thus step size $= h = 0.1$

Predictor: $y_{i+1} = y_i + h(f(x_i, y_i)) = 1 + 0.1*1 = 1.1$

Corrector: $y_{i+1} = y_i + \frac{h}{2}(f(x_i, y_i) + f(x_{i+1}, y_{i+1}))$

$$= 1 + \frac{0.1}{2}(1 + f(1.1, 1.1))$$

But $f(1.1, 1.1) = dy/dx = xy = 1.1*1.1 = 1.21$

Thus $y_{i+1} = 1 + \frac{0.1}{2}(1 + 1.21) = 1.1105.$

(b) at $x_i = 1.1$, $y_i = 1.1105$

then determine y_{i+1} at $x_{i+1} = 1.2$

$$h = 0.1$$

Predictor: $y_{i+1} = y_i + h(f(x_i, y_i))$

$$f(x_i, y_i) = f(1.1, 1.1105) = xy = 1.222$$

Then $$y_{i+1} = 1.1105 + 0.1*1.22 = 1.2327$$

Corrector: $$y_{i+1} = y_i + \frac{h}{2}(f(x_i, y_i) + f(x_{i+1}, y_{i+1}))$$

$$f(x_{i+1}, y_{i+1}) = f(1.2, 1.232)$$
$$= dy/dx = xy = 1.2* 1.232 = 1.478$$

$$y_{i+1} = 1.1105 + \frac{0.1}{2}(1.22 + 1.478) = 1.2455$$

Example 5.4: Use classical Runga Kutta method of fourth order for solving following initial value problem.

$$10\, dy/dx = x^2 + y^2, \text{ i.e., } y(0) = 1; \text{ at } x = 0.1, 0.2$$

Solution: $$y_{i+1} = y_i + \frac{h}{6}(K_1 + 2K_2 + 2K_3 + K_4)$$

$$K_1 = f(x_i, y_i)$$
$$K_2 = f(x_i + h/2, y_i + (h/2)\, K_1)$$
$$K_3 = f(x_i + \frac{h}{2}, y_i + \frac{h}{2} K_2)$$
$$K_4 = f(x_i + h, y_i + hK_3)$$
$$10dy/dx = x^2 + y^2$$

or $$dy/dx = \frac{1}{10}(x^2 + y^2) \text{ i.e. } y(0) = 1 \text{ at } x_o = 0, y = 1\,(x_0)$$

(a) at $x = 0.1$

$$K_1 = f(x_i, y_i) = \frac{dy}{dx} = \frac{1}{10}(x^2 + y^2)$$
$$= \frac{1}{10}(0 + 1) = 1/10 = 0.1$$

$$K_2 = f(x_i + \frac{h}{2}, y_i + \frac{h}{2} K_1)$$

Step size $h = 0.1$

and hence $h/2 = 0.05$

$$x_1 + \frac{h}{2} = 0 + 0.05 = 0.05$$

$$y_i + \frac{h}{2} K_1 = 1 + 0.05*0.1 = 1.005$$

$$K_2 = f(0.05, 1.005) = \frac{1}{10}((0.05)^2 + (1.005)^2)$$
$$= \frac{1}{10}(0.0025 + 1.010025)$$
$$= \frac{1}{10} * 1.0125 = 0.10125$$

$$K_3 = f(x + \frac{h}{2}, y_i + \frac{h}{2}K_2)$$
$$= f(0.05, 1 + 0.05 * 0.10125)$$
$$= f(0.05, 1.005)$$
$$= \frac{1}{10}((0.05)^2 + (1.005)^2)$$
$$= \frac{1}{10}(0.0025 + 1.010025)$$
$$= \frac{1}{10} * 1.0125 = 0.10125$$

$$K_4 = f(x_i + h, y_i + hK_3)$$
$$= f(0 + 0.1, 1 + 0.1 * 0.10125)$$
$$= f(0.1, 1.010125)$$
$$= \frac{1}{10}((0.1)^2 + (1.010125)^2)$$
$$= \frac{1}{10}(0.01 + 1.02)$$
$$= \frac{1}{10}(1.03) = 0.103$$

$$y_{i+1} = y_i + \frac{h}{6}(K_1 + 2K_2 + 2K_3 + K_4)$$
$$= 1 + \frac{0.1}{6}(0.1 + 0.202 + 0.202 + 0.103)$$
$$= 1 + \frac{0.1}{6}(0.607) = 1 + 0.1 * 0.101$$
$$= 1.0101 \text{ Ans.}$$

(b) At $x = 0.2$

$$y_{i+1} = y_i + \frac{h}{6}(K_1 + 2K_2 + 2K_3 + K_4)$$

at $x_i = 0.1$, $y_i = 1.0101$

$$K_1 = f(x_i, y_i) = \frac{1}{10}((0.1)^2 + (1.01)^2) = \frac{1}{10}(0.01 + 1.02)$$
$$= 0.103$$

$h = 0.1$

$$K_2 = f(x_i + \frac{h}{2}, y_i + \frac{h}{2} * K_1)$$
$$= f(0.1 + 0.05, 1.01 + 0.05 * 0.103)$$
$$= f(0.15, 1.015)$$

$$= \frac{1}{10}((0.15)^2 + (1.015)^2)$$

$$= \frac{1}{10}(0.0225 + 1.025) = \frac{1}{10}(1.047) = 0.104$$

$$K_3 = f(x_i + \frac{h}{2}, y_i + \frac{h}{2}K_2)$$

$$= f(0.1 + 0.05, 1.01 + 0.05*0.104)$$

$$= f(0.15, 1.0 + 0.00520) = f(0.15, 1.005)$$

$$= \frac{1}{10}((0.15)^2 + (1.005)^2)$$

$$= \frac{1}{10}(0.0225 + 1.01) = \frac{1}{10}(1.0325) = 0.10325$$

$$K_4 = f(x_i + h, y_i + hK_3)$$

$$= f(0.1 + 0.1, 1.01 + 0.1*0.103)$$

$$= f(0.2, 1.02)$$

$$= \frac{1}{10}((0.2)^2 + (1.02)^2)$$

$$= \frac{1}{10}(0.04 + 1.04) = \frac{1}{10}(1.08) = 0.108$$

$$y_{i+1} = y_i + \frac{h}{6}(K_1 + 2K_2 + 2K_3 + K_4)$$

$$= 1.01 + \frac{0.1}{6}(0.103 + 0.208 + 0.206 + 0.108)$$

$$= 1.01 + \frac{0.1}{6}(0.625) = 1.01 + 0.1*0.104$$

$$= 1.0204$$ Ans.

Example 5.5: Solve the following initial value problem by Milne's Predictor Corrector method at $x = 0.4, 0.5$ when their values are given at four points $x = 0.0, 0.1, 0.2, 0.3$.

$$\dot{y} + y = 2e^x;\ y_0 = 2,\ y_1 = 2.010,\ y_2 = 2.040,\ y_3 = 2.090$$

Solution: Milne Method : $\dot{y} + y = 2e^x$ or $\dot{y} = 2e^x - y$,

i.e. $dy/dx = 2e^x - y$

Predictor: $y_{i+1} = y_{i-3} + \frac{4h}{3}(2f_i - f_{i-1} + 2f_{i-2})$

$$x = 0, 0.1, 0.2, 0.3;\ y = 2, 2.010, 2.040, 2.090$$

f_i at $x = 0.3$ and $y = 2.090 = 2e^{0.3} - 2.090 = 2*1.35 - 2.090$

$$= 2.70 - 2.090 = 0.610$$

f_{i-1} at $x = 0.2$ and $y = 2.040 = 2e^{0.2} - 2.020$

$$= 2*1.22 - 2.040 = 2.44 - 2.040 = 0.40$$

f_{i-2} at $x = 0.1$ and $y = 2.010 = 2e^{0.1} - 2.010$

$$= 2*1.105 - 2.010 = 2.210 - 2.010 = 0.200$$

$$= 0.200$$

at $i - 2$ i.e. $y(0) = 2$

$$y_{i+1} = y_{i-3} + \frac{4h}{3}(2f_i - f_{i-1} + 2f_{i-2})$$

$$= 2 + \frac{4*0.1}{3}(1.22 - 0.40 + 0.40)$$

$$= 2 + \frac{0.4}{3}*1.22 = 2 + 0.4*0.41$$

$$= 2 + 0.16$$

$$= 2.16$$

$y_{i+1} = 2.16$ at $x = 0.4$ for predictor.

Corrector :

$$y_{i+1} = y_{i-1} + \frac{h}{3}(f_{i+1} + 4f_i - f_{i-1})$$

$$f_{i+1} = f(x_{i+1}, y_{i+1}) = f(0.4, 2.16)$$

$$= 2e^{0.4} - 2.16 = 2*1.49 - 2.16$$

$$= 2.98 - 2.16 = 0.82$$

$$y_{i+1} = y_{i-1} + \frac{h}{3}(f_{i+1} + 4f_i + f_{i-1})$$

$$= 2.040 + \frac{0.1}{3}(0.82 + 2.44 + 0.4)$$

$$= 2.04 + \frac{0.1}{3} * 3.66 = 2.04 + 0.1*1.22$$

$$= 2.162 \text{ Ans.}$$

at $x = 0.5$ i.e. x_{i+1} find y_{i+1}

at $x_i = 0.4$ $\quad x_{i-1} = 0.3, \quad x_{i-2} = 0.2$

$y_i = 2.162 \quad y_{i-1} = 2.090, y_{i-2} = 2.040$

Predictor :

$$y_{i+1} = y_{i-3} + \frac{4h}{3}(2f_i - f_{i-1} + 2f_{i-2})$$

$$f_i = dy/dx = 2e^x - y = 2e^{0.4} - 2.162$$

$$= 2*1.49 - 2.162$$

$$= 2.98 - 2.162 = 0.818$$

$$f_{i-1} = 2e^{0.3} - 2.090 = 0.610$$

$$f_{i-2} = 2e^{0.2} - 2.040 = 0.40$$

$$y_{i+1} = 2.010 + \frac{4*0.1}{3}(2*0.818 - 0.610 + 2*0.40)$$

$$= 2\ 010 + \frac{0.4}{3}(1.636 - 0.610 + 0.80)$$

$$= 2.010 + \frac{0.4}{3} * 0.742$$

$$= 2.010 + 2.968*0.1$$

$$= 2.010 + 0.2968$$

$$= 2.306$$

Corrector :

$$y_{i-1} = y_{i-1} + \frac{h}{3}(f_{i+1} + 4f_i + f_{i-1})$$

$$f_{i+1} = dy/dx = 2e^x - y$$

$$= 2e^{0.5} - 2.306$$

$$= 2*1.65 - 2.306$$

$$= 3.30 - 2.306 = 0.994$$

Thus $y_{i+1} = y_{i-1} + \frac{h}{3}(f_{i+1} + 4f_i + f_{i-1})$

$$= 2.090 + 0.01\ (0.994 + 4*0.818 + 0.610)$$

$$= 2.090 + \frac{0.1}{3}(0.994 + 3.272 + 0.610)$$

$$= 2.090 + \frac{0.1}{3} * 4.876$$

$$= 2.090 + 0.1 * 1.625$$

$$= 2.090 + 0.1625 = 2.2525 \quad \textbf{Ans.}$$

Problems

5.1 Solve the following system by the Gaussian Elimination method. (a) Without maximum pivot strategy and (b) With maximum pivot strategy

$$6x_1 + 3x_2 + 2x_3 = 6$$
$$6x_1 + 4x_2 + 3x_3 = 0$$
$$20x_1 + 15x_2 + 12x_3 = 0$$

5.2 State Gauss-Seidel iterative method for solving a system of linear algebraic equations. Solve the following system of equations with the help of Gauss-Seidel iterative technique by doing iteration thrice with initial approximations;

$x_1 = x_2 = x_3 = 0$

$$10x_1 + 2x_2 + x_3 = 9$$
$$2x_1 + 20x_2 - 2x_3 = -44$$
$$-2x_1 + 3x_2 + 10x_3 = 22$$

5.3 Solve the following system using Gauss-Jordan reduction scheme with maximum pivot strategy

$$2x_1 + 4x_2 + 16x_3 = -6$$
$$x_1 - 2x_2 - 3x_3 = 0$$
$$-x_1 + 2x_2 + x_3 = 2$$

5.4 Solve the following initial value problem by modified Euler's method

$$y' = xy, \; y(1) = 1.5 \quad \text{at} \quad x = 1.1, 1.2, 1.3, 1.4$$

5.5 Solve the following initial value problem at $x = 0.4, 0.5$ by Milne's predictor corrector method, given their values at four points $x = 0\,(0.1)\,0.3$

$$y' = 4y, \; y_0 = 1, \; y_1 = 1.492, \; y_2 = 2.226, \; y_3 = 3.320$$

5.6 Use the classical Runga Kutta method of 4th order for solving following initial value problem

$$y' = x - y^2, \; y(0) = 1 \quad \text{at} \quad x = 0.1, 0.2$$

References

1. Ratson, A., *A first Course in Numerical Analysis*, McGraw-Hill 1965.
2. Carnahan, B., Luther, H.A. and Wilkes, J.O., *Applied Numerical Methods*, Wiley 1969.
3. Hildebrand, F.B. *Introduction to Numerical Analysis*, McGraw-Hill 1956.
4. Henrice, P., *Elements of Numerical Analysis*, Wiley, New York, 1964.
5. Krishnamurthy, E.V. and Sen, S.K., *Computer Based Numerical Algorithms*, East-West Press, 1976.
6. Demidovich, B.P. and Maron, I.A. *Computational Mathematics*, MIR Publishers, Moscow, 1973.
7. Jain, M.K. and Chawla, M.M., *Numerical Analysis for Scientists and Engineers*, S.W.S. Publishers, Delhi, 1971.

Chapter 6

Load Flow Studies

6.1 Introduction

Planning the operation of power systems under existing conditions, its improvement and also future expansion requires the load flow studies, short circuit studies and stability studies. However the load flow studies are very important for planning, control and operations of existing systems as well as planning its future expansion as the satisfactory operation of the system depends upon knowing the effects of interconnections, new loads, new generating stations or new transmission lines etc. before they are installed. With the help of load flow studies we can also determine the best size and as well as the most favourable locations for the power capacitors both for the improvement of the p.f. and also raising the bus voltages of the electrical network. The load flow studies also help us to determine the best locality as well as optimal capacity of the proposed generating stations, substations or new lines. This is why the load flow studies are really important for planning the existing system as well as its future expansion.

The information obtained from the load flow studies are usually the magnitude and phase angle of voltages at each bus and active and reactive power flow in each line.

The extensive calculations required both for power flow as well as for voltage determination, necessiated the use of some type of automatic calculators or computer. This led to the design of special purpose analog computer, called a.c. network analyzer sometimes in the year 1929. The operation of the power system under existing conditions as well as the proposed future expansion could be simulated by this device. Digital computer for the load flow studies gained importance during the beginning of 1950 and the first planning studies on the digital computer was completed by the year 1956. Today most of the load flow studies are carried out on the digital computers. This change from the network analyzer to the digital computer has resulted in greater flexibility, economy accuracy and quicker operation.

6.2 Formulation of Load Flow Problem

Actually the load flow problem consists of calculations of voltage magnitude and its phase angle at the buses, and also the active and reactive line flows for the specified terminal or bus conditions. Associated with each bus of the power system network, there are four quantities such as:

(i) magnitude of voltage, i.e. $|V|$
(ii) phase angle of the voltage i.e. ϕ
(iii) real power i.e. $|P|$
(iv) reactive power i.e. $|Q|$

6.2.1 *Type of Buses*

All the buses of the power system network are generally classified into three categories viz. generation bus, load bus and the slack bus and two of the four quantities as mentioned above, are specified at each of the buses as shown below:

(*i*) *Generation Bus* (or voltage controlled bus): This is also called *p-v* bus and here the voltage magnitude $|V|$ and the real power $|P|$ are specified.

(*ii*) *Load Bus:* This is also called *p-q* bus and here real power $|P|$ and reactive power $|Q|$ are specified.

(*iii*) *Slack or Swing Bus:* This is also known as the reference bus and the voltage magnitude $|V|$ and phase angle ϕ are specified here. This bus is selected to provide additional real and reactive power to supply the transmission losses since these are unknown until final solution is obtained. If slack bus is not specified, than a generation bus usually with maximum real power $|P|$ is taken as a slack bus. There can be more than one slack bus in a given scheme.

6.2.2 *Techniques of Solving Load flow Problems*

The development of any method for the load flow studies on the digital computer, requires the following main considerations:

(i) The mathematical formulation of the load flow problem.
(ii) Application of numerical technique to solve these problems.

The mathematical formulation of the load flow problem is a system of nonlinear algebraic equations and since the digital computer can only perform the four basic arithmatic operations viz. additions, subtractions, multiplications and division (actually only additions and subtractions since the remaining operations can also be reduced to these two basic operations as mentioned above) numerical techniques are needed, such that the equations can be solved by the digital computer.

The basic equations for the purpose of load flow studies are a set of network equations which can be established either in the loop frame of reference or bus frame reference. The coefficient of these equations (which is actually nonlinear algebraic equations) depend upon the selection of independent variables viz. voltages or currents. Thus either the impedance or the admittance matrix can be used.

The early approach to the load flow problem employed the loop frame of reference in the admittance form but later on this method was discarded because of the tedious data preparations required to specify the network loops. Later approaches used the bus frame of reference both in the admittance form and also in the impedance form. This method gained importance due to various reasons such as the simplicity of data preparation, ease by which the bus admittance matrix (i.e. Y_{BUS}) can be obtained from a given power system network (i.e. primitive network matrix) and various other advantages both using Y_{BUS} or Z_{BUS} technique. This is why most of the load flow problems are formulated today in the bus frame of reference using Y_{BUS} or Z_{BUS} matrices. Already we have seen that the Y_{BUS} matrix can be obtained directly by inspection from a given primitive network (matrix) and hence the Y_{BUS} matrix is usually sparse i.e. it contains mostly zero elements (nearly 80 to 85 elements are zero) at the off diagonal locations and hence the sparsity technique to conserve the sparsity structure of the Y_{BUS} matrix, are employed. This aspect will be discussed later in detail. This method is followed by W.F. Tinney and his associates at B.P.A. (PORTLAND—ORE). However the inverse of Y_{BUS} i.e. Z_{BUS} is a full matrix and hence sparsity techniques (i.e. optimal ordering to exploit sparsity) cannot be used in this case. Thus the formulation of load flow problem using Z_{BUS} (i.e. Bus impedance matrix which is a full matrix) employs Diakoptics techniques which is actually the piecewise solution of the power system problem by using tearing off technique. This method is followed by Happ and his associates at G.E. (U.S.A). Both these techniques try to economize the computer memory and hence the computer time and thus the cost of computations.

We shall discuss the formulation of the load flow problems and also its solution in the bus frame of reference using both Y_{BUS} and Z_{BUS} which is basically the solution of nonlinear algebraic equations. We shall discuss first of all the solution techniques of load flow problem using Y_{BUS}. Actually the different solution techniques are nothing but the solution techniques of solving nonlinear algebraic equations because the mathematical model of the power system network (i.e. network equations) are a set of nonlinear algebraic equations. For the purpose of load flow studies we shall assume the balanced excitation and hence the formulation will be based upon the positive sequence network only (i.e. single phase circuit consisting of only positive sequence parameters)

6.3 Solution Technique Using Y_{BUS} in the Bus Frame of Reference

For the purpose of load flow studies, not only we work with the positive sequence power system network but we also neglect the mutual coupling between the elements of the power system network, i.e. we assume that the primitive network matrix is a diagonal matrix. If any bus in the power system network has both the load and generation, then either we split this bus into two buses by creating a fictitious bus or we take the load as the negative generation. By this arrangement we will finally obtain a set of buses known as *generation buses* where $|P|$ and $|V|$ are known, a set of buses known as *load buses* where $|P|$ and $|Q|$ are known and a *slack bus* which is actually the reference bus where $|V|$ and ϕ are specified.

Moreover with this assumption, Y_{BUS} matrix can be found directly by inspection. Normally ground is included in the development of Y_{BUS} matrix and at the same time ground is taken as a reference node. With this development, the elements of Y_{BUS} matrix include also the shunt connections between the buses and the ground such as transformer magnetizing impedance, static load and line charging etc.

The performance equation using Y_{BUS} in the bus frame of reference is:

$$\bar{I}_{BUS} = [Y_{BUS}]\, E_{BUS} \tag{6.1}$$

And if the power at any bus K is known, then it can be expressed by the following basic equation viz :

$$P_K - jQ_K = E_K^* I_K \tag{6.2}$$

where P_K and Q_K are real and reactive power at the bus K. From the above equation (i.e. (eqn. 6.2)), we get,

$$I_K = \frac{P_K - jQ_K}{E_K^*} \tag{6.3}$$

As explained, these nonlinear algebraic equations can be solved by any solution techniques used to solve the nonlinear algebraic equations such as Gauss iterative techniques, Gauss-Seidel iterative techniques and Newton's method etc. We shall discuss all these techniques including their merits and demerits.

6.3.1 Gauss and Gauss Seidel Iterative Technique

Load Buses: Here real power P and reactive power Q is given. Let us assume n-node system including a slack bus "S" where both V and ϕ are are specified and they remain fixed throughout. Since $|P|$ and $|Q|$ are given for all the buses except the slack bus, we have for any bus K (from equation (6.3))

$$I_K = \frac{P_K - jQ_K}{E_K^*} \quad \text{for} \quad K = 1, 2, \ldots, n \;\; \neq S \tag{6.4}$$

where S is a slack bus.

Now the performance equation in the bus frame of reference using Y_{BUS} where ground is included as reference node will be (from equation (6.1)):

$$Y_{BUS} = [Y_{BUS}]\, \bar{E}_{BUS} \tag{6.5}$$

Here for n-node system, there will be $n-1$ linear independent equations to be solved. Expanding equation (6.5), we get:

$$I_1 = Y_{11}E_1 + Y_{12}E_2 + \ldots$$

and hence for Kth bus

$$I_K = Y_{K1}E_1 + Y_{K2}E_2 + \ldots = \sum_{q=1}^{n} Y_{Kq}E_q \tag{6.6}$$

i.e.

$$I_K = \sum_{q=1}^{n} Y_{Kq}E_q, \; K = 1, \ldots, n; \; K \neq S$$

$$= Y_{KK}E_K + \sum_{\substack{q=1 \\ q \neq K}}^{n} Y_{Kq}E_q \tag{6.7}$$

Thus $$E_K = \frac{1}{Y_{KK}}[I_K - \sum_{\substack{q=1 \\ q \neq K}}^{n} Y_{Kq}E_q] \quad \text{for} \quad K = 1, \ldots, n \quad K \neq S$$

Substituting for I_K from equation (6.4), we have

$$E_K = \frac{1}{Y_{KK}}\left[\frac{P_K - jQ_K}{E_K^*} - \sum_{\substack{q=1 \\ q \neq K}}^{n} Y_{Kq}E_q\right] \tag{6.8}$$

where $K = 1, \ldots, n.$

and $K \neq S$

In the *Gauss method*, we assume the voltage for all the buses except the slack bus where the voltage magnitude and phase angle are specified and remain fixed. Normally, we set i.e. assume the voltage magnitude and phase angle of these buses equal to that of the slack bus and work in per unit system (i.e. we may take in p.u., the voltage magnitude and phase angle as $1 \angle 0$). The assumed bus voltage and the slack bus attach along with P and Q are substituted in the right hand side of the equation (6.8) to obtain new set of bus voltages. After the entire iteration is complete, the new set of bus voltages are again substituted along with the specified slack bus voltage in the right hand side of equation (6.8) to obtain a new set of bus voltages. The process is continued till

$$| E_K^{c+1} - E_K^c | \leqslant \varepsilon \tag{6.9}$$

where c is an iteration count and ε is a very small number which depends upon the system accuracy and is normally equal to 0.0001 etc.

Thus the process is continued till the mod of the bus voltage obtained at the current iteration minus the value of the bus voltage at the previous iteration is less than a chosen very small number and in this way we obtain the solution viz. $|V|$ and ϕ.

Gauss-Seidel Iterative Technique: In the case of Gauss-Seidel method, the value of bus voltages calculated for any bus immediately replace the previous values in the next step while in the case of Gauss method as stated earlier, the calculated bus voltages replace the earlier value only at the end of the iteration. Due to this Gauss-Seidel method converges much faster than that of Gauss method, i.e. number of iterations needed to obtain solution is much less in the Gauss-Seidel method compared to the Gauss method. This is why Gauss-Seidel method is very popular

Generation Bus: This is the voltage controlled bus where $|P|$ and $|E|$ are specified, i.e. given. However, usually limit of reactive power i.e. Q_{max} and Q_{min} to hold the generation voltage within limits are also given. For any bus p we have,

$$P_p - jQ_p = E_p^* I_p \tag{6.10}$$

and $$I_p = \sum_{q=1}^{n} Y_{pq} E_q \tag{6.11}$$

Then substituting eqn. (6.11) in eqn. (6.10), we get

$$P_p - jQ_p = E_p^* \sum_{q=1}^{n} Y_{pq} E_q \tag{6.12}$$

Thus,

$$Q_p = \text{Imaginary } E_p^* \sum_{q=1}^{n} Y_{pq} E_q \tag{6.13}$$

The process starts by assuming the bus voltages and its phase angles say $|E|$ and ϕ, from which we determine the real and the imaginary components (say e and f) of the bus voltages. Say for any bus p

$$e_p = E_p \cos \phi_p$$
$$f_p = E_p \sin \phi_p \tag{6.14}$$

Since here we are given voltage magnitude i.e. for any bus p, E_p is specified and therefore selected value of e_p and f_p must satisfy the following equation approximately.

$$e_p^2 + f_p^2 \simeq [|E_p| \text{ scheduled}]^2 \tag{6.15}$$

where E_p (scheduled) is the specified bus voltage for any bus p.

Then substituting this assumed value of e_p and f_p in the reactive power Q_p (in equation (6.13)), the reactive power Q is calculated. If the reactive power calculated exceeds the Q_{max} or Q_{min}, then we do the following,

$$\text{if} \quad Q \geqslant Q_{max}$$
$$\text{put} \quad Q = Q_{max}$$
and
$$\text{if} \quad Q \leqslant Q_{min}$$
$$\text{put} \quad Q = Q_{min} \tag{6.16}$$

and treat this bus as a load bus to find the voltage solution. If the above i.e. (eqn. (6.16)) is not true then use the phase angle of this assumed bus voltage i.e. ϕ_p to recalculate e_p and f_p for any bus p.

We know that $\phi_p^K = \tan^{-1}(f_p^K / e_p^K)$, where K is the iteration count. Assuming this phase angle ϕ_p also to be that of the scheduled bus voltage E_p (for this bus, ϕ_p is unknown but E_p is given), we get

$$e_p^K \text{ (new)} = |E_p \text{ scheduled}| \cos \phi_p^K$$
$$f_p^K \text{ (new)} = |E_p \text{ scheduled}| \sin \phi_p^K \tag{6.17}$$

We substitute the e_p^K and f_p^K from eqn. (6.17) in the eqn. (6.8) to recalculate bus voltages and the process is continued till the solution converges. However, at every step the reactive power is also calculated to check whether Q calculated is within the limit.

Acceleration factor: Some times we use *acceleration factor* to increase the rate of convergence. The acceleration factor chosen depends upon the system and its value normally lies within 1.4 to 1.6. Then after calculating e_p^{K+1} and f_p^{K+1} at $K+1$ st. iteration and knowing the acceleration factors say α and β, we calculate the new estimate for the bus voltages say

$$e_p^{K+1} \text{ (accelerated)} = e_p^K + \alpha\,[e_p^{K+1} - e_p^K]$$

$$f_p^{K+1} \text{ (accelerated)} = f_p^K + \beta\,[f_p^{K+1} - f_p^K] \quad (6.18)$$

and this new estimate replaces the calculated value e_p^{K+1} and f_p^{K+1}.

After calculating bus voltages and their phase angles for all the buses, the line flow and line losses are calculated.

Line Flow and Losses: Here we assume the normal π representation of transmission line, as shown in Fig. 6.1.

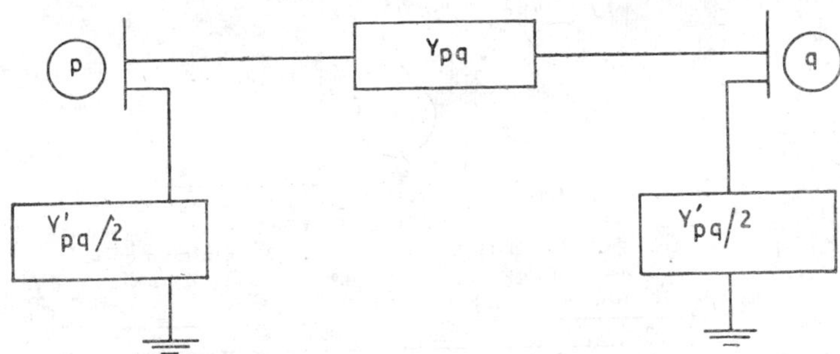

Y_{pq}—series admittance between any bus p and q and Y'_{pq}—line charging.

Fig. 6.1

As we have already found the solution of bus voltages (say both $|V|$ and ϕ for load buses and only ϕ for generation bus), we calculate the line flows between any buses p and q with the help of the above nominal π representation (Fig. 6.1).

Let i_{pq} = be current flow from bus p towards q, then

$$i_{pq} = [E_p - E_q]\,Y_{pq} + E_p \frac{Y'_{pq}}{2} \quad (6.16)$$

where E_p and E_q are the bus voltages at the buses p and q which are already calculated from the load flow studies.

The power flow in the line $p-q$ at the bus p is given by

$$P_{pq} - jQ_{pq} = E_p^*\, i_{pq}$$

$$= E_p^*\,[E_p - E_q]\,Y_{pq} + E_p^*\,E_p \frac{Y'_{pq}}{2} \quad (6.20)$$

Similarly the line flow in the line $p-q$ at the bus q is given by

$$P_{pq} - jQ_{pq} = E_q^*\,[E_q - E_p]\,Y_{pq} + E_q^*\,E_q \frac{Y''_{pq}}{2} \quad (6.21)$$

The algebraic sum of the above equations (eqns. 6.20 and 6.21) is the line losses in the element $p-q$.

6.3.2 *Newton-Raphson Method*

Load Bus: Here P and Q are given and $|V|$ and ϕ are to be calculated. For any bus p, the real and the reactive power is given by the equation:

$$P_p - jQ_p = E_p^* I_p \tag{6.22}$$

But $\bar{I}_{BUS} = [Y_{BUS}]\,\bar{E}_{BUS}$ and hence expanding this equation for any bus p, we get,

$$I_p = Y_{p1} E_1 + Y_{p2} E_2 + \ldots$$

$$= \sum_{q=1}^{n} Y_{pq} E_q \tag{6.23}$$

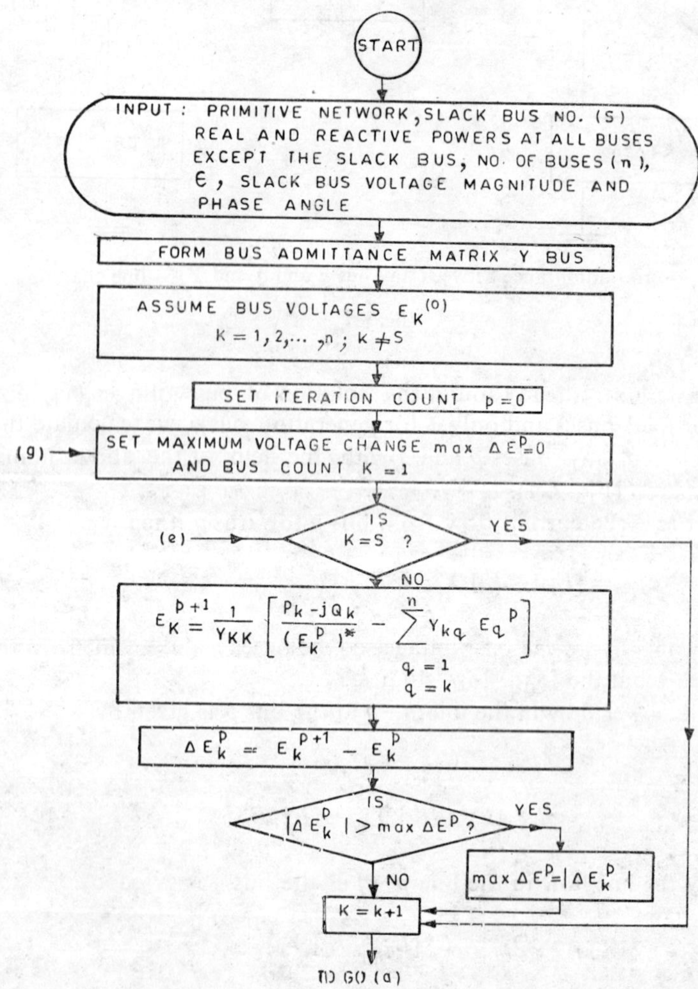

(a)

Substituting equation (6.23) in the equation for power (eqn. 6.22), we obtain for any bus ⓟ

$$P_p - jQ_p = E_p^* \sum_{q=1}^{n} Y_{pq} E_q \tag{6.24}$$

If we want formulation in rectangular coordinates, we express

$$E_p = e_p + jf_p$$

where e_p and f_p are the real and imaginary components of the bus voltage E_p, and hence

$$E_p^* = e_p - jf_p \tag{6.25}$$

Similarly,

$$E_q = e_q + jf_q \tag{6.26}$$

and

$$Y_{pq} = G_{pq} - jB_{pq} \tag{6.27}$$

where G_{pq} and B_{pq} are conductance and suceptance respectively. Substituting eqns. (6.25), (6.26) and (6.27) in eqn. (6.24) for power, we get

$$P_p - jQ_p = [e_p - jf_p] \sum_{q=1}^{n} [(G_{pq} - jB_{pq})(e_q + jf_q)] \tag{6.28}$$

From above eqn. (6.28), we determine for any bus p,

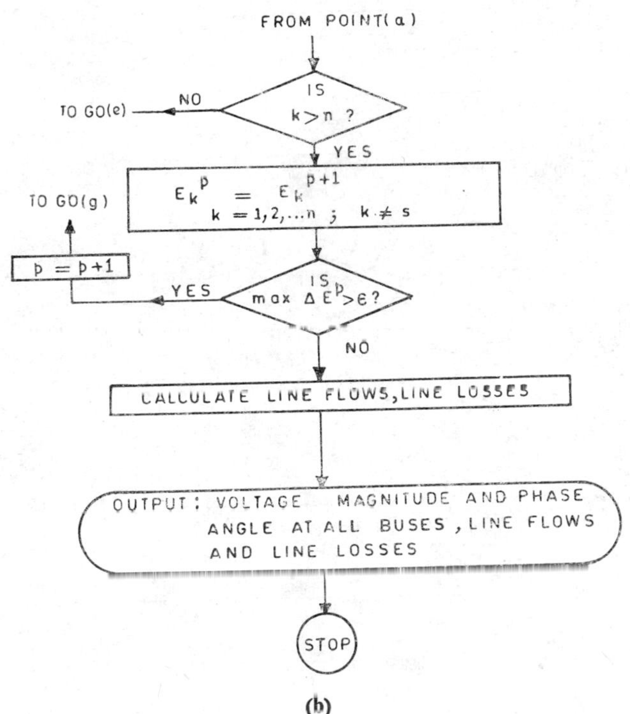

(b)

Fig. 6.2 Gauss method.

$$P_p = \text{Real part of } E_p^* \sum_{q=1}^{n} Y_{pq} E_q$$

$$= \sum_{q=1}^{n} [e_p (e_q G_{pq} + f_q B_{pq}) + f_p (f_q G_{pq} - e_q B_{pq})] \tag{6.29}$$

and

$$Q_p = \text{Imaginary } E_p^* \sum_{q=1}^{n} Y_{pq} E_q$$

$$= \sum_{q=1}^{n} [f_p (e_q G_{pq} + f_q B_{pq}) - e_p (f_q G_{pq} - e_q B_{pq})] \tag{6.30}$$

Separating for pth bus, the power equations (6.29) and (6.30) becomes

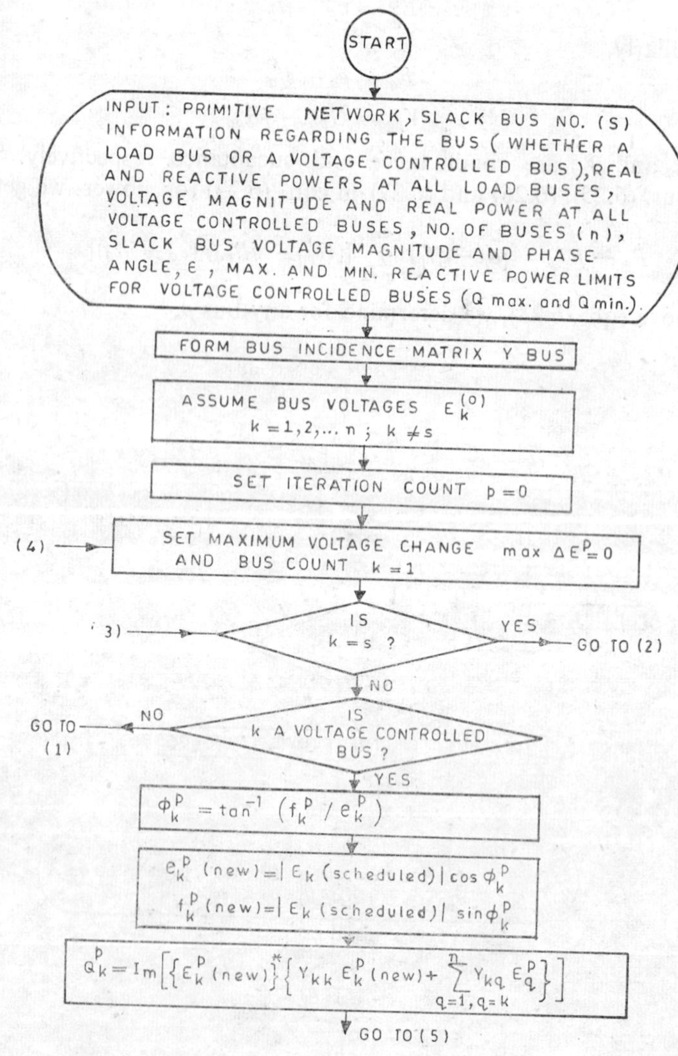

(a)

$$P_p = e_p (e_p G_{pp} + f_p B_{pp}) + f_p (f_p G_{pp} - e_p B_{pp})$$

$$+ \sum_{\substack{q=1 \\ q \neq p}}^{n} [e_p(e_q G_{pq} + f_q B_{pq}) + f_p (f_q G_{pq} - e_q B_{pq})] \quad (6.31)$$

and

$$Q_p = f_p (e_p G_{pp} + f_p B_{pp}) - e_p (f_p G_{pp} - e_p B_{pp})$$

$$+ \sum_{\substack{q=1 \\ q \neq p}}^{n} [f_p (e_q G_{pq} + f_q B_{pq}) - e_p (f_q G_{pq} - e_q B_{pq})] \quad (6.32)$$

Thus the above formulation results in a system of nonlinear algebraic equations, two equation (one for P_p and the other for Q_p) at each bus. So excluding the slack bus where $|V|$ and ϕ are specified and remains

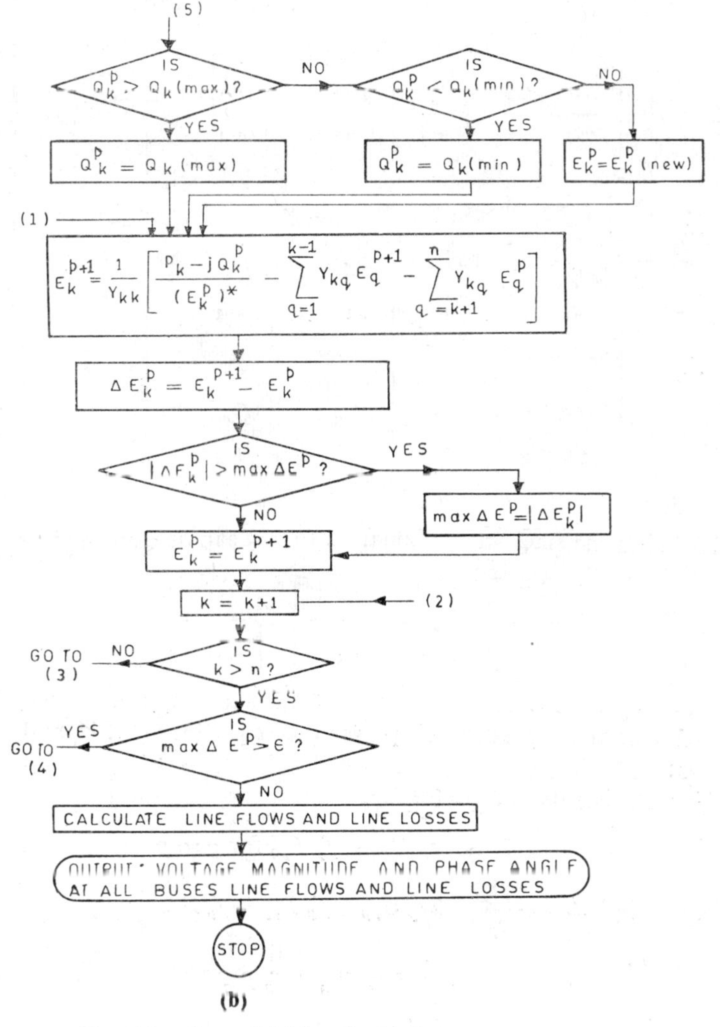

(b)

Fig. 6.3 Gauss Seidel method.

fixed throughout, the total number of equations to be solved for n bus system will be 2 $(n-1)$ equations.

With the help of the Newton Raphson's method, the above non-linear algebraic equations of power is transformed into a set of linear algebraic equations inter-relating the changes in power (i.e. error in power) with the change in real and reactive components of bus voltages with the help of Jacobian matrix. This is actually done by expanding the function by Taylor's series and neglecting higher order derivatives and higher power. Of course it is necessary that the initial guess is near the solution point otherwise there will be quite a large truncation error. Thus we get from eqn. (6.29) and (6.30)

$$\begin{bmatrix} \Delta P_1 \\ \vdots \\ \Delta P_{n-1} \\ \cdots \\ \Delta Q_1 \\ \cdots \\ \Delta Q_{n-1} \end{bmatrix} = \left[\begin{array}{cc|cc} \dfrac{\partial P_1}{\partial e_1} & \dfrac{\partial P_1}{\partial e_{n-1}} & \dfrac{\partial P_1}{\partial f_1} & \dfrac{\partial P_1}{\partial f_{n-1}} \\ \dfrac{\partial P_{n-1}}{\partial e_1} & \dfrac{\partial P_{n-1}}{\partial e_{n-1}} & \dfrac{\partial P_{n-1}}{\partial f_1} & \dfrac{\partial P_{n-1}}{\partial f_{n-1}} \\ \hline \dfrac{\partial Q_1}{\partial e_1} & \dfrac{\partial Q_1}{\partial e_{n-1}} & \dfrac{\partial Q_1}{\partial f_1} & \dfrac{\partial Q_1}{\partial f_{n-1}} \\ \dfrac{\partial Q_{n-1}}{\partial e_1} & \dfrac{\partial Q_{n-1}}{\partial e_{n-1}} & \dfrac{\partial Q_{n-1}}{\partial f_1} & \dfrac{\partial Q_{n-1}}{\partial f_{n-1}} \end{array}\right] \begin{bmatrix} \Delta e_1 \\ \vdots \\ \Delta e_{n-1} \\ \vdots \\ \Delta f_1 \\ \vdots \\ \Delta f_{n-1} \end{bmatrix} \tag{6.33}$$

Here nth bus is the slack bus. Equation (6.33) can be expressed as

$$\begin{bmatrix} \Delta P \\ \hline \Delta Q \end{bmatrix} \left[\begin{array}{c|c} J_1 & J_2 \\ \hline J_3 & J_4 \end{array}\right] \begin{bmatrix} \Delta e \\ \hline \Delta f \end{bmatrix}$$

where J_1, J_2, J_3 and J_4 are the elements of the Jacobian matrix which are calculated from the expression of power (i.e. equations (6.31) and (6.32) as follows:

Off diagonal elements of $J_1 = \partial P_p/\partial e_q$

$$= e_p\, G_{pq} - f_p\, B_{pq} \text{ for } q \neq p \tag{6.35}$$

$$\text{Dia. elements of } J_1 = \frac{\partial P_p}{\partial e_p} = 2e_p\, G_{pp} + f_p\, B_{pp} - f_p\, B_{pp}$$

$$+ \sum_{\substack{q=1 \\ q \neq p}}^{n} [(e_q\, G_{pq} + f_q\, B_{pq})] \tag{6.36}$$

Elements of J_2:

Off dia. element $= \partial P_p/\partial f_q$

$$= e_p B_{pq} + f_p G_{pq} \quad q \neq p \tag{6.37}$$

Dia. elements $= \partial P_p/\partial f_p$

$$= e_p B_{pp} + 2f_p G_{pp} - e_p B_{pp} + \sum_{\substack{q=1 \\ q\neq p}}^{n} [f_q G_{pq} - e_q B_{pq}] \tag{6.38}$$

Elements of J_3:

Off dia. element $= \partial Q_p/\partial e_q$

$$= f_p G_{pq} + e_p B_{pq} \text{ for } q \neq p \tag{6.39}$$

Dia. elements $= \partial Q_p/\partial e_p$

$$= f_p G_{pp} - f_p G_{pp} + 2e_p B_{pp} - \sum_{\substack{q=1 \\ q\neq p}}^{n} [f_p G_{pq} - e_q B_{pq}] \tag{6.40}$$

Elements of J_4:

Off. dia. elements $= \partial Q_p/\partial f_q$

$$= f_p B_{pq} - e_p G_{pq} \text{ for } q \neq p \tag{6.41}$$

and

Dia. elements $= \partial Q_p/\partial f_p$

$$= e_p G_{pp} + 2f_p B_{pp} - e_p G_{pp} + \sum_{\substack{q=1 \\ q\neq p}}^{n} (e_q G_{pq} + f_q B_{pq}) \tag{6.42}$$

Algorithm: Following steps in sequence are followed to obtain load flow solution by *N-R* Method.

1. So, for the load buses where P and Q are given, we assume the bus voltages magnitude and phase angle for all the buses except the slack bus where $|V|$ and ϕ are specified. Normally we have the flat voltage start i.e. we set the assumed bus voltage magnitude and its phase angle (in other words the real and imaginary component e and f of the bus voltages) equal to the slack bus quantities.

2. Substituting this assumed bus voltages (i.e. e and f) in eqns. (6.31) and (6.32), we calculate the real and reactive components of power, i.e. P_p and Q_p for all the buses $p = 1, \ldots, n-1$ except the slack bus.

3. Since P_p and Q_p for any bus p is given, i.e. scheduled, the error in the power will be

$$\Delta P_p^K = P_p \text{ (scheduled)} - P_p^K \tag{6.43}$$

$$\Delta Q_p^K = Q_p \text{ (scheduled)} - Q_p^K$$

where K is an iteration count

Here P_p^K and Q_p^K are the power calculated with the latest value of bus voltages at any iteration K.

4. Then the elements of Jacobian matrix (J_1, J_2, J_3 and J_4) are

calculated with the latest bus voltages and calculated power eqns. (6.31) and (6.32).

5. After this we solve the linear set of equation (6.33) by either iterative technique or by the method of elimination (normally by Gaussian elimination method) to determine the voltage correction, i.e. Δe_p and Δf_p at any bus p.

6. This value of voltage correction is used to determine the new estimate of bus voltages as follows:

$$e_p^{K+1} = e_p^K + \Delta e_p^K$$

$$f_p^{K+1} = f_p^K + \Delta f_p^K \tag{6.44}$$

where K is an iteration count.

7. Now this new estimate of the bus voltage i.e. e_p^{K+1} and f_p^{K+1} is used in equations (6.31) and (6.32) for power to recalculate the error in power and thus the entire algorithm starting from step 3 as listed above is repeated.

Here in each iteration, the elements of Jacobian is calculated since it depends upon the latest voltage estimate and calculated power. The process is continued till the error in power becomes very small i.e.

$$|\Delta P| < \varepsilon$$

and

$$|\Delta Q| < \varepsilon \tag{6.45}$$

where ε is very small number.

This method converges faster than the Gauss-Seidel method because of quadratic convergence. Moreover, while the number of iterations to obtain the nominal solutions increases with the problem size in the case of Gauss-Seidel method, the number of iterations to obtain solution is nearly constant in the Newton Raphsons method. We shall see later that with the modified Newton's method, nearly 5 to 6 iterations are needed to obtain solution though the time taken to complete an iteration is nearly seven times more than that of the Gauss Seidel method.

Formulation in Polar Coordinates: We can also formulate the load flow problems using Newton-Raphson method in polar coordinates, say for any bus p we have

$$E_p = |E_p| e^{j\delta p}, \quad \text{then} \quad E_p^* = |E_p| e^{-j\delta p}$$

$$E_q = |E_q| e^{j\delta q} \quad \text{and} \quad Y_{pq} = |Y_{pq}| e^{-j\theta pq} \tag{6.46}$$

where δ is the phase angle of the bus voltages and θ_{pq} is an admittance angle.

Then for any bus p,

$$P_p - jQ_p = E_p^* \sum_{q=1}^{n} Y_{pq} E_q \tag{6.47}$$

Substituting relations (6.46) in eqn. (6.47), we have

$$P_p - jQ_p = \sum_{q=1}^{n} |E_p E_q Y_{pq}| e^{-j(\theta_{pq}+\delta_p-\delta_q)} \tag{6.48}$$

Thus

$$P_p = \text{Real } E_p^* \sum_{q=1}^{n} Y_{pq} E_q$$

$$= \sum_{q=1}^{n} | E_p E_q Y_{pq} | \cos (\theta_{pq} + \delta_p - \delta_q)$$

$$= | E_p E_p Y_{pp} | \cos (\theta_{pp}) + \sum_{\substack{q=1 \\ q \neq p}}^{n} | E_p E_q Y_{pq} | \cos (\theta_{pq} + \delta_p - \delta_q) \quad (6.49)$$

and

$$Q_p = \text{Imaginary } E_p^* \sum_{q=1}^{n} Y_{pq} E_q$$

$$= \sum_{q=1}^{n} | E_p E_q Y_{pq} | \sin (Q_{pq} + \delta_p - \delta_q)$$

$$= | E_p E_p Y_{pp} | \sin \theta_{pp} + \sum_{\substack{q=1 \\ q \neq p}}^{n} | E_p E_q Y_{pq} | \sin (\theta_{pq} + \delta_p - \delta_q) \quad (6.50)$$

for $p = 1, \ldots, n - 1$ as the nth bus is a slack bus.

Now the linear equation in the polar form becomes:

$$\begin{bmatrix} \Delta P \\ \Delta Q \end{bmatrix} = \begin{bmatrix} J_1 & J_2 \\ J_3 & J_4 \end{bmatrix} \begin{bmatrix} \Delta \delta \\ \Delta | E | \end{bmatrix} \quad (6.51)$$

where J_1, J_2, J_3 and J_4 are the elements of Jacobian which can be calculated from the power eqns. (6.49) and (6.50) as follows:

J_1: Off-dia. elements

$$\frac{\partial P_p}{\partial \delta_q} = | E_p E_q Y_{pq} | \sin (\theta_{pq} + \delta_p - \delta_q) \text{ for } q \neq p$$

Dia. element

$$\frac{\partial P_p}{\partial \delta_p} = - \sum_{\substack{q=1 \\ q \neq p}}^{n} | E_p E_q Y_{pq} | \sin (\theta_{pq} + \delta_p - \delta_q)$$

J_2: Off-dia. elements

$$\frac{\partial P_p}{\partial | E_q |} = | E_p Y_{pq} | \cos (\theta_{pq} + \delta_p - \delta_q) \text{ for } q \neq p$$

Dia. element

$$\frac{\partial P_p}{\partial | E_p |} = 2 | E_p Y_{pp} | \cos \theta_{pp} + \sum_{\substack{q \neq 1 \\ q \neq p}}^{n} | E_q Y_{pq} | \cos (\theta_{pq} + \delta_p - \delta_q)$$

J_3: Off-dia. elements

$$\frac{\partial Q_p}{\partial \delta_q} = - | E_p E_q Y_{pq} | \cos (\theta_{pq} + \delta_p - \delta_q)$$

Dia. elements

$$\partial Q_p / \partial \delta_p = \sum_{\substack{q=1 \\ q \neq p}}^{n} | E_p E_q P_{pq} | \cos (\theta_{pq} + \delta_p - \delta_q)$$

and

J_4: Off. dia elements

$$\frac{\partial Q_p}{\partial | E_q |} - | E_p Y_{pq} | \sin (\theta_{pq} + \delta_p - \delta_q) \text{ for } q \neq p$$

and diagonal element

$$\frac{\partial Q_p}{\partial |E_p|} = 2\,|E_p\,Y_{pp}|\sin\theta_{pp} + \sum_{\substack{q=1\\q\neq p}}^{n} |E_q\,Y_{pq}|\sin(\theta_{pq}+\delta_p-\delta_q) \quad (6.52)$$

The elements of Jacobian are calculated with the latest voltage estimate and calculated power. However, the procedure (i.e. algorithm) here, is the same as that of the rectangular coordinates. The formulation in the polar coordinates takes less computational efforts and also requires less memory space.

Since the real power P is less sensitive to changes in the voltage magnitude $\Delta\,|E|$ and similarly the reactive power Q is less sensitive to the changes in the phase angle δ, we can approximate the above equation (i.e. equation (6.51)), as shown below

$$\begin{bmatrix} \Delta P \\ \Delta Q \end{bmatrix} = \left[\begin{array}{c|c} J_1 & 0 \\ \hline 0 & J_4 \end{array}\right] \begin{bmatrix} \Delta\delta \\ \Delta\,|E| \end{bmatrix} \quad (6.53)$$

Generation Bus: Here P and $|V|$ are given. Now the real power P for any bus p is given by

$$P_p = \text{Real}\; E_p^* \sum_{q=1}^{n} Y_{pq}\,E_q \quad (6.54)$$

and also for bus p, we have,

$$|E_p|^2 = e_p^2 + f_p^2 \quad (6.55)$$

where E_p is the voltage magnitude and e_p and f_p are its real and imaginary components.

The matrix equations inter-relating the changes in bus powers and square of the bus voltage magnitude to the changes in the real and imaginary components of voltages are,

$$\begin{bmatrix} \Delta P \\ \hline \Delta Q \\ \hline \Delta\,|E|^2 \end{bmatrix} = \left[\begin{array}{c|c} J_1 & J_2 \\ \hline J_3 & J_4 \\ \hline J_5 & J_6 \end{array}\right] \begin{bmatrix} \Delta e \\ \hline \Delta f \end{bmatrix} \quad (6.56)$$

where $\Delta\,|(E_p^K|^2 = [|E_p\ (\text{scheduled})|^2 - |E_p^K|^2]$ (E_p^K is the calculated bus voltage after the kth iteration).

Elements of Jacobian are calculated as follows:

Off. diagonal elements of $J_5 = \dfrac{\partial |E_p|^2}{\partial e_q} = 0$ for $q \neq p$

and

Diagonal elements of $J_5 = \dfrac{\partial |E_p|^2}{\partial e_p} = 2e_p$

Similarly

$$\text{off diagonal elements of } J_6 = \frac{\partial |E_p|^2}{\partial f_q} = 0 \text{ for } q \neq p$$

and

$$\text{diagonal elements of } J_6 = \frac{\partial |E_p|^2}{\partial f_p} = 2f_p \tag{6.57}$$

Here E_p^K is the bus voltage calculated at the Kth iteration and E_p (scheduled) is the voltage given (i.e. specified) at any bus p as it is the generation bus. Calculations for the elements J_1, J_2, J_3 and J_4 are discussed earlier.

After obtaining bus voltages, power flow and line losses are calculated using equations (6.20) and (6.21).

We now give a modified Newton's method. This method has several merits over the conventional Newton-Raphsons method which we shall discuss later.

6.4 Power Flow Solution by Modified Newton's Method

This method is duc to W.F. Tinney. Here the load and generation

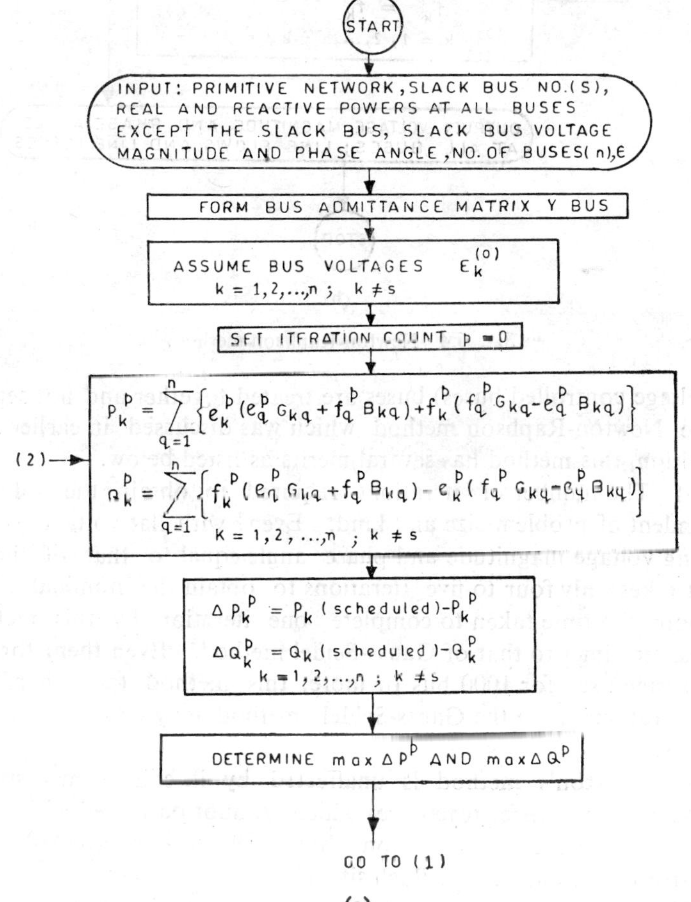

(a)

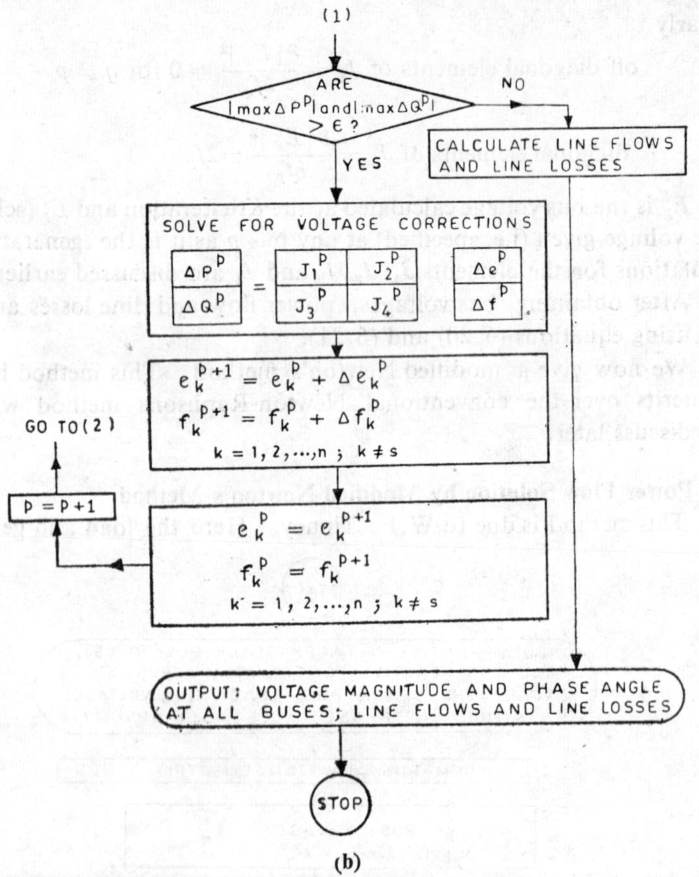

(b)

Fig. 6.4 Newton-Raphson method.

(i.e. voltage controlled buses) buses are treated together and not separately like the Newton-Raphson method which was discussed in earlier section. In addition, this method has several merits as listed below.

(i) The number of iterations required to obtain the solution is independent of problem size and kind. Even with flat voltage start (i.e assuming voltage magnitude and phase angle equal to that of the slack bus), it takes only four to five iterations to obtain the nominal solution. No doubt, the time taken to complete one iteration by this method is nearly seven times to that of Gauss-Seidel method. Even then, for a large power system, say for 1000 bus or more, this method takes hardly four to five iterations while the Gauss-Seidel method may take 100 or more iterations.

(ii) Newton's method is unaffected by ill-conditioned situations such as negative transfer reactances which cannot be netted with positive reactances of the line or a situation where high and low impedance branch terminate at the same node. Such situations do not effect this method.

(iii) The programme can easily be extended with slight modifications to include in the same Jacobian matrix, all the factors, such as, remote reactance control, area exchange and tie line control, and also inclusion of fixed off nominal tap settings or tap changing under load or phase shifting transformers.

6.4.1 *Basic Method*

At any node K, the relationship between node current I_K and node-to-datum voltage E_K in a network of N-nodes is given by the linear equation:

$$\bar{I}_K = \sum_{m=1}^{N} [Y_{Km}]\,\bar{E}_m \tag{6.58}$$

where $[Y_{Km}]$ is a bus admittance matrix. Complex power at the node K is given by

$$[P_K + JQ_K] = \bar{E}_K \sum_{m=1}^{N} [Y^*_{Km}]\,\bar{E}_m^* \tag{6.59}$$

where P_K and Q_K are the real and the reactive power at the node K.

In a power flow problem, the above equations (i.e. eqn. 6.59) which are $N-1$ nonlinear equations, are to be solved by some iterative scheme. By the Newton's method, these nonlinear algebraic equations are transformed into a system of linear equations provided a corresponding Jacobian matrix can be evaluated and a sufficient good starting condition is possible so as to avoid truncation error. Both these conditions are fulfilled in this method. Let us take a 7 node (i.e. 6 bus) problem as shown in Fig. 6.5.

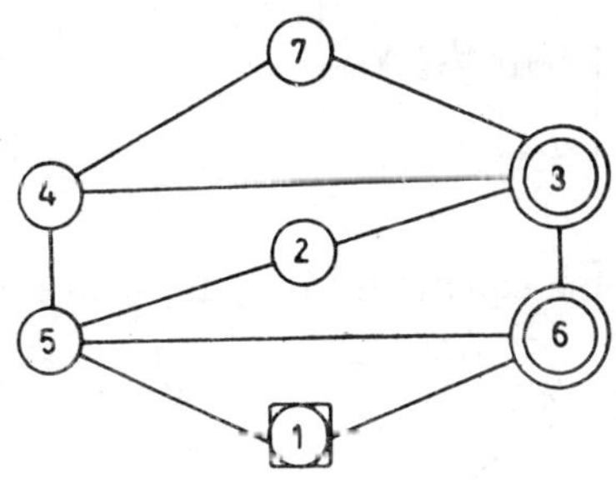

Fig. 6.5

Here,

[1] is a slack node where $|E|$ and δ are given

○ are type A nodes where P and Q given. These are called load buses

and, ◎ are type B nodes where P and E are given. These are called generation buses.

Newton's method involves repeated direct solutions of a system of linear equations derived from eqn. (6.59). Jacobian matrix of the equation (6.59) gives the linearized relationship between small change in voltage angle $\Delta\delta_K$ and voltage magnitude $\Delta E_K/E_K$ with the small changes in real and reactive power ΔP_K and ΔQ_K. For the sample network as shown in Fig. 6.5, the linear equations will be as shown below :

$$\begin{bmatrix} \Delta P_2 \\ \Delta Q_2 \\ \hline \Delta P_3 \\ \hline \Delta P_4 \\ \Delta Q_4 \\ \hline \Delta P_5 \\ \Delta Q_5 \\ \hline \Delta P_6 \\ \hline \Delta P_7 \\ \Delta Q_7 \end{bmatrix} = \left[\begin{array}{cc|c|cc|cc|c|cc} H_{22} & N_{22} & H_{23} & & & H_{25} & N_{25} & & & \\ J_{22} & L_{22} & J_{23} & & & J_{25} & L_{25} & & & \\ \hline H_{32} & N_{32} & H_{33} & H_{34} & N_{34} & & & H_{36} & H_{37} & N_{37} \\ \hline & & H_{43} & H_{44} & N_{44} & H_{45} & N_{45} & & H_{47} & N_{47} \\ & & J_{43} & J_{44} & L_{45} & J_{46} & L_{45} & & J_{47} & L_{47} \\ \hline H_{52} & N_{52} & & H_{54} & N_{54} & H_{55} & N_{55} & H_{56} & & \\ J_{52} & L_{52} & & J_{54} & L_{54} & J_{65} & N_{55} & J_{56} & & \\ \hline & & H_{63} & & & H_{65} & N_{65} & H_{66} & & \\ \hline & & H_{73} & H_{74} & N_{74} & & & & H_{77} & N_{77} \\ & & J_{73} & J_{74} & L_{74} & & & & J_{77} & L_{77} \end{array}\right] \begin{bmatrix} \Delta\delta_2 \\ \Delta E_2/E_2 \\ \hline \Delta\delta_3 \\ \hline \Delta\delta_4 \\ \Delta E_4/E_4 \\ \hline \Delta\delta_5 \\ \Delta E_5/E_5 \\ \hline \Delta\delta_6 \\ \hline \Delta\delta_7 \\ \Delta E_7/E \end{bmatrix} \tag{6.60}$$

The elements of the equations (6.60) are defined below :

$$H_{Km} = \partial P_K/\partial\delta_m, \qquad N_{Km} = \frac{\partial P_K\,|E_m|}{\partial E_m}$$

$$J_{Km} = \frac{\partial Q_K}{\partial\delta_m} \quad \text{and} \quad L_{Km} = \frac{\partial Q_K\,|E_m|}{\partial E_m} \tag{6.61}$$

The partial derivatives defined in equation (6.61), are real functions of the admittance matrix and node voltages. Problems have been formulated in the polar form because it takes less memory space in the polar form. Problem can also be formulated in the rectangular form. The term ΔP_K and ΔQ_K are residuals of equations and these are defined below :

$$\Delta P_K = P_K \text{ (scheduled)} - P_K \text{ (actual)}$$

$$\Delta Q_K = Q_K \text{ (scheduled)} - Q_K \text{ (actual)} \tag{6.62}$$

For a system of N nodes including a slack node and having S nodes (type B node) with fixed voltage magnitude (i.e. Generation bus where

P and $|E|$ are given), there are $(2N-S-2)$ linear equations similar to eqn. (6.60) which are to be solved. There are no equations for the slack bus where $|E|$ and δ are given and they remain fixed. The pattern of non-zero elements of Jacobian matrix is similar to that of the system admittance matrix. This fact will be utilized in Chapter 8—Sparsity Techniques, both for the optimization of the computational efforts and also memory space in the solution of Jacobian.

6.4.2 *Algorithm*

(i) An initial approximation to the voltage solution of eqn. (6.59) is assigned. One way is to set the voltage magnitudes, where given, to their given values and to set the other voltage magnitudes equal that of the slack node. The use of per unit system is assumed. All angles are set equal to the slack node angles. This is referred to as flat voltage start.

(ii) One cycle is performed without over correction to assure a favourable start.

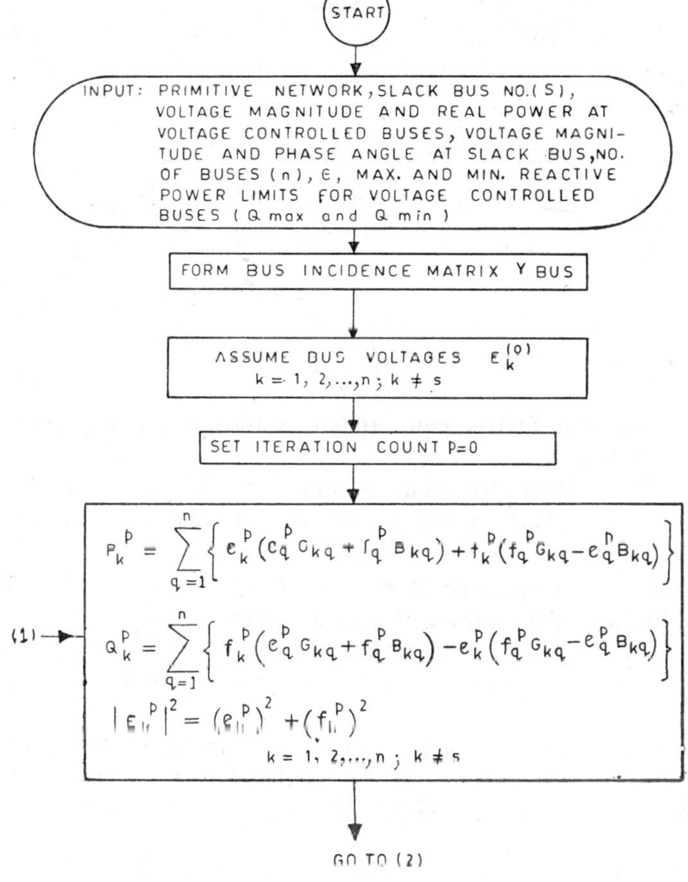

(a)

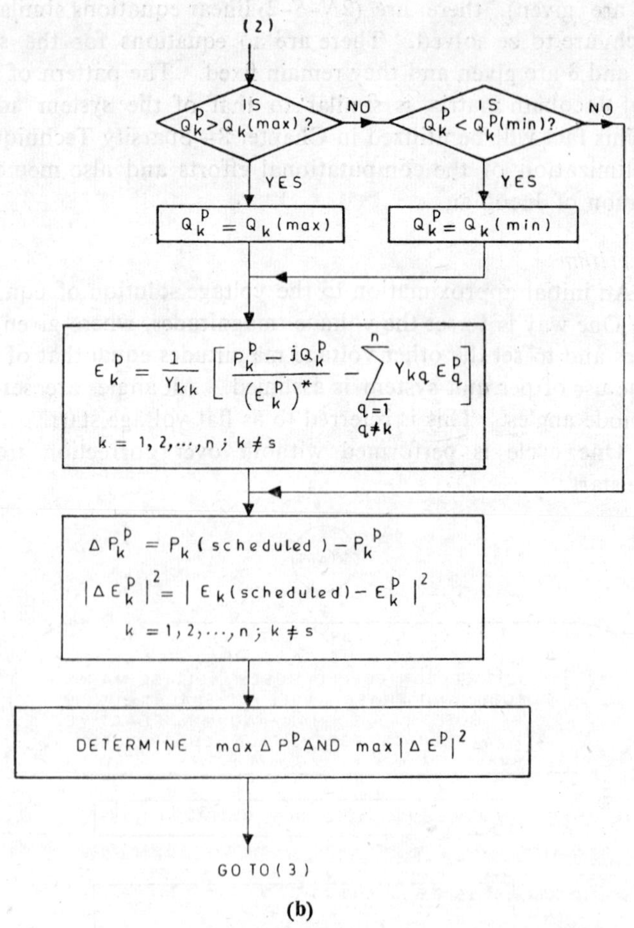

(b)

(iii) Jacobian matrix eqn. (6.60) is formed and augmented with columns of residuals.

(iv) The voltage corrections are solved by Gaussian elimination and back substitution. This operation transforms the Jacobian into an upper triangular matrix and augmented columns of residuals becomes a column of voltage angles and magnitude corrections in polar form. This correction is applied to the earlier estimate of node voltages.

(v) The residuals ΔP and ΔQ are checked. If they are sufficiently small, the problem is solved. If not the procedure is repeated, starting with step 3.

After the solution of voltage magnitude and phase angle at different buses, the line flows and line losses are calculated by the method described earlier (as in the case of Gauss-Seidel method).

6.5 Solution Technique Using Z_{BUS} in the Bus Frame of Reference

As discussed earlier, while Y_{BUS} matrix can be directly formed by

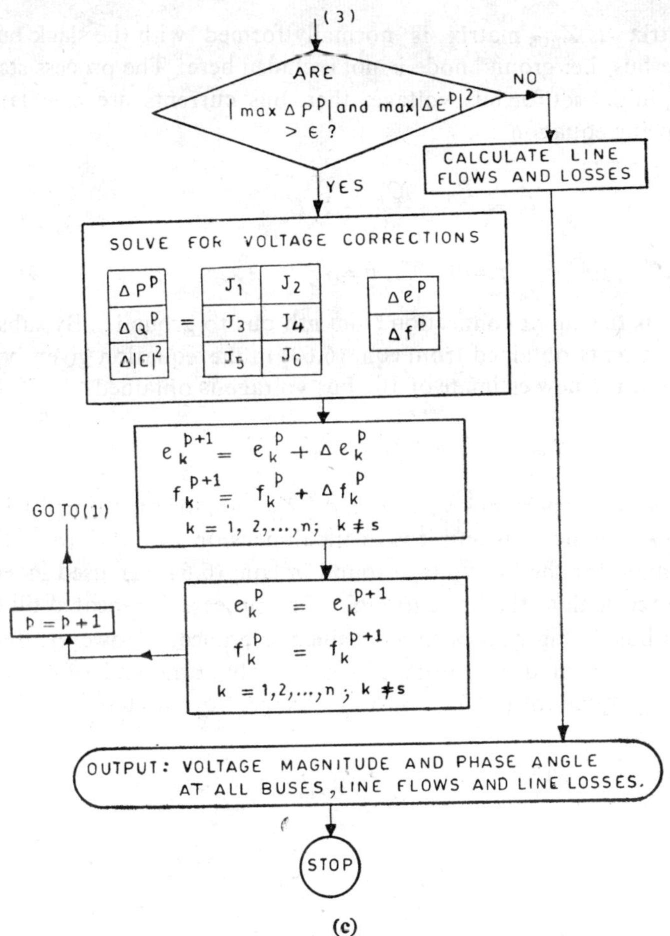

(c)

Fig. 6.6 *N—R* method for Voltage controlled bus.

inspection from the given power system network (i.e. primitive network matrix), there is no direct relation between the elements of Z_{BUS} matrix and the given power system network. The Z_{BUS} matrix, though can be found by finding inverse of Y_{BUS} matrix (i.e. $Z_{BUS} = Y_{BUS}^{-1}$), is normally determined by building algorithm starting with a slack bus, with the result ground node and the shunt connections between buses and the ground node, such as static capacitor and loads etc. are not included in the elements of Z_{BUS} matrix. These shunt connections between buses and the ground are treated as current sources. Moreover since Z_{BUS} is a full matrix, and hence sparsity techniques cannot be used here. In the case of load flow problem using Z_{BUS}, diakoptics techniques, which is actually piecewise solution of load flow problem (i.e. tearing of techniques) are used.

6.5.1 Gauss Iterative Method Using Z_{BUS}

As discussed, the shunt connections between buses and the ground is taken as current sources since these are not included in the elements of

Z_{BUS} matrix as Z_{BUS} matrix is normally formed with the slack bus as the reference bus, i.e. ground node is not included here. The process starts with selecting initial set for bus voltage, then bus currents are calculated from the following equation :

$$I_p = \frac{P_p - jQ_p}{E_p^*} - Y_p E_p \tag{6.63}$$

$$p = 1, \ldots, n \neq S$$

where Y_p is the shunt connection from pth bus to ground. By substituting the bus currents obtained from eqn. (6.63) in the equation given below i.e. eqn. (6.64), the new estimate of the bus voltage is obtained.

$$\overline{E}_{BUS} - E_s = [Z_{BUS}]\, \overline{I}_{BUS} \tag{6.64}$$

Here E_s is the slack bus voltage and $[Z_{BUS}]$ matrix is formed with slack bus as the reference bus and it is of the dimension $(n-1)X\,(n-1)$. This new estimate for the bus voltage found in eqn. (6.64), is used in equation (6.63) to recalculate the bus currents. The process is repeated till changes in all the bus voltages are within specified tolerance. However, the substitution for calculated bus voltage is made only at the end of the iteration. Combining equations (6.63) and (6.64), we get for any bus p,

$$E_p^{K+1} - E_S = [Z_{BUS}]\, \overline{I}_{BUS} = \sum_{\substack{q=1 \\ q \neq S}}^{n} Z_{pq}\, I_q^K \tag{6.65}$$

$$p = 1, 2, \ldots, p \neq S$$

where $I_q^K = \dfrac{P_q - jQ_q}{[E_q^K]^*} - Y_q\, E_q^K$.

6.5.2 *Gauss-Seidel Method Using* Z_{BUS}

In this case the bus voltage equations (eqn. 6.64) are solved one at a time in sequence. After each equation is solved to obtain a new estimate for the bus voltages, the corresponding bus currents are calculated i.e. the substitution is made immediately which is unlike Gauss-method where substitution is made only at the end of the iteration. With the result, Gauss-Seidel method converges faster than the Gauss method. Formulation for Gauss-Seidel method for any bus p when calculations have been already completed upto $p-1$ bus is as follows:

$$E_p^{K+1} - E_S = \sum_{q=1}^{p-1} Z_{pq}\, I_q^{K+1} + \sum_{\substack{q=p \\ q \neq S}}^{n} Z_{pq}\, I_q^K \tag{6.66}$$

$$p = 1, \ldots, n$$

$$p \neq S$$

where $$I_q^{K+1} = \frac{P_q - jQ_q}{[E_q^{K+1}]^*} - Y_q E_q^{K+1}$$

After this, as usual, line flows and losses are calculated.

6.6 Representation of Transformer

We shall discuss here the modelling of different types of transformers with off nominal taps for the purpose of load flow studies.

6.6.1 *Fixed Tap Setting Transformer*

A transformer with the fixed tap setting is represented by its impedance or admittance Y_{pq} in series with an ideal autotransformer, say, for such a transformer at the bus p in the line $p—q$, the one line diagram will be as shown in Fig. 6.7.

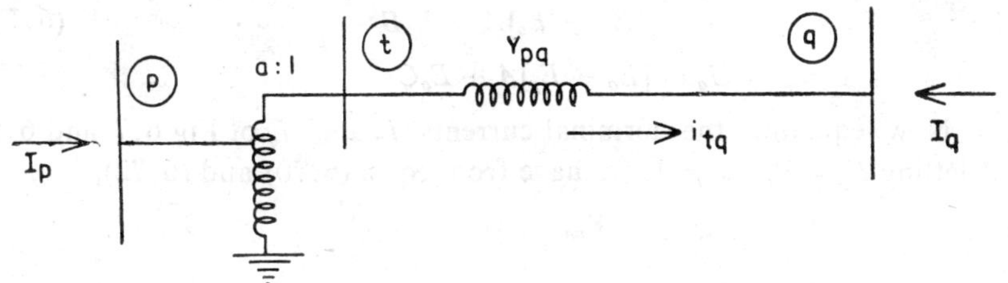

E_p, E_t and E_q are bus voltages at buses p, t and q.

Fig. 6.7 Transformer with fixed tap setting having ratio 'a'

We have from Fig. 6.7

$$\frac{E_p}{E_t} = \frac{i_{tq}}{I_p} = a \tag{6.67}$$

Thus, $$I_p = i_{tq}/a$$

but $$I_{tq} = (E_t - E_q)Y_{pq}$$

Hence $$I_p = \frac{i_{tq}}{a} = (E_t - E_q)\frac{Y_{pq}}{a} \tag{6.68}$$

But from eqn. (6.67), we have,

$$E_t = E_p/a \tag{6.69}$$

Substituting equation (6.69), in equation (6.68), we have

$$I_p = (E_p - aE_q)\frac{Y_{pq}}{a^2} \tag{6.70}$$

And also $$I_q = (E_q - E_t)Y_{pq}$$

$$= (aE_q - E_p)\frac{Y_{pq}}{a} \tag{6.71}$$

Such transformer at the bus p in the line $p—q$ can be represented by the equivalent π-circuit as shown in Fig. 6.8.

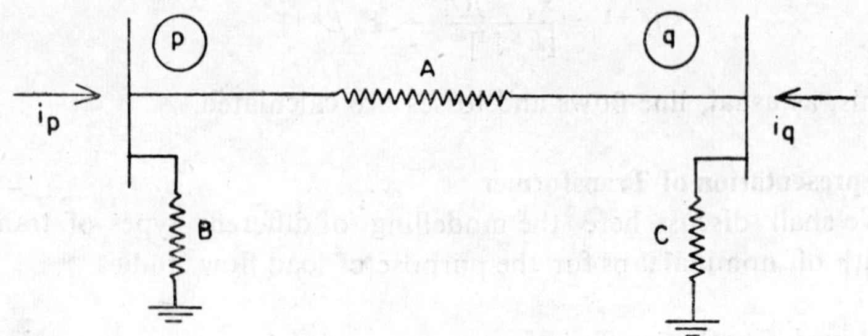

Fig. 6.8 Equivalent π circuit.

The corresponding currents of the equivalent circuit (Fig. 6.8) will be,

$$I_p = [E_p - E_q]A + E_pB \tag{6.72}$$

$$I_q = [E_q - E_p]A + E_qC \tag{6.73}$$

Now equating the terminal currents I_p and I_q of Fig 6.7 and 6.8, and letting $E_p = 0$, $E_q = 1$, we have from eqns. (6.70) and (6.72),

$$I_p = -\frac{Y_{pq}}{a} = -A$$

i.e.

$$A = \frac{Y_{pq}}{a} \tag{6.74}$$

Similarly from eqns. (6.71) and (6.73), we get

$$I_q = Y_{pq} = A + C$$

Hence

$$C = Y_{pq} - A = Y_{pq} - \frac{Y_{pq}}{a}$$

$$= \left(1 - \frac{1}{a}\right)y_{pq} \tag{6.75}$$

Similarly for B, we equate terminal current I_p for eqns. (6.70) and (6.72). Thus we get,

$$(E_p - aEq)\frac{Y_{pq}}{a^2} = (E_p - E_q)A + E_pB \tag{6.76}$$

Substituting for $A = Y_{pq}/a$, and letting $E_p = 1$ and $E_q = 0$, we get from equation (6.76)

$$B = \frac{Y_{pq}}{a^2} - \frac{y_{pq}}{a} = \left[\frac{1}{a^2} - \frac{1}{a}\right]y_{pq}$$

$$= \frac{1}{a}\left[\frac{1}{a} - 1\right]y_{pq} \tag{6.77}$$

With the substitutions of A, B and C from eqns. (6.74), (6.75) and (6.77), Fig. 6.8 takes the form of Fig. 6.9.

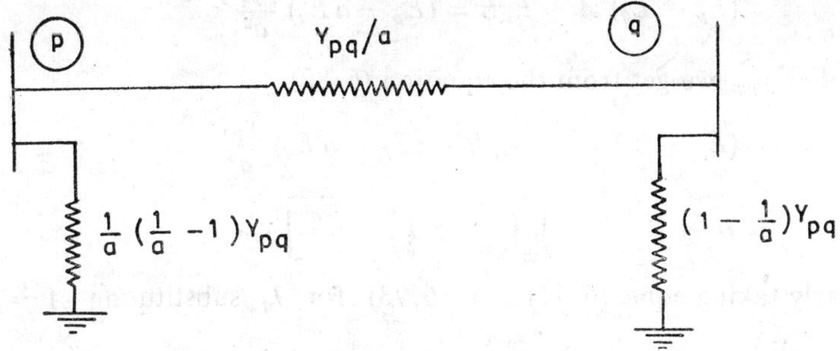

Fig. 6.9 Equivalent π-circuit of the transformed.

Thus to represent this transformer in the elements of Y_{BUS} matrix, we have from the Fig. 6.9.

$$Y_{pp} = y_{p1} + \dots + \frac{y_{pq}}{a} + \dots + y_{pn} + \frac{1}{a}\left[\frac{1}{a} - 1\right] y_{pq}$$

$$= y_{p1} + \dots + \frac{y_{pq}}{a^2} + \dots + y_{pn} \tag{6.78}$$

i.e. this element is changed.

and

$$Y_{pq} = Y_{qp} = -\frac{y_{pq}}{a} \tag{6.79}$$

and also,

$$Y_{qq} = y_{q1} + \dots + y_{pq}/a + \dots + y_{qn} \dots \left(1 - \frac{1}{a}\right) y_{pq}$$

$$= y_{q1} + \dots + y_{qp} + \dots + y_{qn} \tag{6.80}$$

i.e. no change in this element.

6.6.2 *Tap Changing Under Load Transformer (i.e. TCUL)*

In this case, the tappings of the transformer is changed to maintain voltage magnitude within specified tolerance. Normally, tappings are changed once in two iterations and corresponding values of Y_{pp}, Y_{pq} and Y_{qq} are calculated, such that for any bus (p), we have

$$|E_p^K - E_p\,(\text{scheduled})| < \rho \tag{6.81}$$

However, to avoid extensive calculations, the series impedance of the equivalent π-circuit is set equal to the series impedance of the transformer and the shunt parameters B and C are changed to simulate the changes in tappings. Thus taking eqns. (6.70) and (6.72), for I_p, we have

$$(E_p - E_q)\,A + E_pB = (E_p - a\,E_q)\,\frac{y_{pq}}{a^2} \tag{6.82}$$

with $A = y_{pq}$, we get from the equation (6.82),

$$(E_p - E_q)\,y_{pq} + E_pB = (E_p - a\,E_q)\,\frac{y_{pq}}{a^2}$$

i.e.
$$B = \left(\frac{1}{a} - 1\right)\left[\left(\frac{1}{a} + 1\right) - \frac{E_q}{E_p}\right] y_{pq} \tag{6.83}$$

Similarly taking eqns. (6.71) and (6.73) for I_q substituting $A = y_{pq}$, we have

$$[E_q - E_p]\,y_{pq} + E_qC = [aE_q - E_p]\,\frac{y_{pq}}{a} \tag{6.84}$$

From equation (6.84), we obtain

$$C = \left(1 - \frac{1}{a}\right) y_{pq}\,\frac{E_p}{E_q} \tag{6.85}$$

Afterwards, only the elements Y_{pp} and Y_{qq} are calculated with the change in tappings.

6.6.3 *Phase Shifting Transformer*

A phase shifting transformer is used to advance the phase angle of the bus voltages and is represented by an ideal autotransformer with complex turn ratio $(a + jb)$ in series with impedance or admittancc y_{pq} as shown in Fig. 6.10.

We have from Fig. 6.10, $E_p = E_r$

and
$$E_p/E_S = E_r/E_S = a_S + jb_S \tag{6.86}$$

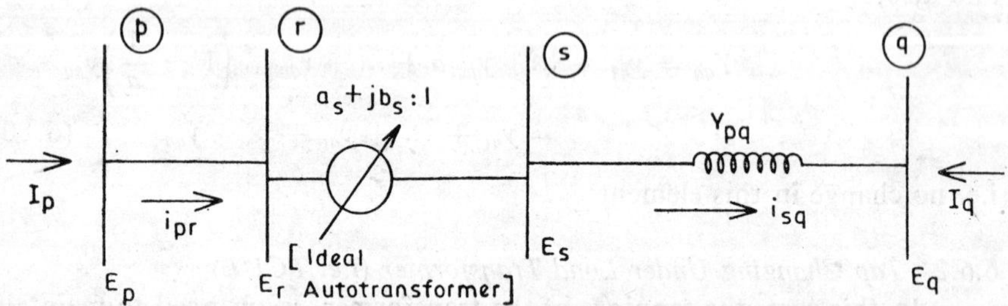

Fig. 6.10 Phase shifting transformer

Since the power loss in the ideal autotransformer in negligible, we have from Fig. 6.10,

$$E_p^*\,i_{pr} = E_S^*\,i_{sq}$$

i.e.
$$i_{sq}/i_{pr} = E_p^*/E_S^* = a_S - jb_S \tag{6.87}$$

Now
$$i_{sq} = (E_S - E_q)\,y_{pq}$$

then
$$i_{pr} = \frac{i_{sq}}{a_S - jb_S} = [(E_S - E_q)]\,y_{pq}/(a_S - jb_S) \tag{6.88}$$

but since from eqns. (6.86),

$$E_S = \frac{E_p}{a_S + jb_S} \tag{6.89}$$

By substituting eqn. (6.89) in eqn. (6.88), we obtain,

$$i_{pr} = [(E_p - (a_S + jb_S) E_q] (y_{pq}/a_S^2 + b_S^2) \tag{6.90}$$

Similarly $i_{qs} = (E_q - E_s) y_{pq}$

$$= [(a_S + jb_S) E_q - E_p] \frac{y_{pq}}{a_S + jb_S} \tag{6.91}$$

Now we calculate the elements of Y_{BUS} matrix by conducting short-circuit test. The diagonal element Y_{pp} is found by connecting a unit voltage source at the pth bus, i.e. $E_p = 1$ p.u. and short circuiting remaining buses, i.e. putting other buses equal to zero, then we get,

$$\begin{aligned} I_p = Y_{pp} &= I_{p1} + I_{p2} + \ldots + i_{pr} \\ &= [E_p - E_1] y_{p1} + [E_p - E_2] y_{p2} + \ldots + i_{pr} \\ &= y_{p1} + y_{p2} + \ldots \\ &+ [E_p - (a_S + jb_S) E_q] y_{pq}/a_S^2 + b_S^2 \end{aligned} \tag{6.92}$$

as $E_P = 1$ p.u. and $E_i = 0$ for $i = 1, \ldots n; i \neq p$.

We have, from the above eqn. (6.92),

$$I_p = Y_{pp} = y_{p1} + y_{p2} + \ldots + (y_{pq}/(a_S^2 + b_S^2)) + \ldots \tag{6.93}$$

i.e. this element is changed.

Similarly for Y_{qq}, we keep $E_q = 1$ p.u. and rest bus voltages are zero, we get

$$\begin{aligned} I_q = Y_{qq} &= I_{q1} + I_{q2} + \ldots + I_{qS} \\ &= [E_q - E_1] y_{q1} + [E_q - E_2] y_{q2} + ((a_S + jb_S) E_q - E_p) y_{qq}/a_S + jb_S \end{aligned} \tag{6.94}$$

since $E_q = 1$ p.u. and the rest of the bus voltages viz. $E_1 = E_2 \ldots = 0$, we get

$$I_q = Y_{qq} = Y_{q1} + Y_{q2} + \ldots y_{pq} \tag{6.95}$$

i.e. no change in this element.

For mutual admittances, say Y_{qp}, we apply $E_p = 1$ p.u. and measure I_q keeping other bus voltages equal to zero, i.e. $I_q = Y_{qp} E_p$ but since $E_p = 1$ p.u. we get from eqn. (6.91)

$$I_q = Y_{qp} = -\frac{y_{pq}}{a_S + jb_S} \tag{6.96}$$

i.e. this element is also changed.

For Y_{pq} we keep $E_q = 1$ and the rest of the buses are short circuited, then from eqn. (6.90) we get,

$$I_p = Y_{pq} = -\frac{y_{pq}}{a_S - jb_S} \tag{6.97}$$

i.e. this element is also changed. Moreover, $Y_{pq} \neq Y_{qp}$. Now the complex turn ratio $a_S + jb_S = a\,[\cos\theta + j\sin\theta]$ where

$$|E_p| = a|E_S| \tag{6.98}$$

However, if θ is positive, the phase of $|E_p|$ is advanced, i.e. leading with respect to that of $|E_S|$ or E_q.

6.7 Fast Decoupled Load Flow Method

The fast decoupled load flow method is a very fast method of obtaining load flow solution. In this method, both, the speed as well as the sparsity are exploited. This is actually an extension of Newton's method formulated in polar coordinates with certain approximations which will be discussed here.

The earlier equation of Load flow studies (i.e. equation 6.51) using Newtons-Raphson's method can be expressed in polar coordinates as

$$\begin{bmatrix} \Delta P \\ \Delta Q \end{bmatrix} = \begin{bmatrix} H & N \\ M & L \end{bmatrix} \begin{bmatrix} \Delta\delta \\ \dfrac{\Delta|E|}{E} \end{bmatrix} \tag{6.99}$$

where H, N, M and L are the elements (viz., J_1, J_2, J_3 and J_4) of the Jacobian matrix. Since changes in real power (i.e. ΔP) is less sensitive to the changes in voltage magnetude (i.e. ΔE) and changes in reactive power (i.e. ΔQ) is less sensitive to the changes in angle (i.e. $\Delta\delta$), equation (6.99) takes the form

$$\begin{bmatrix} \Delta P \\ \Delta Q \end{bmatrix} = \begin{bmatrix} H & O \\ O & L \end{bmatrix} \begin{bmatrix} \Delta\delta \\ \dfrac{\Delta|E|}{E} \end{bmatrix} \tag{6.100}$$

Equations (6.100) are decoupled equations which cen be expanded as,

$$[\Delta P] = [H][\Delta\delta] \tag{6.101}$$

and

$$[\Delta Q] = [L]\frac{[\Delta|E|]}{E} \tag{6.102}$$

The elements of Jacobian (eqn. 6.100) are defined again (eqn. 6.61)

$$H_{pq} = \frac{\delta P_p}{\delta\delta_q} \quad \text{and} \quad L_{pq} = \frac{\delta Q_p}{\delta E_q}|E_q|$$

Equations (6.49) and (6.50) for power are again expressed below, for calculating elements of Jacobian (i.e. H and L)

$$P_p = \sum_{q=1}^{n} |E_p E_q Y_{pq}| \cos[\theta_{pq} + \delta_p - \delta_q]$$

$$= |E_p E_p Y_{pp}| \cos\theta_{pp} + \sum_{\substack{q=1 \\ \neq p}}^{n} |E_p E_q Y_{pq}| \cos[\theta_{pq} + \delta_p - \delta_q] \tag{6.103}$$

and

$$Q_p = |E_p E_p Y_{pp}| \sin\theta_{pp} + \sum_{\substack{q=1 \\ \neq p}}^{n} |E_p E_q Y_{pq}| \sin[\theta_{pq} + \delta_p - \delta_q] \tag{6.104}$$

Therefore, the elements of Jacobian (i.e., H and L) can be calculated as shown from the equations (eqns. 6.103 and 6.104) of power. Off diagonal element of H is

$$H_{pq} = \frac{\delta P_p}{\delta \delta_q} = |E_p E_q Y_{pq}| \sin [\theta_{pq} + \delta_p - \delta_q]$$

$$= |E_p E_q Y_{pq}| [\sin [\theta_{pq}] \cos [\delta_p - \delta_q] + \cos \theta_{pq} \sin [\delta_p - \delta_q]]$$

$$= |E_p E_q| [Y_{pq} \sin \theta_{pq} \cos [\delta_p - \delta_q] + Y_{pq} \cos \theta_{pq} \sin [\delta_p - \delta_q]]$$

$$= |E_p E_q| [-B_{pq} \cos [\delta_p - \delta_q] + G_{pq} \sin [\delta_p - \delta_q]] \tag{6.105}$$

Similarly off-diagonal element of L is

$$L_{pq} = \frac{\delta Q_p |E_q|}{\delta E_q} = |E_p E_q Y_{pq}| \sin [\theta_{pq} + \delta_p - \delta_q]$$

$$= |E_p E_q| [G_{pq} \sin [\delta_p - \delta_q] - B_{pq} \cos [\delta_p - \delta_q]] \tag{6.106}$$

From equations (6.105) and (6.106), it is evident that

$$H_{pq} = L_{pq} = |E_p E_q| [G_{pq} \sin [\delta_p - \delta_q] - B_{pq} \cos [\delta_p - \delta_q]] \tag{6.107}$$

Diagonal elements of H is given by

$$H_{pp} = \frac{\delta P_p}{\delta \delta_p} = -\sum_{\substack{q=1 \\ q \neq p}}^{n} |E_p E_q Y_{pq}| \sin [\theta_{pq} + \delta_p - \delta_q]$$

$$= -[\sum_{q=1}^{n} |E_p E_q Y_{pq}| \sin [\theta_{pq} + \delta_p - \delta_q] - |E_p E_p Y_{pp}| \sin \theta_{pp}]$$

$$= -[+Q_p + E_p^2 B_{pp}]$$

$$= -E_p^2 B_{pp} - Q_p \tag{6.108}$$

and of L is given by

$$L_{pp} = \frac{\delta Q_p |E_p|}{\delta E_p}$$

$$= |2E_p^2 Y_{pp}| \sin \theta_{pp} + [\sum_{\substack{q=1 \\ \neq p}}^{n} |E_p E_q Y_{pq}| \sin [\theta_{pq} + \delta_p - \delta_q]$$

$$= |2E_p^2 Y_{pp}| \sin \theta_{pp} + Q_p - |E_p^2 Y_{pp}| \sin \theta_{pp}$$

$$= Q_p + |E_p^2 Y_{pp}| \sin \theta_{pp}$$

$$= Q_p - E_p^2 B_{pp}$$

$$= -E_p^2 B_{pp} + Q_p \tag{6.109}$$

In the case of fast decoupled load flow problem, following approximations are made:

$$\cos [\delta_p - \delta_q] \simeq 1$$

$$G_{pq} \sin [\delta_p - \delta_q] \ll B_{pq}$$

and $$Q_p \ll B_{pp} E_p^2 \tag{6.110}$$

These approximations are made for calculations of elements of Jacobian H and L. With these assumptions the elements of Jacobian become

$$H_{pq} = L_{pq} = -|E_p||E_q|B_{pq} \quad \text{for } q \neq p$$

and $$H_{pp} = L_{pp} = -B_{pp}|E_p^2| \tag{6.111}$$

Equations (6.101) and (6.102) with the substitutions of elements of Jacobian eqn. (6.111), take the form

$$\Delta P = H\Delta\delta$$

i.e. $$[\Delta P_p] = [E_p][E_q][B'_{pq}][\Delta\delta_q] \tag{6.112}$$

and similarly $$\Delta Q = L\frac{\Delta|E|}{E}$$

i.e. $$[\Delta Q_p] = [E_p][E_q][B''_{pq}]\frac{\Delta|E_q|}{E_q} \tag{6.113}$$

where B'_{pq} and B''_{pq} are elements of $[-B]$ matrix. Further decoupling and final algorithm for the fast decoupled load flow studies are obtained by:

(1) Omitting from B', the representation of those network elements that affects MVAR flows i.e. shunt reactances and off nominal in phase transformer taps.

(2) Omitting from B'', the angle shifting affects of phase shifters.

(3) Dividing eqns. (6.112) and (6.113) by E_p and setting $E_q = 1$ p.u. and also neglecting the series resistance in calculating the elements of B'.

With these assumptions, equations (6.112) and (6.113) take the following final form,

$$\left[\frac{\Delta P_p}{E_p}\right] = [B'][\Delta\delta] \tag{6.114}$$

and $$\left[\frac{\Delta Q_p}{E_p}\right] = [B''][\Delta E] \tag{6.115}$$

Here both $[B']$ and $[B'']$ are real and sparse and have structure of H and L respectively. Since, they contain only network admittances, they are constant and need to be triangularized only once at the beginning of the iteration. This algorithm, which results in a very fast solution of $\Delta\delta$ and ΔE, is known as fast decoupled load flow formulation of load flow studies.

6.7.1 Load Flow Studies by Phase Coordinate Method (i.e. 3-phase load flow studies)

The solution of load flow problem constitutes an important aspect of the power system planning, opration and control. Traditionally, the electrical power systems are assumed to be balanced and are represented by one line diagram. The load flow studies, which is carried out based upon one line diagram, give the operating parameters of one phase and that of remaining two phases are obtained on the basis of balanced system concepts. This assumption is true for most of the steady state analysis where the imbalance of a power system can be ignored and a single phase analysis is adequate. However, there are numerous situations, such as very long untransposed transmission lines, large single phase loads like 1-φ induction furnaces and 1-φ traction motors, single pole switching, one, two or four-phase supply, and bundled conductors etc., under which power systems becomes unbalanced and therefore, more detailed analysis become desirable. Under such imbalances, the system cannot be represented by a single line diagram and all available formulations of load flow problem become incapable of analysing the system.

Unbalanced currents can cause serious problems. Negative sequence currents may cause over heating of machines. Ground (i.e. zero sequence) currents can cause improper operation of protective relaying zero sequence current greatly increase the effect of inductive coupling between transmission lines. Consequently, in order to solve load-flow problem of the unbalanced power system, a method using phase coordinate has been developed. The load flow studies in phase coordinate (i.e. 3-phase load flow studies) is handled much like a single phase load flow study where each single phase voltage, power and current becomes a three element vector each single phase admittance element is replaced with three by three admittance matrix. In addition, some components such as synchronous generators, induction motors and transformers etc. are represented in more detail than would be used for a single-phase study system representation.

Representations: The three-phase load is decoupled and one third of it, is assigned to each phase. The three phase transmission-line model follows as extensions of the nominal or distributed π representation. Here one half of the transmission line equivalent capacitance is connected

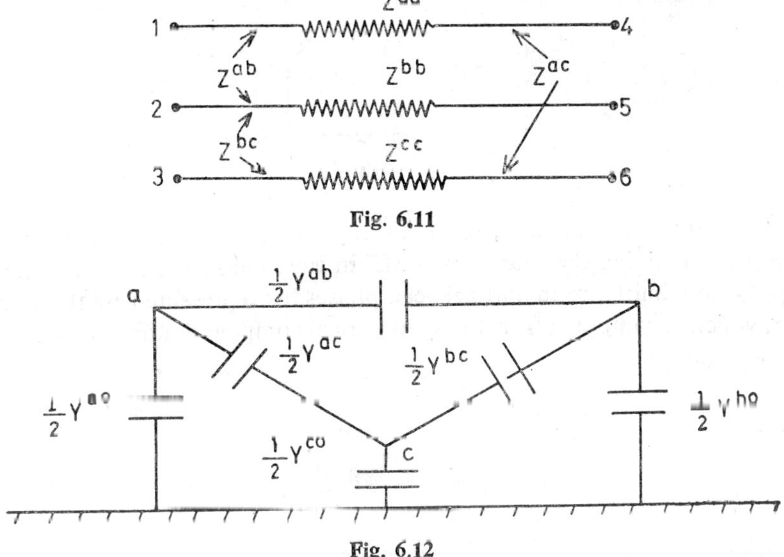

Fig. 6.11

Fig. 6.12

to each end of the line. Figure 6.11 shows schematically the series impedance Z^{abc} of a three phase transmission line between bus pairs 1, 2, 3 and 4, 5, 6. Figure 6.12 shows the equivalent capacitive susceptances in $\frac{1}{2}Y_{sh}$ at each set of bus bars 1, 2, 3, and 4, 5, 6.

The nodal i.e., bus voltages and currents (note, in this case, every phase becomes a node i.e. a bus) into the bus bars for this one three-phase element alone, are then related by,

$$\begin{bmatrix} I_1 \\ I_2 \\ I_3 \\ \hline I_4 \\ I_5 \\ I_6 \end{bmatrix} = \left[\begin{array}{c|c} Y^{abc}+\frac{1}{2}Y_{sh} & -Y^{abc} \\ \hline -Y^{abc} & Y^{abc}+\frac{1}{2}Y_{sh} \end{array}\right] \begin{bmatrix} V_1 \\ V_2 \\ V_3 \\ \hline V_4 \\ V_5 \\ V_6 \end{bmatrix} \quad (6.116)$$

where $Y^{abc} = Z^{abc-1}$.

The similarity between the three phase and single phase admittance matrix is evident, each element in the single phase matrix is replaced by three by three admittance submatrices.

Synchronous and induction machines are represented as shown in Figure 6.13.

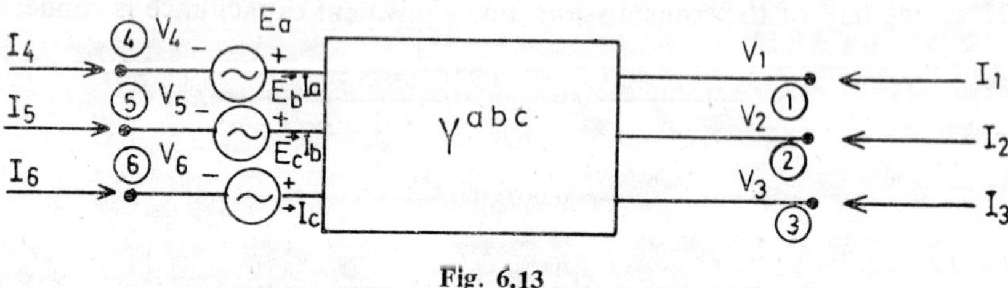

Fig. 6.13

Figure 6.13 illustrates, schematically, a general three phase circuit element containing the sources of EMF in each phase and self and mutual impedance/admittance in and between phases, as represented by the matrix Y^{abc}, which is computed from the symmetrical compound admittances, using the relation,

$$Y^{abc} = \frac{1}{3}TY^{0,1,2}T^* \quad (6.117)$$

where

$$T = \begin{bmatrix} 1 & 1 & 1 \\ 1 & \alpha^2 & \alpha \\ 1 & \alpha & \alpha^2 \end{bmatrix}$$

and $\alpha = 3\sqrt{1} = 1\angle 120 = -0.5 + j0.866$.

Here Y^0, Y^1 and Y^2 are zero, positive and negative sequence admittances. The nodal equations (note, here every phase becomes a node) for the element shown in Figure 6.13 are,

$$\begin{bmatrix} I_1 \\ I_2 \\ I_3 \\ I_4 \\ I_5 \\ I_6 \end{bmatrix} = \begin{bmatrix} Y_{11} & Y_{12} & Y_{13} & -Y_{11} & -Y_{12} & -Y_{13} & -Y_{11} & -Y_{12} & -Y_{13} \\ Y_{21} & Y_{22} & Y_{23} & -Y_{21} & -Y_{22} & -Y_{23} & -Y_{21} & -Y_{22} & -Y_{23} \\ Y_{31} & Y_{32} & Y_{33} & -Y_{31} & -Y_{32} & -Y_{33} & -Y_{31} & -Y_{32} & -Y_{33} \\ -Y_{11} & -Y_{12} & -Y_{13} & Y_{11} & Y_{12} & Y_{13} & Y_{11} & Y_{12} & Y_{13} \\ -Y_{21} & -Y_{22} & -Y_{23} & Y_{21} & Y_{22} & Y_{23} & Y_{21} & Y_{22} & Y_{23} \\ -Y_{31} & -Y_{32} & -Y_{33} & Y_{31} & Y_{32} & Y_{33} & Y_{31} & Y_{32} & Y_{33} \end{bmatrix} \begin{bmatrix} V_1 \\ V_2 \\ V_3 \\ V_4 \\ V_5 \\ V_6 \\ E_1 \\ E_2 \\ E_3 \end{bmatrix} \tag{6.118}$$

where $E_1 = E_a$, $E_2 = E_b$ and $E_3 = E_c$ are internal machine voltages of phases a, b and c respectively, and

$$Y^{abc} = \begin{bmatrix} Y_{11} & Y_{12} & Y_{13} \\ Y_{21} & Y_{22} & Y_{23} \\ Y_{31} & Y_{32} & Y_{33} \end{bmatrix}$$

$$= \begin{bmatrix} Y_0 + Y_1 + Y_2 & Y_0 + \alpha Y_1 + \alpha^2 Y_2 & Y_0 + \alpha^2 Y_1 + \alpha Y_2 \\ Y_0 + \alpha^2 Y_1 + \alpha Y_2 & Y_0 + Y_1 + Y_2 & Y_0 + \alpha Y_1 + \alpha^2 Y_2 \\ Y_0 + \alpha Y_1 + \alpha^2 Y_2 & Y_0 + \alpha^2 Y_1 + \alpha Y_2 & Y_0 + Y_1 + Y_2 \end{bmatrix}$$

Adding last three equations of (6.118), generates one extra equation for I_N (since $I_N = I_4 + I_5 + I_6$). However, adding equations 4th, 5th and 6th of (6.118) implies joining the nodes 4, 5 and 6 to form a neutral point N. Further, summation of columns for $V_N = V_4 = V_5 = V_6$ will further reduce equation (6.118). We also have,

$$I_4 = I_a = S_1^*/E_1^* \; ; \quad I_5 = I_b = S_2^*/E_2^* \text{ and } I_6 = I_C = S_3^*/E_3^*$$

Here S_1, S_2 and S_3 are the per phase power of the machine. We also know,

$$E_2 = \alpha^2 E_1; \; E_3 = \alpha E_1; \; E_2^* = \alpha E_1^* \text{ and } E_3^* = \alpha^2 E_1^*$$

since the machine is balanced.

With the above substitutions and also summing the rows 4, 5 and 6 and columns 4, 5 and 6, equation (6.118) reduces to,

$$\begin{bmatrix} I_1 \\ I_2 \\ I_3 \\ \dfrac{S_1^* + S_2^* + S_3^*}{E_1^*} \\ I_N \end{bmatrix} = \begin{bmatrix} Y_{11} & Y_{12} & Y_{13} & -Y_1 & -Y_0 \\ Y_{21} & Y_{22} & Y_{23} & -\alpha^2 Y_1 & -Y_0 \\ Y_{31} & Y_{32} & Y_{33} & -\alpha Y_1 & -Y_0 \\ -Y_1 & -\alpha Y_1 & -\alpha^2 Y_1 & 3Y_1 & 0 \\ -Y_0 & -Y_0 & -Y_0 & 0 & 3Y_0 \end{bmatrix} \begin{bmatrix} V_1 \\ V_2 \\ V_3 \\ E_1 \\ V_N \end{bmatrix} \tag{6.119}$$

Equation (6.119) is the final model capable of representing imbalancing in machines. The unbalancing may be either due to machine inductances or external circuit. In case of any unbalancing in the system, the machine currents I_a, I_b and I_c are not balanced and also each phase power is not the one third of total machine power and hence these are unknown. But the total power assigned is known. I_1, I_2 and I_3 are the injected machine currents and I_N is the neutral current. These injected currents are zero for induction motor. They are also zero for synchronous generator if there is no local load. Since left hand side column vector and matrix Y are known, equation (6.119) is solved for all the five voltages, i.e. terminal voltages V_1, V_2 and V_3 of the machine and only the internal machine voltage (i.e. induced e.m.f.) E_1 because the machine has symmetrical structure and, therefore, the induced e.m.f.s for all the three-phases are equal in magnitude, they only differ in phase by 120° as shown earlier. If the neutral is solidly earthed, V_N is equal to zero.

The third type of network element is the transformer in the three-phase circuit with all possible variations of construction and connections. The transformers may be connected in different ways such as $Y-Y$, $Y-\Delta$, $\Delta-Y$ and $\Delta-\Delta$ for 3-phase power transfer. Transformer in p.u., is represented by an ideal transformer with off-nominal tappings in series with an equivalent leakage impedance/admittance. The nodal admittance

Table 6.1 Connection Table for star-star transformer
Here $\alpha = 1 + t_\alpha$ and $\beta = 1 + t_\beta$ in p.u.

Admittances	Between nodes
y/α^2	$N-A$, $N-B$, $N-C$
y/β^2	$n-a$, $n-b$, $n-c$
$y/\alpha\beta$	$A-a$, $B-b$, $C-c$
$-y/\alpha\beta$	$n-A$, $n-B$, $n-C$; $N-a$, $N-b$, $N-c$
$3y/\alpha\beta$	$N-n$

(i.e. Y_{BUS}) matrix of transformer which may be developed directly by inspection, depends upon the type of connections. Following connection table will be helpful in order to assemble the nodal admittance matrix of the three phase transformer (or 3 single phase transformer for three phase power transformer). Here y is the leakage admittance per phase in p.u. and α and β are tappings on primary and secondary side respectively in p.u.

Table 6.2 Connection Table for $\Delta-\Delta$ transformer
Here $\alpha = \sqrt{3}\,[1 + t_\alpha]$ and $\beta = \sqrt{3}\,[1 + t_\beta]$ in p.u.

Admittance	Between nodes
y/α^2	$A-B,\ B-C,\ C-A$
y/β^2	$a-b,\ b-c,\ c-a$
$2y/\alpha\beta$	$A-a,\ B-b,\ C-c$
$-y/\alpha\beta$	$A-b,\ B-c,\ C-a;\ a-B,\ b-C,\ c-A$

Table 6.3 Connection Table for star-delta transformer
Here $\alpha = 1 + t_\alpha$, $\beta = \sqrt{3}\,[a + t_\beta]$ in p.u.

Admittance	Between nodes
y/α^2	$A-N,\ B-N,\ C-N$
y/β^2	$a-b,\ b-c,\ c-a$
$y/\alpha\beta$	$A-c,\ B-a,\ C-b$
$-y/\alpha\beta$	$A-b,\ B-c,\ C-a$

Three phases and neutral on primary side is referred as A, B, C and N and corresponding values on secondary side as a, b, c and n. It is clear from the above that every phase becomes a node. Nodal admittance matrix in phase coordinates is now assembled with the help of connection table given above. The nodal admittance matrix Y_{ABC} of the transformer relates the nodal currents with nodal voltages as shown below:

$$\begin{bmatrix} I_1 \\ I_2 \\ I_3 \\ I_4 \\ I_5 \\ I_6 \end{bmatrix} = [Y_{ABC}]_{6\times 6} \begin{bmatrix} V_1 \\ V_2 \\ V_3 \\ V_4 \\ V_5 \\ V_6 \end{bmatrix} \qquad (6.120)$$

The column vector $[I_1\ I_2\ I_3\ I_4\ I_5\ I_6]^T$ are the nodal currents representing the three phase currents on primary side and also the three phase currents on secondary side of the transformer. Similarly the column vector $[V_1\ V_2\ V_3\ V_4\ V_5\ V_6]^T$ are the corresponding nodal voltages.

Solution Technique: The analysis of unbalanced load flow solution involves the formulation of the system nodal admittance matrix and solution of nodal performance equation.

The nodal admittance matrix of the unbalanced 3-phase network is assembled taking one element at time and modifying the matrix of the partial network to reflect the addition. The process is continued till all the elements such as machines, transmission line, transformer and their tappings and other shunt parameters are exhausted.

Final nodal equation is of the form,

$$[Y]\,\overline{V} = \overline{I} \tag{6.121}$$

Here $\overline{V}$ and $\overline{I}$ are the nodal voltages and currents respectively and Y is the polyphase nodal admittance matrix containing all sorts of unbalances of the system. For the solution of equation (6.121), each conventional busbar is represented by 3 nodes, one each representing a phase. Each 3-phase static load is assumed to be decoupled in three parts, and each part is assigned to a node. Each phase is a node. Similarly, each neutral is a node if not solidly grounded. The above nodal eqn. (eqn. 6.121) is solved by any iterative technique such as Gauss-Seidel, Netwon-Raphsons etc. sparsity technique to optimize computer memory as well as computational efforts can also be used since nodal admittance matrix is highly sparse.

The following are the solution steps:

1. Each phase of each bus bar including the neutral is assigned a node number.
2. Nodal polyphase admittance matrix is formed.
3. Equations (6.119) and (6.121) are solved iteratively by any numerical method exploiting sparsity till convergence criterion is satisfied.
4. Nodal voltage magnitude, phase angle, line flows, line losses and power at swing bus is calculated.
5. Desired results are computed and printed.

6.8 Numerical Example

With bus 1 in the figure shown below as slack bus, use the following methods to obtain load flow solution (a) Gauss-Seidel method using Y_{BUS} with an acceleration factor of 1.4 and tolerance of 0.0001 p.u. both for real and imaginary components of voltages.

(b) Newton-Raphson's method using Y_{BUS} with a tolerance of 0.01 p.u.

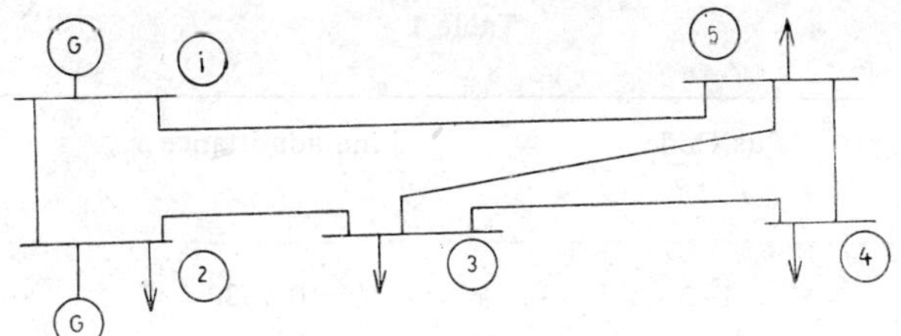

Fig. 1

Bus code p–q	Impedance Z_{pq}	Line charging $Y'_{pq}/2$
1–2	$0.02 + j0.04$	$j0\ 020$
2–3	$0.04 + j0.2$	$j0.020$
3–5	$0.15 + j0.4$	$j0.025$
3–4	$0.02 + j0.06$	$j0.01$
4–5	$0.02 + j0.04$	$j0.01$
1–5	$0.08 + j0.2$	$j0.02$

Using per unit at the base of 100 MVA, the scheduled generation and loads are :

Bus code 'p'	Generation		Load	
	MW	MVAR	MW	MVAR
1 (slack bus)	0	0	0	0
2	50	25	15	10
3	0	0	45	20
4	0	0	40	15
5	0	0	50	25

p.u. voltage at bus 1 = $1\angle 0$ (slack bus).

Solution: *By Gauss-Siedel Method*

Since bus 1 is a slack bus, we will start with assuming initially the voltage mode and angles equal to slack bus quality i.e. $1\angle 0$.

Determination of Y_{BUS} Matrix

In order to determine the elements of bus admittance matrix, the transmission line and line charging admittance taking ground as reference is calculated. The transmission line admittance is obtained by taking the reciprocal of line impedance and is given in Table 1.

Table 1

Bus Code $p-q$	Line admittance y_{pq}
1–2	$10-j20$
2–3	$0.96154-j4.80769$
3–5	$0.82192-j2.19178$
3–4	$5-j15$
4–5	$10-j20$
1–5	$1.72414-j4.31034$

The total line charging admittance to ground at each bus is shown in Table 2.

Table 2

Bus Code p	Admittance to ground y_p
1	$0.0+j0.04$
2	$0.0+j0.04$
3	$0.0+j0.055$
4	$0.0+j0.02$
5	$0.0+j0.03$

Since there is no mutual coupling in the representation of the system shown, the diagonal elements of the bus admittance matrix for bus 1 by inspection is

$$Y_{11} = y_{12} + y_{15} + y_1$$
$$= 10 - j20 + 1.72414 - j4.31034 + 0.0 + j0.04$$
$$= 11.72414 - j24.27034$$

The off-diagonal element of bus 1 by inspection are

$$Y_{12} = Y_{21} = -y_{12} = -10 + j20$$
$$Y_{13} = Y_{31} = -y_{13} = 0$$

Proceeding in the above manner, the elements of Y_{BUS} matrix are calculated and are given in Table 3.

Table 3

	①	②	③	④	⑤
①	11.72414 −j24.27034	−10.+j20	0+j0	0+j0	−1.72414 +j4.31034
②	−10+j20	10.96154 −j24.76769	−0.96154 +j4.80769	0+j0	0+j0
③	0+j0	− 0.96154 +j4.80769	6.78346 −j21.9445	−5+j15	−0.82192 0+j2.19178
④	0+j0	0+j0	−5+j15	15−j34.98	−10+j20
⑤	−1.72414 +j4.31034	0+j0	−0.82192 +j2.19178	−10+j20	12.54606 −j26.44713

The equations for the Gauss-Siedel iterative solution for the given network are as:

$$E_2^{k+1}=\frac{1}{Y_{22}}\left[\frac{P_2-jQ_2}{E_2^*}-Y_{21}E_1-Y_{23}E_3^k\right]$$

$$E_3^{k+1}=\frac{1}{Y_{33}}\left[\frac{P_3-jQ_3}{(E_3^k)^*}-Y_{32}E_2^{k+1}-Y_{34}E_4^k-Y_{35}E_5^k\right]$$

$$E_4^{k+1}=\frac{1}{Y_{44}}\left[\frac{P_4-jQ_4}{(E_4^k)^*}-Y_{43}E_3^{k+1}-Y_{45}E_5^k\right]$$

$$E_5^{k+1}=\frac{1}{Y_{55}}\left[\frac{P_5-jQ_5}{(E_5^k)^*}-Y_{51}E_1-Y_{53}E_3^{k+1}-Y_{54}E_4^{k+1}\right]$$

The first step in the iterative solution is to calculate new estimate of voltage for bus 2, the equations is

$$E_2^{(1)}=\frac{1}{Y_{22}}\left[\frac{P_2-jQ_2}{(E_2^{(0)})^*}-Y_{21}E_1-Y_{23}E_3^{(0)}\right]$$

The change in voltage is

$$E_2^{(1)}=E_2^{(1)}-E_2^{(0)}$$

The accelerated value of bus voltage is given by the equation

$$E_{2(\text{accelerated})}^{(1)}=E_2^{(0)}+\alpha E_2^{(1)}$$

which replaces the original assumed values of bus 2 and is used in subsequent calculation of voltage for remaining buses.

The process is continued for remaining buses to complete one iteration. This process is continued till $| E_p^{k+1} - E_p^k | \leqslant 0.0001$ which is achieved in 28 iteration, the voltage at various buses obtained on computer are given below in Table 4.

Table 4 Bus voltages from Gauss-Siedel iterative tech.

Bus No.	Complex voltage	Voltage Magnitude	Voltage angle
1	$1 + j0$	1.00	0.0
2	0.98442767 $-j0.00807412$	0.98446078	−0.46973124
3	0.87243819 $-j0.12531753$	0.88129250	−8.1707929
4	0.8587696 $-j0$ 13333714	0.86916535	−8.8209417
5	0.86609264 $-j0.1286197$	0.87504685	−8.2003366

Line Flows:

The line flows have been calculated with the final bus voltages and the given line admittance and line charging admittance. The flow in line 1−2 at bus 1 from the equation.

Table 5

Bus code p–q	Complex Power
1–2	$0.31720569 - j0.21070545$
2–1	$-0.3141288 + j0.24393403$
2–3	$0.66433486 - j0.39424428$
3–2	$-0.63905826 + j0.30278162$
3–5	$0.00546896 + j0.00748921$
5–3	$-0.00543569 + j0.03098607$
3–4	$0.18457288 - j0.11114618$
4–3	$-0.18333177 + j0.122744587$
4–5	$-0.21549821 + j0.02648868$
5–4	$0.21673715 - j0.01375502$
1–5	$0.76907292 - j0.34190762$
5–1	$-0.71127690 + j0.23273176$

$$P_{pq} - jQ_{pq} = E_p^* (E_p - E_q) y_{pq} + E_p^* E_p \frac{y'_{pq}}{2}$$

From the above equation, the flow in line 1—2 is

$$P_{12} - jQ_{12} = E_1^*(E_1 - E_2) y_{12} + E_1^* E_1 \frac{y'_{12}}{2}$$

The flow in line 1—2 at bus 2 is

$$P_{21} - jQ_{21} = E_2^*(E_2 - E_1) y_{21} + E_2^* E_2 \frac{y'_{21}}{2}$$

Similarly all line flow have been calculated on computer and are given in Table 5. The slack bus power has been determined by summing the flow on the line terminating at the slack bus. The real and reactive powers are given in Table 6.

Table 6

Bus No.	Real Power	Reactive Power
1	1.0862787	—0.5526132
2	0.35	0.15
3	—0.45	—0.20
4	—0.40	—0.15
5	—0.50	—0.25

(*b*) *Newton Raphson method:* The matrix equation for the solution of load flow by N.R. method is

$$\begin{bmatrix} \Delta P^k \\ \Delta Q^k \end{bmatrix} = \begin{bmatrix} J_1^k & J_2^k \\ J_3^k & J_4^k \end{bmatrix} \begin{bmatrix} \Delta e^k \\ \Delta f^k \end{bmatrix}$$

The changes in the bus power obtained from:

$$\Delta P_p^k = P_{p(\text{scheduled})} - P_p^k \quad \text{(a)}$$

$$\Delta Q_p^k = Q_{p(\text{scheduled})} - Q_p^k \quad \text{(b)}$$

$P_{p(\text{scheduled})}$ and $Q_{p(\text{scheduled})}$ are given in Data.

The calculated bus power are obtained from the equations

$$P_p^k = \sum_{q=1}^{n} [e_p^k(e_q^k G_{pq} + f_q^k B_{pq}) + f_p^k(f_q^k G_{pq} - e_q^k B_{pq})]$$

$$Q_p^k = \sum_{q=1}^{n} [f_p^k(e_q^k G_{pq} + f_q^k B_{pq}) - e_p^k(f_q^k G_{pq} - e_q^k B_{pd})]$$

Assuming flat start i.e. all bus voltages are assumed equal to slack bus voltage of $1\angle 0°$

$$P_2^{(0)} = e_2^{(0)}[e_1^{(0)} G_{21} + e_2^{(0)} G_{22} + e_5^{(0)} G_{25}]$$

$$Q_2^{(0)} = e_2^{(0)}[e_1^{(0)} B_{21} + e_2^{(0)} B_{22} + e_5^{(0)} B_{25}]$$

Similarly power for the remaining buses are calculated. Then the equations (a) and (b) gives change in real and reactive power.

The bus currents are calculated from the equations

$$I_p^k = \frac{P_p^k - jQ_p^k}{(E_p^k)^*}$$

The current for bus 2 is

$$I_2^{(0)} = \frac{P_2^{(0)} - jQ_2^{(0)}}{(E_2^{(0)})^*}$$

Similarly the components of the bus currents, i.e., $c_p^{(0)}$ and $d_p^{(0)}$ for remaining buses are calculated.

The elements of Jacobian are calculated using the voltages and currents and elements of Y_{BUS}. Y_{BUS} is obtained as per the procedure given in Gauss-Seidal method.

The diagonal elements in first row of J_1^k for the equation

$$\frac{\delta P_p}{\delta e_p} = e_p^k\, G_{pp} - f_p^k\, B_{pp} + C_p^k$$

is

$$\frac{\delta P_2}{\delta e_2} = e_2^{(0)}\, G_{22} + C_2^{(0)}$$

And off-diagonal elements for the equation

$$\frac{\delta P_p}{\delta e_q} = e_p^k\, G_{pq} - f_p^k\, B_{pq}$$

$$\frac{\delta P_2}{\delta e_3} = e_2^{(0)}\, G_{23}$$

$$\frac{\delta P_2}{\delta e_4} = e_2^{(0)}\, G_{24}$$

and

$$\frac{\delta P_2}{\delta e_5} = e_2^{(0)}\, G_{25}$$

The diagonal elements in first row of J_2^k from the equation

$$\frac{\delta P_p}{\delta f_p} = e_p^k\, B_{pp} + f_p^k\, G_{pp} + d_p^k$$

is $$\frac{\delta P_2}{\delta f_2} = e_2^{(0)} B_{22} + d_2^{(0)}$$

And off-diagonal elements for the equation

$$\frac{\delta P_p}{\delta f_q} = e_p^k B_{pq} + f_p^k G_{pq}$$

are $$\frac{\delta P_2}{\delta f_3} = e_2^{(0)} B_{23}$$

$$\frac{\delta P_2}{\delta f_4} = e_2^{(0)} B_{24}$$

$$\frac{\delta P_2}{\delta f_5} = e_2^{(0)} B_{25}$$

The diagonal elements in the first row of J_3^k for the equation

$$\frac{\delta Q_p}{\delta e_p} = e_p^k B_{pp} + f_p^k G_{pp}$$

is $$\frac{\delta Q_2}{\delta e_2} = e_2^{(0)} B_{22} - d_2^{(0)}$$

And off-diagonal elements for the equation

$$\frac{\delta Q_p}{\delta e_q} = f_p^k G_{pq} + e_p^k B_{pq};$$

are $$\frac{\delta Q_2}{\delta e_3} = e_2^{(0)} B_{23}$$

$$\frac{\delta Q_2}{\delta e_4} = e_2^{(0)} B_{24}$$

$$\frac{\delta Q_2}{\delta e_5} = e_2^{(0)} B_{25}$$

The diagonal element in the first row of J_4^k for the equation

$$\frac{\delta Q_p}{\delta f_p} = - e_p^k G_{pp} + f_p^k B_{pp} + C_p^k ;$$

is $$\frac{\delta Q_2}{\delta f_2} = - e_2^{(0)} G_{22} + C_2^{(0)}$$

And off-diagonal elements for the equation

$$\frac{\delta Q_p}{\delta f_q} = f_p^k B_{pq} - e_p^k G_{pq};$$

are $$\frac{\delta Q_2}{\delta f_3} = - e_2^{(0)} G_{23}$$

$$\frac{\delta Q_2}{\delta f_4} = -e_2^{(0)} G_{24}$$

$$\frac{\delta Q_2}{\delta f_5} = -e_2^{(0)} G_{25}$$

The process is repeated to obtain the elements of the remaining rows of Jacobian for $k = 0$.

The solution of the matrix equation for Δe_p and Δf_p, $p = 2, 3, 4, 5$ is obtained from

$$\begin{bmatrix} \Delta e^k \\ \hline \Delta f^k \end{bmatrix} = J^{-1} \begin{bmatrix} \Delta P^k \\ \hline \Delta Q^k \end{bmatrix}$$

The new bus voltages are obtained from the equation

$$E_p^{k+1} = E_p^k + \Delta E_p^k$$

$$E_2^{(1)} = E_2^{(0)} + \Delta E_2^{(0)}$$

$$E_3^{(1)} = E_3^{(0)} + \Delta E_3^{(0)}$$

$$E_4^{(1)} = E_4^{(0)} + \Delta E_4^{(0)}$$

$$E_5^{(1)} = E_5^{(0)} + \Delta E_5^{(0)}$$

These values are used to compute bus power and currents and elements of Jacobian for next iteration. The process is terminated when change in both real and reactive power at each bus are less than 0.01. The line flows are calculated as indicated in load flow solution by Gauss-Siedel method.

Load flow solution is obtained on computer in 2 iterations, the results obtained are given below.

Bus Voltages

Bus No.	Voltage magnitude	Voltage angle in degrees
1	1.0	0
2	0.985	−0.47468
3	0.882	−8.14892
4	0.870	−8.79050
5	0.876	−8.17651

Line Flows

Bus code	Complex power
1–2	0.3175695 −*J*0.20728906
2–1	−0.31451928+*J*0.24057702
2–3	0.66211276−*J*0.3916638
3–2	−0.63705210+*J*0.30131070
3–5	0.00537251 +*J*0.007605
5–3	−0.00533988+*J*0.03093636
3–4	0.18285697−*J*0.1111761
4–3	−0.1863379 +*J*0.1228549
4–5	−0.2141475 +*J*0.027647
5–4	0.21537020−*J*0.0148550
1–5	0.766369 −*J*0.3389190
5–1	−0.709077 +*J*0.231031

Problems

6.1 With bus 1 in Fig. 1 as slack bus, use the following methods to obtain load flow solution. (a) Gauss-Seidel method using Y_{BUS} with an acceleration factor of 1.4 and tolerance of 0.0001 p.u. both for real and imaginary components of voltages, (b) Newton-Raphson's methods using Y_{BUS} with a tolerance of 0.01 p.u. for changes in the real and reactive bus powers, and also by (c) Modified Newton's method with a tolerance of 0.01 p.u.

Bus code p–q	Impedance Z_{pq}	Line charging $Y'_{pq}/2$
1–2	0.01+*j*0.025	*j*0.025
2–3	0.02+*j*0.2	*j*0.020
3–5	0.20+*j*0.3	*j*0.025
3–4	0.02+*j*0.04	*j*0.025
4–5	0.02+*j*0.05	*j*0.015
1–5	0.15+*j*0.4	*j*0.025

Values are given in p.u. at the base of 100 MVA. The scheduled generation and loads are:

Bus Code 'p'	Generation MW—MVARS		Load MW—MVARS	
1 (slack bus)	0	0	0	0
2	75	40	20	15
3	0	0	50	25
4	0	0	40	15
5	0	0	60	25

p.u. voltage at bus No. 1=1 ∠0 slack bus.

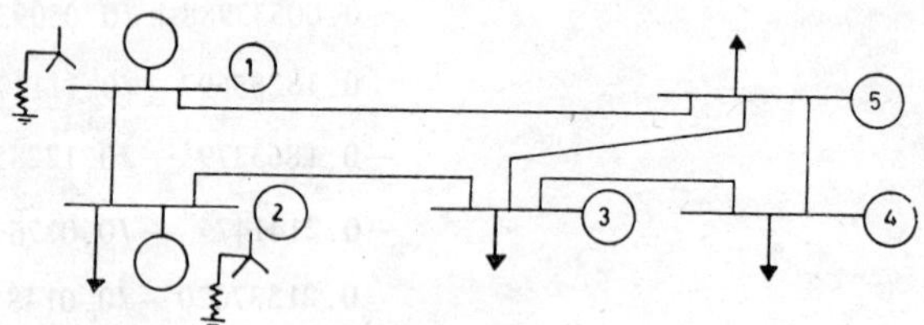

Fig. 1

6.2 With bus 1 in Fig. 2 as slack bus, use the following methods to obtain load flow solutions: (a) Gauss-Seidel method using Y_{BUS} with an acceleration factor of 1.35 and tolerance of 0.0001 p.u. (b) Newton-Raphson's method using Y_{BUS} with a tolerance of 0.0015 p.u. and (c) Fast decoupled load flow method. Use data not given by suitable assumptions.

Impedance of sample network of Fig. 2.

Bus code p–q	Imp. Z_{pq}	Line charging $Y'_{pq}/2$
1–2	$0.12+j0.05$	$j0.025$
2–3	$0.05+j0.02$	$j0.015$
3–4	$0.25+j0.075$	$j0.02$
1–3	$0.45+j0.045$	$j0.015$
1–4	$0.05+j0.015$	$j0.01$

Scheduled generation and loads

Bus code p	Generation MW	Generation MVARS	Load MW	Load MVARS
1	0	0	0	0
2	0	0	50	25
3	100	0	0	0
4	0	0	125	30

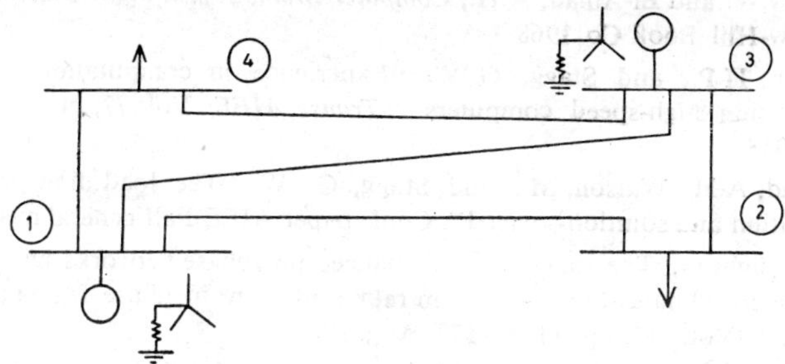

Fig. 2

References

1. J.B. Ward and H.W. Hale, "Digital computer solution of power flow problems", *Trans. AIEE* (Power Apparatus and Systems), Vol. 75, p. 398, June 1956.
2. R.J. Brown and W.F. Tinney, "Digital solutions for large power networks", *Trans. AIEE* (Power Apparatus and Systems), Vol. 76, p. 347, June 1957.
3. A.F. Glimn and G.W. Stagg, "Automatic calculation of load flows", *Trans. AIEE* (Power Apparatus and Systems), Vol. 76, p. 817, October 1957.
4. J.E. Van Ness, "Iteration methods for digital load flow studies", *Trans. AIEE* (Power Apparatus and Systems), Vol. 78A, p. 583, August 1959.
5. J.E. Van Ness, "Convergence of iterative load flow studies", *Trans. AIEE* (Power Apparatus and Systems), Vol. 78B, 1590, February 1959.
6. H.W. Hale and R.W. Goodrich, "Digital computation of power flow—some new aspects", *Trans. AIEE* (Power Apparatus and Systems), Vol. 78A, p. 919, October 1959.
7. J.E. Van Ness ond J.H. Griffin, "Elimination methods for load flow studies", *Trans. AIEE* (Power Apparatus and Systems), Vol. 80, p. 299, June 1961.
8. H.E. Brown, G.K. Carter, H. H, Happ, and C.E. Person, "Power flow solution by matrix iterative method", *Trans. AIEE* (Power Apparatus and Systems), Vol, 82, p. 1, April 1963.
9. M.A. Laughton and M.W. Humphrey Davics, "Numerical techniques in solution of power-system load-flow problems", *Proc. IEE* (London), Vol. III, p. 1575, September 1964.
10. N. Sato and W.F. Tinney, "Techniques for exploiting the sparsity of the network admittance matrix", *IEEE Trans. Power Apparatus and Systems*, Vol. 82, p. 944, December 1963.
11. L. Carpentier, "Ordered climinations", Proc. 1963 Power System Computation Conf.
12. R. Baumann, "Some new aspects of load flow calculation", Proc. 1965 Power Industry Computer—Applications Conf. p. 91.
13. D. Feingold and D. Spohn, "Bounded variables load flow problems", Proc. 1966 Power System Computation Conf., Sec. 4.3.
14. D. Feingold and D. Spohn, "Direct methods for linear networks of load flow computation" Proc. 1966 Power System Computation Conf., Sec. 4.4.
15. J. Peschon, W.F. Tinney, D.S. Piercy and O.J. Tvelt, "Optimum control of reactive power flow", presented at 1967 IEEE Winter Power Meeting, New York, N.Y. January 29—February 3.

16. Stagg, G.W. and El-Abiad, A.H., *Computer Methods in Power System Analysis*", McGraw-Hill Book Co. 1968.

17. St. Clair, H.P., and Stagg, G.W., "Experience in computation of load flow studies using high-speed computers", *Trans. AIEE*, Vol. 77, pt. III, pp. 1275–1282, 1958.

18. El-Abiad, A.H., Watson, M., and Stagg, G. W. "The load flow problems—its formulation and solution—part I", Conf. paper AIEE Fall general meeting, 1961.

19. M.A. Laughton, 'The analysis of unbalanced polyphase networks by the methods of phase coordinates-Part 1—System representation in phase frame of reference' *Proc. IEE*, Vol. 115, pp. 1163–1172, Aug. 1968.

20. L. Roy, B.H. Raw and M.A. Laughton, 'Analysis of unbalanced polyphase networks-Part III—Load flow analysis', IEEE Winter Power Meeting, New York, Paper A79026-6, February 1979.

21. M.A. Laughton and A.O.M. Saleh, 'United phase coordinate load flow and fault analysis of polyphase networks'. *International Journal of Electrical Power and Energy Systems*. Vol. 2, No. 4, pp. 181–192, 1980.

22. K.A. Birt, J.A. Graffy, J.D. McDonald and A.H. El-Abiad, 'Three phase load flows', *IEEE Trans. on PAS*, PAS-95, No. 1, pp. 59-65, 1976.

Chapter 7

Economic Load Scheduling of Power System

7.1. Introductory Remarks

The economic load scheduling of a power system is perhaps the most exciting branch for the power system engineers. Of course, this topic was not very important in the beginning when there were small power generating stations for each locality such as urban power systems. But now, with the growth in the demand of electricity and at the same time, guarantee regarding the continuity of supply to the consumers under normal conditions, have forced the power system engineers to develop grid system, i.e. interconnections of different generating stations located at different places. For such systems, the optimum load scheduling of the different generating plants in the system has become increasingly important.

7.2 Urban (Thermal) Systems

By economic load scheduling we mean to determine the generations of different plants such that the total operating cost is minimum, and at the same time the total demand and the losses at any instant is met by the total generation. In the *urban systems* the generators are close to the load centre, i.e. length of the transmission line is so small that the transmission losses are negligible. At the same time, for the urban power systems, there used to be mostly thermal plants. The operating cost of these thermal plants is mainly the cost of fuel. However, in addition to the fuel cost, there is also cost of labour, supplies, maintenance and water etc. However, since there is no direct method to determine these as a function of generation and therefore, usually these are assumed to vary as a fixed percentage of the fuel cost. Therefore, the operating cost of the thermal plants which is mainly the fuel cost, is given as a function of generation. This cost function is defined as a nonlinear function of plant generations. Normally graph is given between the heat value of fuel in B.T.H.U. and power generation in M.W. and knowing the cost of the fuel, we can definitely determine the fuel cost as a function of generations for each thermal plant.

Hence for the optimal operation, the problem becomes to find the generation of the respective thermal plants, i.e. P_i, $i = 1, \dots, N$ such that the objective function (i.e. total cost of fuel) as defined by the equation:

$$C_t = \sum_{i=1}^{N} C_i(P_i) \qquad (7.1)$$

is minimum, subject to the constraint

$$\sum_{i=1}^{N} P_i = P_L \tag{7.2}$$

Equation (7.2) indicates that the total demand P_L at any instant is met by the total generation as the transmission losses for these schemes (i.e. urban systems) are negligible. As equation (7.1) is the nonlinear objective function with the nonlinear equality constraint (eqn. 7.2), it may be converted by choosing lagranges multiplier λ, into an unconstrained objective function as shown in equation 7.3

$$F = C_t - \lambda\left(\sum_{i=1}^{N} P_i - P_L\right)$$

$$= \sum_{i=1}^{N} C_i(P_i) - \lambda\left(\sum_{i=1}^{N} P_i - P_L\right) \tag{7.3}$$

where P_i is the generation in M.W. of the ith thermal plant and P_L is the total load.

Equation (7.3) can be solved for minimum by simple law of calculus, i.e. by determining the partial derivative of the function 'F' (eqn. 7.3) with respect to variable P_i (plant generation) and equating it equal to zero,

i.e.
$$\frac{\delta F}{\delta P_i} = 0 = \frac{\delta C_i}{\delta P_i} - \lambda \tag{7.4}$$

or
$$\lambda = \frac{\delta C_i}{\delta P_i} = \frac{dC_i}{dP_i} \text{ for } i = 1, \ldots, N \tag{7.5}$$

Hence for the most economic operation, all plants must operate at the equal incremental cost. The incremental cost is normally given in B.Th.U./M. [illegible] and knowing the cost of fuel, it can be converted into actual cost function in Rs. per M.W.hr.

7.3 Transmission Losses

With the development of integrated power systems (i.e. grid systems) and also with the interconnections of different power stations located at far places for the purpose of economy of interchange and reliability of operation, it is necessary to consider not only the incremental fuel cost but also the incremental transmission losses for economy. This is, because, no doubt the operational cost, i.e. the generation cost of any thermal plant may be low, but still the plant may be uneconomical because of its location very far from the load centre. Due to this, the transmission losses may be considerably high and hence the plant may be overall uneconomical. A transmission loss formula expressing the total transmission losses as function of plant generations was first presented by E.E. George in 1943. Since then, there is considerable improvement in the method for the determination of the transmission losses as a function of plant generation. The application of digital computor to calculate losses was developed by Kirchmayer, Stagg, Glimn and Habermann in the year 1943. The term of the equation is given below:

$$P_T = \sum_{m=1}^{N} \sum_{n=1}^{N} P_m B_{mn} P_n + \sum_{n=1}^{N} B_{no} P_{no} + B_{oo}. \tag{7.6}$$

7.3.1 *Derivation of Transmission Loss Formula*

In order to develop a general formula for transmission losses in terms of generations P_i, first of all we perform load flow studies. From the load flow studies, the voltage magnitude and phase angle at both, load and generation buses are obtained. The injected bus currents are calculated from the following network equations

$$\bar{E}_{BUS} - E_S = [Z_{BUS}]\, \bar{I}_{BUS} \tag{7.7}$$

Here E_S is the slack bus voltage and $[Z_{BUS}]$ matrix is developed with slack bus as the reference bus. In performing load flow studies, if any, bus has both load and generation, we split this bus into two buses by creating a fictitious bus, so that at the generation buses there are no loads. In this manner, we will eventually obtain a set of buses known as generation buses, where voltage magnitude and real power are known and a set of buses, known as load buses where real power P and reactive power Q are known. One of the generation buses normally with maximum real power capability is taken as slack (i.e. reference) bus. Equation (7.7) can be expressed as follows:

$$\begin{bmatrix} E_{G_1} - E_S \\ E_{G_2} - E_S \\ \vdots \\ E_{L_1} - E_S \\ E_{L_2} - E_S \\ \vdots \end{bmatrix} = \begin{bmatrix} Z_{G_1G_1} Z_{G_1G_2} \cdots Z_{G_1L_1} Z_{G_1L_2} \cdots \\ Z_{G_2G_1} \cdots\cdots\cdots \\ \vdots \\ \vdots \\ \vdots \\ \vdots \end{bmatrix} \begin{bmatrix} I_{G_1} \\ I_{G_2} \\ \vdots \\ I_{L_1} \\ I_{L_2} \\ \vdots \end{bmatrix} \tag{7.8}$$

As mentioned earlier, E_S (i.e. slack bus voltage) is also included in the E_G. Here E_{Gi} for $i = 1, 2, \ldots$ etc. are the generation bus voltages and E_{Li} for $i = 1, 2, \ldots$ are the load buses and I_{Gi} and I_{Li} are the corresponding bus currents (i.e. injected currents at the buses). Following steps are followed in sequence to obtain general formula for transmission loss in terms of generations (i.e. generating power).

Step 1: We write equation in the reference frame 1. We know that

$$I_{L_1} + I_{L_2} + \ldots = I_L = \sum_{j=1}^{k} I_{Lj} \tag{7.9}$$

where I_L is the total load currents and I_{Lj} for $j = 1, \ldots, k$ are the injected bus currents at k load buses. We can write the individual bus currents at these load buses as a linear function of total load current I_L i.e.

$$I_{L1} = l_1 I_L$$
$$I_{L2} = l_2 I_L \ \ldots \text{ etc.} \tag{7.10}$$

i.e. $I_{Li} = l_i I_L$ for $i = l, \ldots, k$. Here l_i is some complex number and is equal to

$$l_i = \frac{I_{Li}}{I_L} \quad \text{for} \quad i = l, \ldots, k \tag{7.11}$$

In equation (7.10), we have assumed that the individual load currents vary in proportion to the total load. However, if some loads do not vary in proportion to the total load, these loads are known as non-conforming loads and these are treated as negative generation.

Step 2: The equation of bus currents (including both generation bus currents I_{Gi} and load bus currents I_{Li}) after substitution of relation (equation (7.10)) for load bus currents will be as shown in equation (7.12) i.e.

$$\begin{bmatrix} I_{G1} \\ I_{G2} \\ \vdots \\ I_{L1} \\ I_{L2} \\ \vdots \end{bmatrix} = \begin{bmatrix} l & 0 & \cdot & \cdot & \cdot & \cdot & \cdot & \cdot & 0 \\ 0 & l & 0 & \cdot & \cdot & \cdot & \cdot & \cdot & 0 \\ & & & & \vdots & & & & \\ 0 & \cdot & \cdot & \cdot & \cdot & \cdot & \cdot & \cdot & l_1 \\ 0 & \cdot & \cdot & \cdot & \cdot & \cdot & \cdot & \cdot & l_2 \\ & & & & \vdots & & & & l \end{bmatrix} \begin{bmatrix} I_{G1} \\ \cdot \\ \cdot \\ \cdot \\ \cdot \\ \cdot \\ \cdot \\ \cdot \\ I_L \end{bmatrix} \tag{7.12}$$

Old bus currents New bus currents

Equation (7.12) can be written symbolically as shown below

$$\underset{(\text{old})}{\bar{I}_{\text{BUS}}} = [C_2^1] \underset{(\text{new})}{\bar{I}_{\text{BUS}}} \tag{7.13}$$

Here $\bar{I}_{\text{BUS (old)}}$ is the column vector of old bus currents, i.e. total bus currents which include both generation as well as load buses and is of dimension nl where n is the number of buses, C_2^1 is a linear transformation which transforms the total bus currents (equation 7.8) in terms of generation bus currents I_G and total load current I_L is of dimension $nX(n_g+1)$ where n_g is the number of generation buses and $\bar{I}_{\text{BUS (new)}}$ is the column vector of bus currents I_G and total load current I_L and is of dimension $(n_g+1)Xl$. Here C_2^1 is a transformation from reference frame 1 to reference 2, i.e. it transforms the old currents to the new currents (equation 7.13) step 2. We know that

$$I_{G1} + I_{G2} + \ldots = -I_L \tag{7.14}$$

i.e.

$$\begin{bmatrix} I_{G1} \\ I_{G2} \\ \\ \\ I_L \end{bmatrix} = \begin{bmatrix} 1 & & & & 0 \\ & 1 & & & \\ & 0 & \ddots & & \\ & & & \ddots & \\ -1 & -1 & \cdots & \cdots & -1 \end{bmatrix} \begin{bmatrix} I_{G1} \\ I_{G2} \\ \cdot \\ \cdot \\ \cdot \end{bmatrix} \tag{7.15}$$

Equation (7.15) can be written symbolically as follows

$$\bar{I}_{BUS\,(new)(ng+1)X1} = [C_3^2]\,\bar{I}_{G\,BUS_{ngX1}} \tag{7.16}$$

Here $\bar{I}_{BUS\,(new)}$ is the column vector of bus currents as defined by equation (7.13), C_3^2 is a linear transforms from reference frame 2 (i.e. new bus currents) to the reference frame 3 (i.e. generation bus currents) and $\bar{I}_{G\,BUS}$ is the column vector of generation bus currents (i.e. injected currents only at the generation buses).

The equation (7.8) because of transformations C_2^1 and C_3^2 as defined in equations (7.12, 7.13, 7.15 and 7.16) takes the following form

$$\begin{bmatrix} E_{G_1} - E_S \\ E_{G_2} - E_S \\ \vdots \\ E_{L_1} - E_S \\ E_{L_2} - E_S \\ \vdots \end{bmatrix} = \begin{bmatrix} Z_{G_1G_1} & Z_{G_1G_2} & \cdots \\ Z_{G_2G_1} & \cdots & \\ & \cdot & \\ & \cdot & \\ & \cdot & \\ & \cdot & \end{bmatrix} C_2^1 C_3^2 \begin{bmatrix} I_{G1} \\ I_{G2} \\ \cdot \\ \cdot \\ \cdot \\ \cdot \\ \cdot \end{bmatrix} \tag{7.17}$$

$$= [Z_{BUS}][C_2^1][C_3^2]\,\bar{I}_{G\,BUS}$$

$$= [Z_{BUS}]\,[C]\,\bar{I}_{G\,BUS} \tag{7.18}$$

where $[C] = [C_2^1][C_3^2]$ is a unitary matrix, i.e. $C^{*T} = C^{-1}$, since $[C]$ is a linear power invariant transformations, we have from eqn (7.17),

$$\bar{E}_{BUS\,(old)} = [C]\,\bar{E}_{BUS\,(new)}$$

i.e.

$$\bar{E}_{BUS\,(new)} = (C^{*T})\,\bar{E}_{BUS\,(old)} \tag{7.19}$$

and also,

$$\bar{I}_{BUS\,(old)} = [C]\,\bar{I}_{BUS\,(new)};$$

and

$$\bar{I}_{BUS\,(new)} = [C^{*T}]\,\bar{I}_{BUS\,(old)} \tag{7.20}$$

Power in the Bus frame of Reference—Power in the old frame of reference is,

$$P_{old} = \bar{E}^T_{BUS\,(old)}\,\bar{I}^*_{BUS\,(old)} \tag{7.21}$$

Power in the new frame of reference, i.e.

$$P_{new} = \bar{E}^T_{BUS\,(new)}\,\bar{I}^*_{BUS\,(new)} \tag{7.22}$$

since the transformation matrix $[C]$ is linear power invariant,

$$P_{old} = P_{new}$$

i.e.

$$\bar{E}^T_{BUS\,(old)}\,\bar{I}^*_{BUS\,(old)} = \bar{E}^T_{BUS(new)}\,I^*_{BUS(new)} \tag{7.23}$$

Substituting relations (eqns. 7.19 and 7.20) in equation (7.23), we obtain,

$$\overline{E}^T_{\text{BUS (old)}} [[C]\, \overline{I}_{\text{BUS (new)}}]^* = \overline{E}^T_{\text{BUS (new)}} \overline{I}^*_{\text{BUS (new)}}$$

i.e. $$\overline{E}^T_{\text{BUS (old)}} [C^*]\, \overline{I}^*_{\text{BUS (new)}} = E^T_{\text{BUS (new)}} I^*_{\text{BUS (new)}} \quad (7.24)$$

From equation (7.24), we get,

$$E^T_{\text{BUS (old)}} [C^*] = \overline{E}^T_{\text{BUS (new)}}$$

i.e. $$\overline{E}^T_{\text{BUS (new)}} = \overline{E}^T_{\text{BUS (old)}} [C^*] \quad (7.25)$$

We know that,

$$\overline{E}_{\text{BUS (old)}} = [Z_{\text{BUS (old)}}]\, \overline{I}_{\text{BUS (old)}} \quad (7.26)$$

Premultiplying equation (7.26) by C^{*T}, we obtain

$$C^{*T}\, \overline{E}_{\text{BUS (old)}} = C^{*T} [Z_{\text{BUS (old)}}]\, \overline{I}_{\text{BUS (old)}} \quad (7.27)$$

Since $$\overline{I}_{\text{BUS (old)}} = [C]\, \overline{I}_{\text{BUS (new)}} \quad (7.28)$$

Substituting equation (7.28) in equation (7.27), we get

$$[C^{*T}]\, \overline{E}_{\text{BUS (old)}} = C^{*T} [Z_{\text{BUS (old)}}][C]\, \overline{I}_{\text{BUS (new)}} \quad (7.29)$$

Taking transpose of equation (7.25), we get

$$\overline{E}_{\text{BUS (new)}} = [C^{*T}]\, \overline{E}_{\text{BUS (old)}} \quad (7.30)$$

Substituting equation (7.30) in equation (7.29), we obtain,

$$\overline{E}_{\text{BUS (new)}} = [C^{*T}] [Z_{\text{BUS (old)}}] [C]\, \overline{I}_{\text{BUS (new)}} \quad (7.31)$$

But $$\overline{E}_{\text{BUS (new)}} = [Z_{\text{BUS (new)}}]\, I_{\text{BUS (new)}} \quad (7.32)$$

Comparing eqns. (7.31) and (7.32), we get,

$$[Z_{\text{BUS (new)}}] = [C^{*T}] [Z_{\text{BUS (old)}}] [C] \quad (7.33)$$

Step 3: We know that the transmission losses are given by

$$P_T = I^{*T}_{\text{BUS (new)}} \overline{E}_{\text{BUS (new)}} \quad (7.34)$$

Here $\overline{E}_{\text{BUS (new)}}$ is the bus voltages and $\overline{I}_{\text{BUS (new)}}$ the bus currents in the reference frame 3, i.e.,

$$\begin{bmatrix} I_{G1} \\ I_{G2} \\ \vdots \\ I_{L1} \\ I_{L2} \\ \vdots \end{bmatrix} = [C]\, \overline{I}_{\text{BUS(new)}} = CI_3 \quad (7.35)$$

Here $I_{\text{BUS (new)}}$ is expressed as I_3 to show that it is the bus current in the reference frame 3. Similarly, we write $\overline{E}_{\text{BUS (new)}}$ as E_3.

From eqn. (7.30) we have

$$\bar{E}_{\text{BUS (new)}} = E_3 = [C^{*T}]\,\bar{E}_{\text{BUS (old)}} = C^{*(T)} \begin{bmatrix} E_{G_1}-E_S \\ E_{G_2}-E_S \\ \vdots \\ E_{L_1}-E_S \\ E_{L_2}-E_S \\ \vdots \end{bmatrix} \tag{7.36}$$

Substituting for $\bar{E}_{\text{BUS (new)}}$, i.e. E_3 from eqn. (7.31) in eqn. (7.36) we obtain

$$E_3 = [C^{*T}] \begin{bmatrix} E_{G1}-E_S \\ E_{G_2}-E_S \\ \vdots \\ E_{L_1}-E_S \\ E_{L_2}-E_S \\ \vdots \end{bmatrix} = [C^{*T}][Z_{\text{BUS (old)}}]\,[C]\,I_{\text{BUS (new)}}$$

$$= [C^{*T}] \begin{bmatrix} E_{G_1}-E_S \\ E_{G2}-E_S \\ \vdots \\ E_{L_1}-E_S \\ E_{L_2}-E_S \\ \vdots \end{bmatrix} = [C^{*T}]\,[Z_{\text{BUS (old)}}]\,[C]\,I_3 \tag{7.37}$$

We have shown earlier (eqn. 7.18) that,

$$C = C_2^1\, C_3^2 = \begin{bmatrix} 1 & 0 & . & . & . & . & 0 \\ 0 & 1 & 0 & . & . & . & 0 \\ & & \vdots & & & & \\ 0 & 0 & . & . & . & 0 & 1_1 \\ 0 & . & . & . & . & 0 & 1_2 \\ & & \vdots & & & & \end{bmatrix} \begin{bmatrix} 1 & 0 & . & . & . & 0 \\ 0 & 1 & 0 & . & . & 0 \\ & & \vdots & & & \\ & & \vdots & & & \\ -1 & -1 & . & . & . & -1 \end{bmatrix}$$

$$= \begin{bmatrix} 1 & 0 & . & . & . & 0 \\ 0 & 1 & 0 & . & . & 0 \\ & & \vdots & \vdots & \vdots & \\ -1_1 & -1_1 & . & . & . & -1_1 \\ -1_2 & -1_2 & . & . & . & -1_2 \\ & & \vdots & & & \end{bmatrix} \tag{7.38}$$

In eqn. (7.38), C_3^1 and C_3^2 have been substituted from eqns. (7.12) and (7.15) respectively.

Equation (7.34), after the substitution of E_3 for $\bar{E}_{\text{BUS (new)}}$ and I_3 for $\bar{I}_{\text{BUS (new)}}$, takes the form,

$$P^T = I_3^{*T} E^3 \tag{7.39}$$

Substituting eqn. (7.37) in eqn. (7.39), we get

$$P_T = I_3^{*T} [C^{*T}] [Z_{\text{BUS (old)}}] \bar{C} I_3 \tag{7.40}$$

Substituting eqn. (7.33) in eqn. (7.40), we get

$$P_T = I_3^{*T} Z_{\text{BUS (new)}} I_3 \tag{7.41}$$

After dropping the subscript 3 and writing Z for $Z_{\text{BUS (new)}}$, equation (7.41) takes the form,

$$P_T = I^{*T} ZI \tag{7.42}$$

Let the complex power at the generator bus 1 is expressed as

$$P_1 + jQ_1 = E_1 I_{G1}^* \tag{7.43}$$

Here P_1 and Q_1 are the real and reactive power at the generation bus 1 and E_1 and I_{G_1} are the bus voltage and injected (i.e. bus) current at the bus 1. From eqn. (7.43), we obtain,

$$I_{G1}^* = \frac{P_1 + jQ_1}{E_1}$$

i.e.

$$I_{G_1} = \left[\frac{P_1 + jQ_1}{E_1}\right]^* = \frac{P_1 - jQ_1}{E_1}$$

Similarly, for any generation bus i, we have

$$I_{Gi} = \left[\frac{P_i + jQ_i}{E_i}\right]^* = \frac{P_i - jQ_i}{E_i} \tag{7.44}$$

for $i = 1, 2, \ldots, n$.

Substituting equation (7.44) in (7.42), we get

$$P_T = \left[\frac{P_1 + jQ}{E_1} \quad \frac{P_2 + jQ_2}{E_2} \quad \cdots \quad \frac{P_n + jQ_n}{E_n}\right][Z]\begin{bmatrix} \dfrac{P_1 - jQ_1}{E_1^*} \\ \dfrac{P_2 - jQ_2}{E_2^*} \\ \vdots \\ \dfrac{P_n - jQ_n}{E_n^*} \end{bmatrix} \tag{7.45}$$

$$= [P_1 + jQ_1 P_2 + Q_2 \cdots P_n + jQ_n]\begin{bmatrix} \dfrac{1}{E_1} & & & 0 \\ & \dfrac{1}{E_2} & & \\ & & \ddots & \\ 0 & & & \dfrac{1}{E_n} \end{bmatrix}[Z]$$

$$\times \begin{bmatrix} \dfrac{1}{E_1^*} & & & 0 \\ & \dfrac{1}{E_2^*} & & \\ & & \ddots & \\ 0 & & & \dfrac{1}{E_n^*} \end{bmatrix}\begin{bmatrix} P_1 - jQ_1 \\ P_2 - jQ_2 \\ \vdots \\ P_n - jQ_n \end{bmatrix} \tag{7.46}$$

Now, as already mentioned in equations (7.40), (7.41) and (7.42)

$$Z = Z_{BUS(new)} = C^{*T} Z_{BUS(old)} C \tag{7.47}$$

Substituting equation (7.47) in equation (7.46), we get

$$P_T = [P_1 + jQ_1 P_2 + jQ_2 \cdots]\begin{bmatrix} \dfrac{1}{E_1} & & 0 \\ & \dfrac{1}{E_2} & \\ & & \ddots \\ 0 & & \end{bmatrix} C^{*T} Z_{BUS(old)} C$$

$$\begin{bmatrix} \frac{1}{E_1^*} & & & 0 \\ & \frac{1}{E_2^*} & & \\ & & \cdot & \\ & & & \cdot \\ 0 & & & \cdot \end{bmatrix} \begin{bmatrix} P_1 - jQ_1 \\ P_2 - jQ_2 \\ \vdots \end{bmatrix}$$

$$= [P_1 + jQ_1 P_2 + jQ_2 \ \cdots\ ] C_1^{*T} Z_{\text{BUS(old)}} C_1 \begin{bmatrix} P_1 - jQ_1 \\ P_2 - jQ_2 \\ \vdots \end{bmatrix} \tag{7.48}$$

Here

$$C_1 = C \begin{bmatrix} \frac{1}{E_1^*} & & & 0 \\ & \frac{1}{E_2^*} & & \\ & & \cdot & \\ & & & \cdot \\ 0 & & & \cdot \end{bmatrix}$$, substituting C from eqn. (7.38),

we get

$$C_1 = \begin{bmatrix} 1 & 0 & \cdot & \cdot & \cdot & \cdot & \cdot & 0 \\ 0 & 1 & 0 & \cdot & \cdot & \cdot & \cdot & 0 \\ & \vdots & & & & & & \\ -1_1 & -1_1 & \cdot & \cdot & \cdot & \cdot & \cdot & -1_1 \\ -1_1 & -1_2 & \cdot & \cdot & \cdot & \cdot & \cdot & -1_2 \\ & \vdots & & & & & & \end{bmatrix} \begin{bmatrix} \frac{1}{E_1^*} & & & & & 0 \\ & \frac{1}{E_2^*} & & & & \\ & & \cdot & & & \\ & & & \cdot & & \\ & & & & \cdot & \\ 0 & & & & & \cdot \end{bmatrix}$$

$$= \begin{bmatrix} \frac{1}{E_1^*} & 0 & \cdot & \cdot & \cdot & \cdot & \cdot & 0 \\ 0 & \frac{1}{E_2^*} & 0 & \cdot & \cdot & \cdot & \cdot & 0 \\ -\frac{1_1}{E_1^*} & -\frac{1_1}{E_2^*} & \cdot & \cdot & \cdot & \cdot & \cdot & -\frac{1_1}{E_n^*} \\ -\frac{1_2}{E_1^*} & -\frac{1_2}{E_2^*} & \cdot & \cdot & \cdot & \cdot & \cdot & -\frac{1_2}{E_n^*} \\ & \vdots & & & & & & \end{bmatrix} \tag{7.49}$$

Let $C_1^{*T} Z_{BUS(old)} C_1 = C_2$, then with this substitution, equation (7.48) takes the form,

$$P_T = [P_1 + jQ_1 P_2 + jQ_2 \dots P_n + jQ_n] C_2 \begin{bmatrix} P_1 - jQ_1 \\ P_2 - jQ_2 \\ \vdots \\ P_n - jQ_n \end{bmatrix} \tag{7.50}$$

The expression for complex power at any generation bus is given by,

$$\begin{aligned} P_i + jQ_i &= P_i + jS_i P_i + jQ_{i0} \\ &= P_i(1 + jS_i) + jQ_{i0} \end{aligned} \tag{7.51}$$

In equation (7.51), it has been assumed that a fraction of the reactive power Q_i is proportional to the real power P_i, i.e. it is equal to $S_i P_i$ where S_i is a proportionality constant for the generation bus i (this is also known as S factor for generator i) and remaining part of the reactive power is constant, i.e. Q_{i0}.

Then by substituting eqn. (7.51) in eqn. (7.50), we obtain,

$$P_T = [P_1(1 + jS_1) + jQ_{10}\ P_2(1 + jS_2) + jQ_{20} \dots P_n(1 + jS_n)$$

$$+ jQ_0]\ [C_2] \begin{bmatrix} P_1(1 - jS_1) - jQ_{10} \\ P_2(1 - jS_2) - jQ_{20} \\ \vdots \\ P_n(1 - jS_n) - jQ_{n0} \end{bmatrix} \tag{7.52}$$

We can split eqn. (7.52) into the following three parts:

Part 1:

$$P_{T1} = [P_1(1 + jS_1)\ P_2(1 + jS_2) \dots]\ [C_2] \begin{bmatrix} P_1(1 - jS_1) \\ P_2(1 - jS_2) \\ \vdots \end{bmatrix}$$

Part 2:

$$P_{T2} = [P_1(1 + jS_1)\ P_2(1 + jS_2) \dots]\ [C_2] \begin{bmatrix} -jQ_{10} \\ -jQ_{20} \\ \vdots \end{bmatrix}$$

Part 3:

$$P_{T3} = [jQ_{10}\ jQ_{20} \dots]\ [C_2] \begin{bmatrix} -jQ_{10} \\ -jQ_{20} \\ \vdots \end{bmatrix} \tag{7.53}$$

Thus the expression for transmission losses P_T as shown in eqn. (7.52) is the sum of P_{T1}, P_{T2} and P_{T3} given in eqn. (7.53), i.e.

$$P_T = P_{T1} + P_{T2} + P_{T3} \tag{7.54}$$

Now P_{T1} from eqn. (7.53) is

$$P_{T1} = [P_1(1+jS_1)\ P_2(1+jS_2)\ \ldots]\ [C_2] \begin{bmatrix} P_1(1-jS_1) \\ P_2(1-jS_2) \\ \vdots \end{bmatrix}$$

$$= [P_1 P_2 \ldots P_n] \begin{bmatrix} 1+jS_1 & & & 0 \\ & 1+jS_2 & & \\ & & \ddots & \\ 0 & & & 1+jS_n \end{bmatrix} [C_2] \begin{bmatrix} 1-jS_1 & & & 0 \\ & 1-jS_2 & & \\ & & \ddots & \\ 0 & & & 1-jS_n \end{bmatrix} \begin{bmatrix} P_1 \\ P_2 \\ \vdots \\ P_n \end{bmatrix} \quad (7.55)$$

Now,

Real part of P_{T1} = Real P_{T1}

$$= [P_1 P_2 \ldots P_n] \text{ Real} \begin{bmatrix} 1+jS_1 & & 0 \\ & 1+jS_2 & \\ & & \ddots \\ 0 & & \end{bmatrix} [C_2] \begin{bmatrix} 1-jS_1 & & 0 \\ & 1-jS_2 & \\ & & \ddots \\ 0 & & \end{bmatrix} \begin{bmatrix} P_1 \\ P_2 \\ \vdots \\ P_n \end{bmatrix}$$

$$= [P]^T \text{ Real} \begin{bmatrix} 1+jS_1 & & 0 \\ & 1+jS_2 & \\ & & \ddots \\ 0 & & \end{bmatrix} [C_2] \begin{bmatrix} 1-jS_1 & & 0 \\ & 1-jS_2 & \\ & & \ddots \\ 0 & & \end{bmatrix} [P]$$

$$= P_{T1} \quad (7.56)$$

Assuming there are N generation buses (i.e. $n=N$), we get from eqn. (7.56)

$$P_{T1} = \sum_{m=1}^{N} \sum_{n=1}^{N} P_m B_{mn} P_n$$

$$= P_1 B_{11} P_1 + P_1 B_{12} P_2 + \ldots \quad (7.57)$$

Similarly assuming N buses, we have from eqn. (7.53),

$$P_{T2} = [P_1(1+jS_1)\ P_2(1+jS_2)\ \ldots]\ [C_2] \begin{bmatrix} -jQ_{10} \\ -jQ_{20} \\ \vdots \end{bmatrix}$$

$$= [P_1 P_2 \ldots P_N] \begin{bmatrix} 1+jS_1 & & & 0 \\ & 1+jS_2 & & \\ & & \ddots & \\ 0 & & & 1+jS_N \end{bmatrix} [C_2] \begin{bmatrix} -jQ_{10} \\ -jQ_{20} \\ \vdots \\ \end{bmatrix} \quad (7.58)$$

Taking real part of $P_{T2} = P_{T2}$, we have from eqn. (7.58),

$$P_{T_2} = [P_1 P_2 \ldots] \text{ Real} \begin{bmatrix} 1+jS_1 & & 0 \\ & 1+jS_2 & \\ & & \ddots \\ 0 & & \end{bmatrix} [C_2] \begin{bmatrix} -jQ_{10} \\ -jQ_{20} \\ \vdots \end{bmatrix}$$

$$= [P_1 P_2 \ldots P_N]\, B_{N0}$$

$$= \sum_{n=1}^{N} P_n B_{n0} = \sum_{n=1}^{N} B_{n0} P_n \qquad (7.59)$$

Similarly, P_{T_3} from eqn. (7.53) will be of the form

$$P_{T_3} = [jQ_{10}\, jQ_{20} \ldots jQ_{N0}]\, [C_2] \begin{bmatrix} jQ_{10} \\ -jQ_{20} \\ \vdots \\ jQ_{N0} \end{bmatrix}$$

$$= B_{00} \qquad (7.60)$$

Hence combining eqns. (7.57), (7.59) and (7.60), we obtain

$$P_T = P_{T1} + P_{T_2} + P_{T_3}$$

$$= \sum_{m=1}^{N} \sum_{n=1}^{N} P_m B_{mn} P_n + \sum_{n=1}^{N} B_{no} P_n + B_{00} \qquad (7.61)$$

Equation (7.61) is the same as eqn. (7.6).

7.4 Optimal Load Scheduling of Thermal Plants Taking Losses into Account

The optimal load scheduling for thermal plants taking transmission losses into considerations will be mathematically formulated as follows.

Determine the plant generations P_i for $i = 1, \ldots, N$ such that the cost function,

$$C_t = \sum_{i=1}^{N} C_i (P_i) \qquad (7.62)$$

is a minimum subject to the constraint that the demand and the transmission losses at any instant is met by the total generation at that instant, i.e.

$$\sum_{i=1}^{N} P_i = P_L + P_T \qquad (7.63)$$

where P_T is transmission loss which is a function of plant generation P_i (eqns. 7.6 and 7.61). In the same way, with the help of lagranges multiplier, the above nonlinear objective function (eqn. 7.62) with non-linear equality constraint (eqn. 7.63), can be converted into unconstrained objective function, i.e.,

$$F = C_t - \lambda\left(\sum_{i=1}^{N} P_i - P_L - P_T \right) \qquad (7.64)$$

Then with law of calculus, i.e. taking the partial derivative of the function F with respect to the variable P_i and equating it to zero for minimum, we have

$$\frac{\delta F}{\delta P_i} = 0 = \frac{dC_i}{dP_i} - \lambda\left(1 - \frac{\delta P_T}{\delta P_i}\right) \tag{7.65}$$

i.e. $$\lambda = \frac{dC_i}{dP_i} + \lambda \frac{\delta P_T}{\delta P_i} \quad \text{or} \quad \lambda\left(1 - \frac{\delta P_T}{\delta P_i}\right) = \frac{dC_i}{dP_i}$$

i.e. $$\lambda = \frac{dC_i}{dP_i} \Big/ \left(1 - \frac{\delta P_T}{\delta P_i}\right) \tag{7.66}$$

These equations are called coordination equation of thermal plant. Similar equations for hydrothermal plants are derived later (eqns. (7.73) and (7.74)). The multiplier λ is in Rs/MW-hr when the fuel cost is in Rs and generation is MW hr. λ is called incremental cost function. The above equations are solved for each system by assuming value of λ.

Here $1 \Big/ \left(1 - \frac{\delta P_T}{\delta P_i}\right) = \alpha$ is a penalty factor which depends upon the

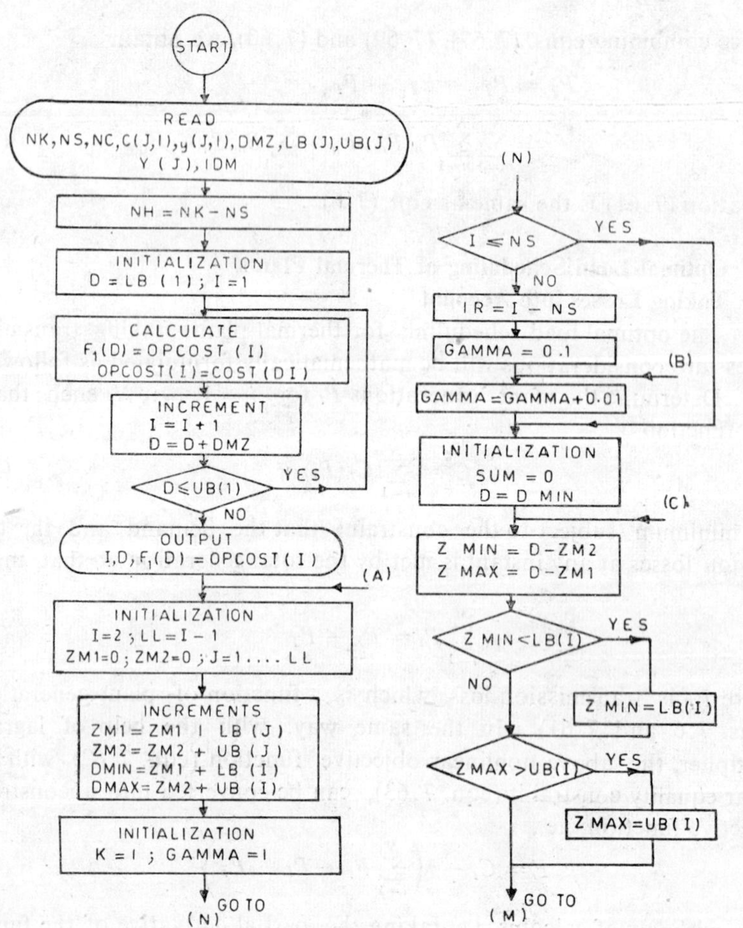

(1)

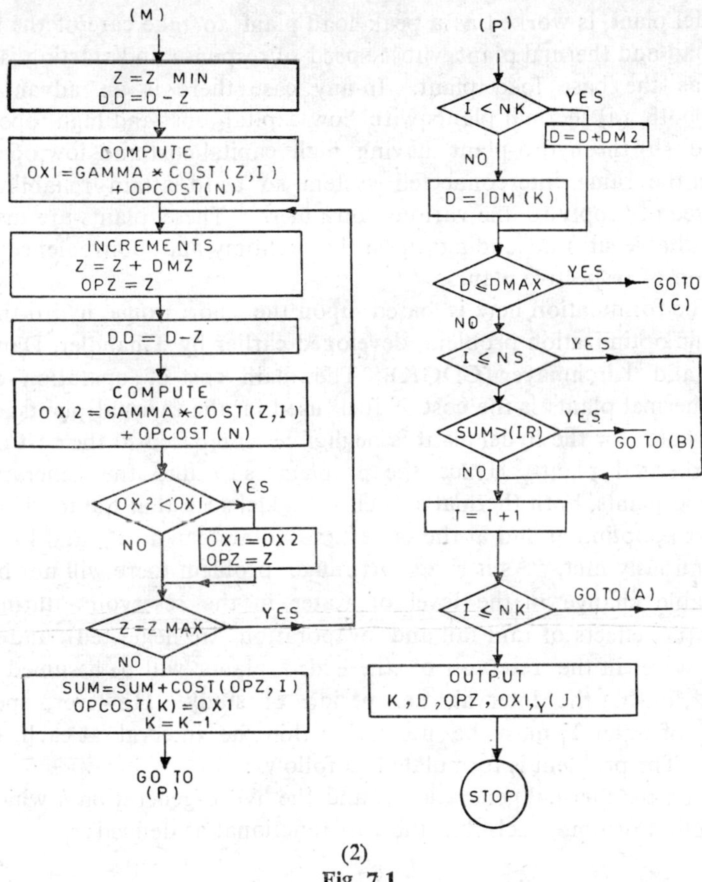

(2)
Fig. 7.1

location of the plant. Hence the effect of transmission losses is to add penalty factor in the solution where the value of penalty factor depends upon the plant location.

7.1 Economic Load Scheduling of Hydro-Thermal Plants

As stated earlier, in the initial stages, there were mostly thermal plants to produce electricity but due to the increase in the demand of electricity for all purposes such as industrial, agricultural, commercial and domestic together with the high cost of fuel as well as its limited reserve, has focused considerable attention upon the need for water power development. As the operating cost of thermal plant is very high and at the same time its capital cost is low compared to the hydro-electric plant whose capital cost is high but operating cost is low, it has become economical as well as convenient to have both thermal as well as hydro plants in the same grid. The thermal plant is run as a *base load plant* and the hydro-electric plant is run as a *peak load plant*. This is because the hydro-plant can be started quickly, has higher reliability and greater speed of response and hence it can take up fluctuating loads. In other words, the hydel plant is worked as a peak load plant to take care of the fluctuating load and thermal plant whose speed of response and starting is slow, is run as the base load plant. In any case there is an advantage of having both (a) thermal plants with low capital cost and high operating cost and (b) the hydro-plant having high capital cost and low operating cost, in the same interconnected system so as to attain reliability and guarantee of supply to the various consumers. These plants are installed at the suitable sites depending upon the economy and convenience in the operation of respective plants.

The formulation here is based upon the short range hydro-thermal economic optimization problem developed earlier by Chandler, Dandeno, Glimn and Kirchmayer (CDGK). The main cost of operation of the hydro-thermal plants is the cost of fuel used in the thermal plants as the cost of water for the hydel plant is negligible compared to the cost of fuel of the thermal plant. Hence the problem is to find the generation of individual plants, both thermal as well as hydel such that the total generation cost is optimum and at the same time total demand P_L and losses P_T is continuously met. As it is a short range problem there will not be any appreciable change in the level of water in the reservoirs during the interval (i.e. effects of rain fall and evaporation are neglected) and hence head of water in the reservoir of the hydro plants will be assumed to be constant during the interval, i.e. period of study. However, specified quantity of water Y_i must be utilized within the interval at each hydro plant i. The problem is formulated as follows:

Find the thermal generation S and the hydro-generation h which are the function of time, such that the cost functional as defined:

$$C_t = \int_0^T \sum_{j=1}^{N} C_j(S_j)\, dt \tag{7.67}$$

is minimum, subject to the equality constraint,

$$\sum_{j=1}^{N} S_j + \sum_{i=1}^{M} h_i = P_L + P_T(S, h) \tag{7.68}$$

i.e. the total demand P_L and the total losses P_T at any instant is equal to the total generations of the plants, and

$$\int_0^T y_i(h_i)\, dt = Y_i \tag{7.69}$$

i.e. in the given interval T, specified quantity of water Y_i is utilized at the hydel plant i.

Here t is the time in hours defining the short range problem, and S_j is the power generation in M.W. of the jth thermal plant, h_i is the power generation in M.W. of the ith hydel plant, N is number of thermal plants and M is number of hydel plants.

$C_j(S_j)$ is the fuel cost in Rs./hr. of the jth thermal plant which is the function of the plant generation.

$y_i(h)$ is the turbine discharge in cubic ft./hr. of the ith hydel plant.

T is the final time defining the short range problem.

$S = (S_1 \ldots S_N)$ = Generations of thermal plants.

$h = (h_1 \ldots h_M)$ = Generations of hydel plants.

P_L = Demand at any time t.

P_T = Transmission losses which are the function of plant generation jth (eqns. 7.6 and 7.61).

The above is the non-linear programming problem with both the objective function (eqn. 7.67) as well as the constraint equations (7.68 and 7.69) being non-linear. The non-linear objective function with the non-linear equality constraints can be converted into the unconstraints objective function F with the proper choice of the lagranges multiplier λ and γ. And hence we obtain

$$F = C_t - \gamma_i \left(\int_0^T y_i(h_i)\, dt - Y_i \right) - \lambda \left(\sum_{j=1}^{N} S_j + \sum_{i=1}^{M} h_i - P_L - P_T \right) \quad (7.70)$$

The above function F can be solved easily by the law of calculus for minimization by taking the partial derivative of F with respect to the plant generation S and h and equating them equal to zero, i.e.

$$\frac{\delta F}{\delta S_j} = 0 = \frac{dC_j}{dS_j} + \lambda \frac{\delta P_T}{\delta S_j} - \lambda \quad (7.71)$$

and
$$\frac{\delta F}{\delta h_i} = 0 = \gamma_i \frac{dy_i}{dh_i} + \lambda \frac{\delta P_T}{\delta h_i} - \lambda \quad (7.72)$$

The above coordination equations (eqns. 7.71 and 7.72) becomes,

$$\frac{dC_j}{dS_j} + \lambda \frac{\delta P_T}{\delta S_j} = \lambda \quad (7.73)$$

and
$$\gamma_i \frac{dy_i}{dh_i} + \lambda \frac{\delta P_T}{\delta h_i} = \lambda \quad (7.74)$$

for $j = 1, \ldots, N$ and $i = 1, \ldots, M$.

The lagranges multipliers γ_i are constant during the interval while the lagranges multiplier λ is the function of time.

The direct application of the above coordination equations (eqns. 7.73 and 7.74) give solution which sometimes dictates generations outside the plant capacity and also negative generation of certain plants since restrictions regarding plant capacities have not been included in the problem formulation Dandeno who has also observed the similar phenomena while working with these coordination equations on the digital computer, has doubted the constancy of γ_i if these restrictions regarding plant capacity is reasonably included in the problem formulation.

Two extensions of the problem considering the situations that exist in practice are studied by Wijeperera and others. The first extension consists of the addition of simple bounds on the operating range of each plant as indicated by the inequalities (eqns. 7.75 and 7.76)

$$\underline{S}_j \leqslant S_j \leqslant \overline{S}_j \tag{7.75}$$

and

$$\underline{h}_i \leqslant h_i \leqslant \overline{h}_i \tag{7.76}$$

where $\underline{S}_j$ and $\underline{h}_i$ are the minimum and $\overline{S}_j$ and $\overline{h}_i$ are the maximum limits of operation of the corresponding plants. This extended problem is solved by the application of Kuhn Tucker conditions for maximizing the hamiltonian. The second extension considers the restrictions (7.75) and (7.76) and in addition recognizes the freedom that any plant may be shut down when conditions permit. It thus encompasses the unit commitment problem and is solved by the use of dynamic programming (Appendix 1) to maximize the hamiltonian.

However, the preliminary in both the extensions is the application of Pontryagin's principle. As far as the part I of the problem is concerned, by applying the Pontryagin's maximum principle followed by the Kuhn Tucker conditions, the problem obtained is to maximize the hamiltonian H at each instant of time, i.e.

$$\text{Maximize } H = -\sum_{j=1}^{N} C_j(S_j) - \sum_{i=1}^{M} \gamma_i y_i(h_i) \tag{7.77}$$

The maximization of H at each instant of time is an auxiliary problem of static type. The constraint γ_i must be so chosen that the boundary conditions given by equation (7.69) is satisfied i.e. specified quantity of water is used at each hydro-plant in the specified period. Since γ_i effectively converts the water consumption in cubic ft/hr to the cost in Rs/hr., it may be termed as water value. The γ_i actually corresponds to the price of water and increase in the value of γ_i results in the lesser water usage at the hydel plant i and vice-versa.

After the application of Kuhn Tucker conditions we obtain

$$\frac{dC_j}{dS_j} - a_j + \beta_j + \lambda \frac{\delta P_T}{\delta S_j} - \lambda = 0 \tag{7.78}$$

and

$$\gamma_i \frac{dy_i}{dh_i} - \mu_i + \eta_i + \lambda \frac{\delta P_T}{\delta h_i} - \lambda = 0 \tag{7.79}$$

where

$$\beta_j = 0 \text{ if } S_j \neq \overline{S}_j\ ;\quad \beta_j \geqslant 0 \text{ if } S_j = \overline{S}_j \tag{7.80}$$

$$a_j = 0 \text{ if } S_j \neq \underline{S}_j;\quad a_j \geqslant 0 \text{ if } S_j = \underline{S}_j \tag{7.81}$$

$$\eta_i = 0 \text{ if } h_i \neq \underline{h}_i\ ;\quad \eta_i \geqslant 0 \text{ if } h_i = \underline{h}_i \tag{7.82}$$

$$\mu_i = 0 \text{ if } h_i \neq \overline{h}_i\ ;\quad \mu_i \geqslant 0 \text{ if } h_i = \overline{h}_i. \tag{7.83}$$

The above equations (7.78) and (7.79) may be considered as the coordination equations of the hydro-thermal dispatch problem when the simple bounds on the operating range of the plant is also taken into considerations. It is observed that due to properties (7.80), (7.81), (7.82) and

(7.83), the above coordination equations (7.78) and (7.79) will be reduced to the original equations (7.73) and (7.74) for the plants where the operation does not fall on the bounds. However, for the plants where the operation falls on the bounds, the incremental cost (or the incremental water rate in the case of the hydro plant) is actually augmented by a_j and β_j (or μ_i/γ_i in η_i/γ_i). Here a_j, β_j, μ_i and η_i are positive and greater than zero if the operation of the corresponding plant falls on the bound and are zero if the operation is within, i.e. inside the bound.

As the minimization of any function is equivalent to the negative of the maximization of the same function and hence the maximization of the Hamiltonian H (equation 7.77) at each instant of time is equivalent to the minimization of the cost function C_T as shown below.

$$\text{Minimize } C_t = -H$$

$$= + \sum_{j=1}^{N} C_j(S_j) + \sum_{i=1}^{M} \gamma_i y_i (h_i) \tag{7.84}$$

The minimization of the cost function C_T corresponds to the minimization of the total operating cost of the thermal as well as the hydel plants. This is because both the terms in the cost function C_T (equation 7.84) represent the operating cost of the combined hydro-thermal system, the first term representing the operating cost of the thermal plant and the second term is that of the hydel plant.

7.6 Formulation of Power System Optimization Problem using Dynamic Programming

The interconnected system which is being considered here have both thermal as well as hydel plants. Let S_j be the generation of jth thermal plant and h_i be the generation of ith hydel plant. Also, let us assume that our system contains N-thermal and M-hydel plants. It will be assumed that the thermal plants are run as a base load plant and therefore depending upon the demand, all the N thermal plants will be put into operation one by one and afterwards when the demand exceeds, hydel plants, which are run as peak load plants to take care fluctuating loads, will be put into operation.

The economic load scheduling of hydro-thermal system can be formulated and studied in a more effective way by dynamic programming (see Appendix-1) because this problem also belongs to the categories of problems known as the multi-state decision process. In the case of economic load scheduling of the hydro-thermal system, we are required to take a sequence of decisions and at any decision point a number of possibilities exist just like any other multi-decision process. The decision that we are required to take in this case is regarding the generation schedule of different plants. The stage corresponds to the number of plants that are in operation and the decision at any stage corresponds to the proper choice of the generation schedule out of several choices of generations. The effect of taking the decision at any stage will result into the change in the state of the system. Here the state corresponds to the demand on the plants which are in operation at that instant. Thus the effect of decision regarding the generation schedule for any plant will naturally change the demand to be handled by the remaining plants. Just like multi-decision process, the state of the system at any stage depends upon the previous stage, decision regarding generation schedule at the stage, and the stage, i.e. plant itself. The change in the state, i.e. the demand on the system will naturally result into some return function which corresponds to the cost function for the hydro-thermal system. Therefore, a sequence of decisions regarding the generation schedule are to be taken so that that the total return function, i.e. total cost function, because of taking such decisions, are optimal.

Thus our problem is to determine the scheduling of generation of hydro-thermal plants, i.e. to determine S_j and h_i such that the total operating cost C_t (see eqn. 7.84),

$$C_t = \sum_{j=1}^{N} C_f(S_j) + \sum_{i=1}^{M} \gamma_i y_i (h_i) \tag{7.85}$$

is minimum, subject to the following constraints (eqns. 7.68, 7.75 and 7.76)

$$\sum_{j=1}^{N} S_j + \sum_{i=1}^{M} h_i = P_L + P_T (S, h) \tag{7.86}$$

$$\underline{h_i} \leqslant h_i \leqslant \overline{h_i};\ \underline{S_j} \leqslant S_j \leqslant \overline{S_j} \tag{7.87}$$

However, the above constraints are only effective if the corresponding plants are in operation. Because if any plant is not in operation, its cost will be zero, i.e.

$$C_j(0) = 0 \text{ and } \gamma_i y_i(0) = 0 \tag{7.88}$$

In order to use dynamic programming, we have to define a function which corresponds to the minimization of the cost function C_t (equation 7.85) and should also take into consideration the constrained equations (7.86, 7.87 and 7.88). In other words, our problem is to determine the optimal decision from amongst multi-decisions, i.e. optimal decisions S_j and h_i out of several choices of decisions such that the total return function C_t (equation 7.85) because of taking decisions S_j and h_i, is minimum subject to the constraints (7.86, 7.87 and 7.88). Actually the effect of taking the decision at any stage will result into the change in the state of the system. Here the state corresponds to demand. Hence it is obvious that the minimum value of C_t depends upon the the total number of stages to be considered which corresponds to the total number of plants (M + N plants) and the initial state of the system which corresponds to the initial demand D. As a matter of fact the initial state of the system takes into account the constraint equations (7.86, 7.87 and 7.88). The initial demand D will be equal to $P_L + P_T$. Let this function be $F_R(D)$.

Hence $F_R(D)$ = minimum value of C_t when there are R stages, i.e. R plants in operation and the initial state is D, where D is the demand on the plants.

The following constraints should also be satisfied by the demand D

$$D : \underline{D} \leqslant D \leqslant \overline{D} \tag{7.89}$$

where $\underline{D}$ is the sum of the lower bounds and $\overline{D}$ is that of the upper bounds of the plants which are in operation at that instant.

Let us assume that out of total number of R plants which supply the total demand D, the Rth plant supplied the demand Z. Let the cost of generating Z by Rth plant be $U_R(Z)$ with the restrictions that:

$$U_R(Z) = C_R(Z) \tag{7.90}$$

if Rth plant is thermal

and

$$U_R(Z) = \gamma_R y_R(Z) \tag{7.91}$$

if Rth plant is hydel.

Hence the remaining demand $D-Z$ will be handled by the remaining number of plants $R-1$.

Thus $F_{R-1}(D-Z)$ = minimum cost of generating $D-Z$ demand from remaining $R-1$ plants which are in operation. However, $D-Z$ must lie between sum of the lower and upper bounds of these $R-1$ plants. Thus we obtain the following equation using the principle of optimality

$$F_R(D) = \underset{Z}{\text{Min}}[U_R(Z) + F_{R-1}(D-Z)] \tag{7.92}$$

where $U_R(Z)$ corresponds to the initial decision that the Rth plant takes demand of Z which is arbitrary and $F_{R-1}(D-Z)$ corresponds to the remaining $R-1$ decisions which are always optimal with regards to the state resulting from the first, i.e. initial decision. Actually by taking the first decision that the Rth plant takes the demand of Z, the state of the system is changed to $D-Z$.

By putting $R = 1$ in equation (7.92), we obtain,

$$\begin{aligned} F_1(D) &= \text{Min } U_1(Z) \\ &= U_1(D) \\ &= C_1(D) \end{aligned} \tag{7.93}$$

if the first is the thermal plant having the cost function $C_j(S_j)$. The demand D should satisfy the constraint equation:

$$\underline{S_1} \leqslant D \leqslant \overline{S_1} \tag{7.94}$$

Similarly by putting $R = 2$ in equation (7.92), we get

$$\begin{aligned} F_2(D) &= \underset{Z}{\text{Min}}\,[U_2(Z) + F_1(D-Z)] \\ &= \underset{Z}{\text{Min}}\,[C_2(Z) + F_1(D-Z)] \\ &= \underset{Z}{\text{Min}}\,[C_2(Z) + C_1(D-Z)] \end{aligned} \tag{7.95}$$

if both the plants are thermal. However, D is varied in small discrete steps. In the same way by putting $R = 3,4,...$ etc, we go on finding $F_3(D)$, $F_4(D)$, ... till we find $F_N(D)$, i.e. when all the N-thermal plants are put into operation. In this way for any demand D, we obtain the optimal cost to generate this demand and the corresponding generation schedule. This is done by varying Z in small discrete steps between the limits, calculating the corresponding cost and comparing them in order to obtain the optimal cost.

However, for the hydel plant which are run as peak load plant, we have from equation 7.92,

$$F_{N+1}(D) = \min_{\underline{h}_1 \leq Z \leq \bar{h}_1} [\gamma_1 y_1(h_1) + C_N(D - Z)] \tag{7.96}$$

where N are the number of thermal plants and $N+1st$ is the first hydel plant which is put into operation with the generation schedule h_1. Here γ_1 is initialized first. The value of γ_1 determines the quantity of water to be used by the first hydel plant. We assign a low value of γ_1 to start with. Thus after determining the generation schedule h_1 for the hydel plant 1 during the entire period, the total discharge in the period is calculated from the given discharge characteristic equation of the plant. The value of total discharge thus obtained is compared with the specified quantity of water to be utilized by this hydel plant (eqn. 7.69),

$$\int_0^T y_i(h_i)dt = Y_i \tag{7.97}$$

If equation (7.97) is not satisfied, γ_1 is incremented by $d\gamma_1$ and again the optimal cost and the corresponding generation schedule is obtained. The process is repeated till equation (7.92) is satisfied. The process is continued till we obtain $F_{N+M}(D)$ i.e. all the thermal as well as the hydel plants are put into operations. The maximum value is restricted by the following equation,

$$D \leq [\sum_{j=1}^{N} S_j + \sum_{i=1}^{M} h_i] \tag{7.98}$$

We observe that at every iteration of the process described, we have to solve the minimization problem with respect to one variable Z only.

7.7 Numerical Example

A short range problem of 24 hours duration has been considered and divided into interval of one hour each (Fig. 7.1 shows the flow diagram and Fig. 7.2 shows the hourly demand on the system). The demand at the generating plant is assumed to be constant during each hourly interval and at the end of each interval, the demand increases in jumps or remains constant. As shown in the example, the system considered consists of three thermal plants and one hydel plants (see Table 7.1).

The optimum cost to generate the demand D and the corresponding

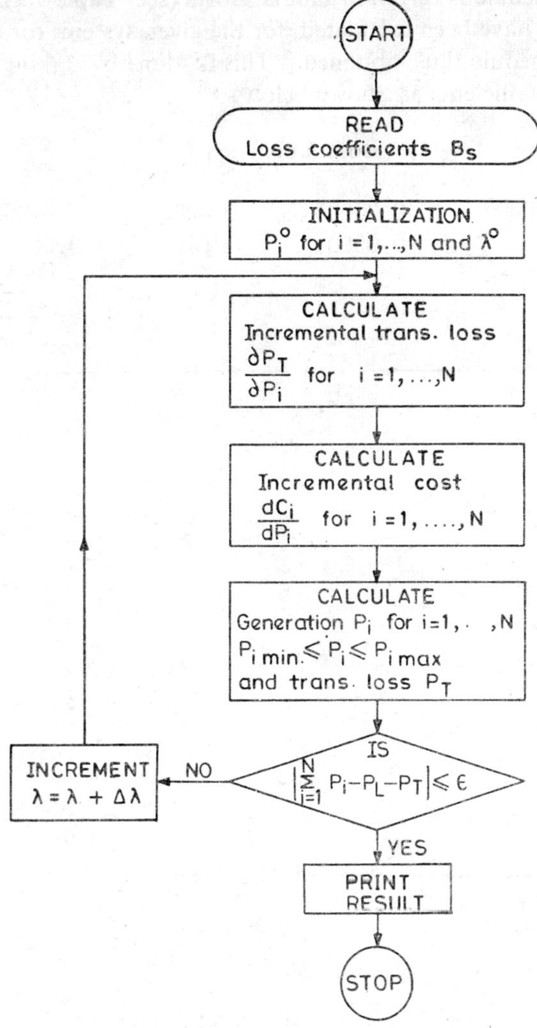

Fig. 7.2 Flow diagram for solving coordination equation.

Table 7.1 Cost characteristics

Lower bound in kW	Upper bound in kW	Cost characteristics in Rs/hr	Quantity of water to be utilized in m³
		(a) Thermal plant	
50	150	$C_1=100+0.1S_1+0.01S_1^2$	
60	150	$C_2=120+0.1S_2+0.02S_2^2$	—
50	200	$C_3=150+0.2S_3+0.01S_3^2$	—
		(b) Hydel plant	
15	65	$y_1=140+20h_1+0.06h_1^2$	509.7

generation schedule is given in tabular form (see Table 7.2). The transmission losses have been calculated for the given system for the values of generation schedule thus obtained. This is done by taking the assumed value of loss coefficients as shown below:

Loss Coefficients

0.0005	0.00005	0.0002	0.00003
	0.00004	0.00018	−0.00011
		0.0005	−0 00012
			0.00023

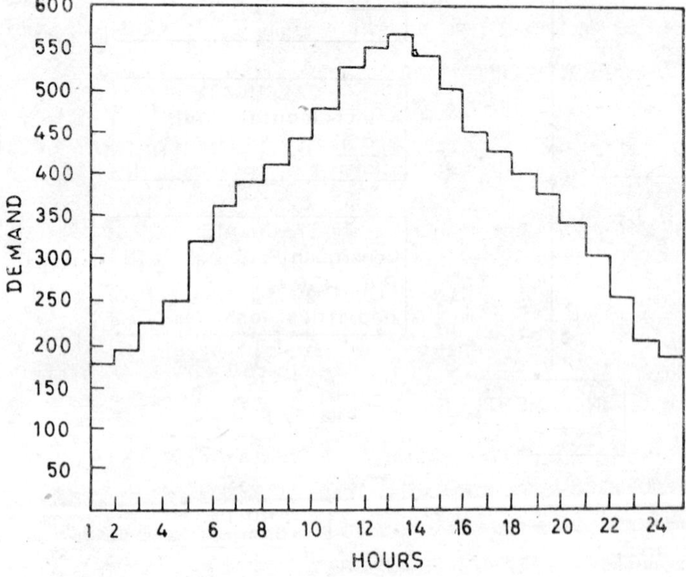

Fig. 7.3

Table 7.2 Results of Generation Schedule

Hours	Demand D	Optimal cost, Rs/hr	P_1	P_2	P_3	h_1	Transmission losses P_T	Load supplied P_L
1–2	175	585.55	100	60	00	15	5.68	169.32
4–5	280	761.30	105	60	100	15	17.23	262.77
7–8	390	1057.59	150	75	150	15	36.29	353.71
10–11	475	1364.75	150	95	190	40	47.05	427.95
13–14	565	1785.95	150	150	200	65	53.49	511.51
16–17	450	1268.03	150	90	180	30	44.34	405.66
19–20	375	1011.30	145	75	140	15	32.95	342.05
22–23	250	699.80	150	85	00	15	12.72	237.28
24–1	180	591.30	105	60	00	15	6.25	173.77

References

1. W.G. Chandler, et al. 'Short Range Economic Operation of a Combined Thermal and Hydroelectric Power System'. *AIEE Transactions*, vol. 72, October 1953, p. 1053.
2. P.L. Dandeno, 'Hydrothermal Economic Scheduling—Computational Experience with the Coordination equations'. *AIEE Transaction*, vol. 80, May 1961, p. 1219.
3. L.S. Pontryagin, et al. *The Mathematical Theory of Optimal Processes.* Interscience Publishers, New York, 1962.
4. K.K. Y. Wije Perera, G.F. Schrack and Frank Noakes. 'Short-Range Economic Optimization of a Hydro-Thermal Power System Considering Operating Range Limitations and Unit Commitment Aspects'. Presented at the IEEE Summer Power Meeting, June 22–27, 1969.
5. R. Bellman, *Dynamic Programming.* Princeton, New Jersey Princeton University Press, 1957.
6. L.K. Kirchmayer, et al. 'Direct Calculation of Transmission Loss Formula'. *AIEE Transactions*, vol. 79, 1960, p. 962.
7. L.K. Kirchmayer, *Economic Operation of Power Systems*, Wiley, New York, 1958.
8. L.P. Singh and R.P. Aggarwal, 'Economic Load Scheduling of Hydro-Thermal Stations Using Dynamic Programming'. *Journal of Inst. of Engineers (India)*, pp. 175–180, vol. 52, No. 8, Part El-4, April 1972.

Chapter 8

Sparsity Technique

8.1 Introduction

There exists a large class of physical problems in practice which give rise to sparse symmetric linear systems, for which the computational efforts required to obtain solution is highly dependent upon ordering the equations. A system of simultaneous algebraic equation viz.

$$[A]\,\overline{x} = \overline{b} \tag{8.1}$$

is called symmetric if the coefficient matrix $[A]$ is symmetric and sparse if $[A]$ has large number of zero elements. In many problems dealing with the power systems, electric networks, structures, finite difference formulation etc., this is precisely the case. It has been pointed out that to solve these problems by first computing the inverse of the system matrix $[A]$ will be highly inefficient and Gaussian elimination, which is in effect one of the Krons techniques, is apparently the most efficient procedure.

It can easily be shown that the matrices associated with the most man-made system are generally sparsely populated in non-zero elements. Recognition of this fact has led to the development of techniques for compact storage and processing of sparse arrays in digital computers. The discovery of sparsity and its exploitation has become a major technological advance in computer programming and numerical analysis.

Since Markowitz observed and exploited sparsity in matrix associated with the linear programming in industrial applications, several others have rediscovered and applied sparsity techniques primarily in electric network calculations. Most of the contributors to the technical literatures deal with the sparsity in terms of zero and non-zero terms. However, sparsity can also be discussed in terms of effective operations or efforts.

The mathematical model of power system network for the purposes of steady state analysis (i.e. studies) such as load flow and planning studies or short circuit studies, are a set of network equations. These network equations can be established either in the bus frame of reference using Y_{BUS} or Z_{BUS} or in the loop frame of reference using Z_{loop} or Y_{loop} or in the branch frame of reference using Y_{BR} or Z_{BR}. The coefficient matrices Y_{BUS}, Z_{loop} or Y_{BR} can be developed in most of the cases directly by inspection from the given power system (i.e. primitive) network and, therefore, usually these coefficient matrices are highly sparse (nearly 90% or more elements at the off diagonal locations are zero). In order to solve network equations of a large scale power system network having coefficient matrices which are largely sparse (this is true in most of the cases), it is necessary to conserve sparsity structure of the coefficient matrices so as to optimize computer memory and also to decrease computational time in order to decrease the cost of computations. This is why sparsity and its exploitations are very important in the formulation and analysis of power system problems. We shall discuss in this chapter sparsity and its consequences as applied to the power systems.

8.2 Sparse Systems

Most of the discrete sample spaces are sparsely populated. Natural occurrences of sparsity is wide ranging. The following is a partial list of sparse systems:

(i) Network of all kinds such as electric power, electronic and communications, hydraulics etc.

(ii) Space trusses and frames of structures.

(iii) Roads, highways and airways connecting all the important cities of the world.

(iv) Street connections among intersections within a city.

(v) Matrices associated with algebraic equations resulting from different methods in the solution of differential equations.

(vi) Matrices arising in the discrete analysis of continuous functions.

(vii) Links in the chemical formulas of complex chains.

(viii) Transition matrix associated with the employment of an individual in the space of all positions in the job market.

(ix) Houses or building containing employed individuals in a country.

(x) The population of a punch card by useful codes.

(xi) Communication path of a single employee in a large organization considering all the employees he could talk to, etc. etc.

From the above partial list, it should be evident that the sparsity occurs in some form in almost all major areas of human operations. Sparsity has been discovered and explored to some extent in certain linear programming problems, large scale electric network calculations, structural analysis and in information transfer along communication channels such as between central processors and remote computer terminals. However, its potential scope of application has hardly even begun to be realized.

8.3 Theorems of Sparse Matrix Method

Matrices are usually considered as multi-dimensional arrays, each of whose elements must explicitly retain its identity, even though it is quite apparent that the matrix associated with most man-made problems or systems are generally sparsely populated in non-zero elements. The discovery of sparsity and its exploitation has become technological advance in computer programming and numerical analysis. Mathematically, we can define:

Given a finite discrete sample space Ω and a non-empty set of sample S such that the cardinality $|S|$ of S is small compared to cardinality $|\Omega|$ of Ω i.e. $|S| \ll |\Omega|$ is said to sparse w.r.t. S. Let

$$n = |S| \tag{8.2}$$

$$N = |\Omega| \tag{8.3}$$

Since each element of S induces some sequence of operations, let t_i be the time it takes to perform the operations induced by element i of s.

Let r_i be the additional time taken to retrieve an element of S in a compact storage scheme.

Let s_i be the additional time taken to store an element of S in a compact storage scheme.

Here the total processing time without compact storage is:

$$\sum_{i=1}^{N} t_i \tag{8.4}$$

and the total processing time with compact storage is:

$$\sum_{i=1}^{n} t_i + r_i + s_i \tag{8.5}$$

Let m be the number of storage cells needed to store and access then elements of S in compact form. Usually $m > n$. Exploitation is possible in storage space, computation time or both. There is a gain in storage space if $m < N$. There is a gain in computation time if

$$\sum_{i=1}^{n} \gamma_i + s_i < \sum_{i=n+1}^{N} t_i \tag{8.6}$$

For a given array of number, the set Ω corresponds to the entire array while the set S corresponds to the active elements of array.

Let Ω be a sample space and let S be the set of active samples in Ω. A sample is active if its identity and considerations ultimately contribute to or effect the final results contained. A sample is not active otherwise. In numerical evaluation, zero terms are non-active. Thus the zero and non-zero pattern of an array can be used to judge sparsity. But there are also non-zero terms which by virtue of the problem being considered make no contribution to final results. By definition these nonzero terms must be treated as though they were zero. *Sparsity exploitation is an art rather than a science.*

The ultimate potential of sparsity concept is enormous. Practically all large problems are sparse in some sense that can be exploited in their computer solution, but this fact is still largely unrecognized. Where sparsity exploitation is possible, an improvement in speed and or storage of the order 10:1 can always be recognized. In many cases improvement of as much as 100:1 is possible. Sparsity techniques cannot be generalized beyond a certain point and almost any of generalization results in a significant loss of efficiency in comparison with what can be done by attacking the problem individually. Each kind of problem has its own peculiar sparsity structure which requires specialized strategy for its optimum exploitation. A much more challenging task may be to develop a generalized sparsity subroutine package for the solution of the sparse system of linear algebraic equations by Gaussian elimination.

8.4 Strategies for Reducing Bandwidth of Matrices

System of linear equations involving sparse, symmetric, positive definite coefficient matrices arise in many applications areas. Our immediate interests in the solution of such systems has arisen from the use of finite element displacement method for analyting a wide variety of structures and the need of automatizing the data generation for such analysis. A proper ordering of system equations with sparse coefficient matrix is essential for their efficient solution.

Proper ordering as used here depends upon many factors including, for example, the strategy used in the computer programmes for solving the equations, the pattern of zero and non-zero elements in the coefficient matrix and the number of times a system with a particular pattern of zero and nonzero elements is to be solved.

Here we assume that a direct method of solution based upon Gaussian elimination is to be used. We also assume that very large system of equations are to be solved so that the partition of matrix must be accessed from relatively slow auxiliary storage during the calculations.

Types of storage schemes for sparse symmetric matrix:

(1) Diagonal band storage: All elements on the diagonal and in as many bands as are needed, parallel to diagonal, are stored in a two dimensional array.

(2) Profile or wave front storage: For each row all elements from first nonzero element to the diagonal are stored or the corresponding elements in the transpose of the matrix are stored. In this case the elements are usually stored in a linear array with pointer array to identify the rows and the columns.

All these storage schemes can lend themselves to efficient data handling and to overlay procedures to conserve memory and to minimize input/output. A number of algorithms have been developed for sequencing the equations and unknowns for efficient solution when these approaches are used.

8.5 Various Application Areas and Sparsity

8.5.1 Load Flow Problem in Power Systems

The load flow problem consists of determination of voltages at the buses and power flows in the lines of the power system network for the specified terminal or bus conditions. The mathematical formulation of the load flow problem is a set of nonlinear algebraic equations which can be established in the following ref. frames:

1. *Bus frame of ref.*

$$\bar{E}_{BUS} = [Z_{BUS}]\, \bar{I}_{BUS} \tag{8.7}$$

and

$$\bar{I}_{BUS} = [Y_{BUS}]\, \bar{E}_{BUS} \tag{8.8}$$

Here we obtain $n-1$ linear independent equations where n is the number of nodes. Y_{BUS} matrix (eqn. 8.8) can be developed by inspection and is sparse while Z_{BUS} is a full matrix. Hence exploitation of sparsity is possible only in Y_{BUS} i.e. formulation of load flow problems in the bus frame of ref. using Y_{BUS}. Enough of work has been done in this area by the system engineering group led by W.F. Tinney at B.P.A. (Oregon), U.S.A. For a given power system network as shown in Fig. 8.1, the complex power at any bus K is given by

$$P_K + jQ_K = \bar{E}_K \bar{I}_K^* = \bar{E}_K \sum_{m=1}^{N} \bar{Y}_{Km}^* \bar{E}_m^* \tag{8.9}$$

where P_K and Q_K are the real and reactive power at the bus K.

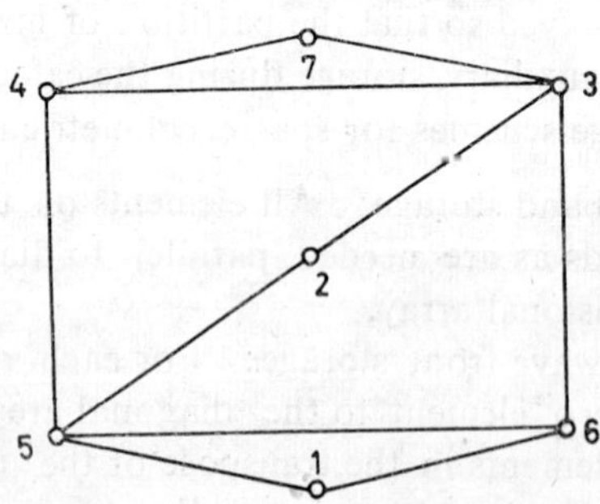

Fig. 8.1 Power system network

Voltage controlled bus where P and E are specified are 3 and 6, load bus where P and Q specified are 2, 4, 5, 7 and slack bus is 1.

By the Newton's method, the above nonlinear equations can be transferred into a set of linear algebraic equations in the following form:

$$
\begin{bmatrix} \Delta P_2 \\ \Delta Q_2 \\ \hline \Delta P_3 \\ \hline \Delta P_4 \\ \Delta Q_4 \\ \hline \Delta P_5 \\ \Delta Q_5 \\ \hline \Delta P_6 \\ \hline \Delta P_7 \\ \Delta Q_7 \end{bmatrix}
=
\left[\begin{array}{cc|c|cc|cc|c|cc}
H_{22} & N_{22} & H_{23} & & & H_{25} & N_{25} & & & \\
J_{22} & L_{22} & J_{23} & & & J_{25} & L_{25} & & & \\
\hline
H_{32} & N_{32} & H_{33} & H_{34} & N_{34} & & & H_{36} & H_{37} & N_{37} \\
\hline
 & & H_{43} & H_{44} & N_{44} & H_{45} & N_{45} & & H_{47} & N_{47} \\
 & & J_{43} & J_{44} & L_{44} & J_{45} & L_{45} & & J_{47} & L_{47} \\
\hline
H_{52} & N_{52} & & H_{54} & N_{54} & H_{55} & N_{55} & H_{56} & & \\
J_{52} & L_{52} & & J_{44} & L_{53} & J_{55} & L_{55} & J_{56} & & \\
\hline
 & & H_{63} & & & H_{65} & N_{65} & H_{66} & & \\
\hline
 & & H_{72} & H_{74} & N_{74} & & & & H_{77} & N_{77} \\
 & & J_{73} & J_{74} & L_{74} & & & & J_{77} & L_{77}
\end{array}\right]
\begin{bmatrix} \Delta\delta_2 \\ \Delta E_2/E_2 \\ \hline \Delta\delta_3 \\ \hline \Delta\delta_4 \\ \Delta E_4/E_4 \\ \hline \Delta\delta_5 \\ \Delta E_5/E_5 \\ \hline \Delta\delta_6 \\ \hline \Delta\delta_7 \\ \Delta E_7/E_7 \end{bmatrix}
\tag{8.10}
$$

JACOBIAN MATRIX

where

$$H_{km} = \delta P_k / \delta\delta_m$$

$$N_{km} = \delta P_k \mid E_m \mid / \delta E_m$$

$$J_{km} = \delta Q_k / \delta\delta_m$$

$$L_{km} = \delta Q_k \mid E_m \mid / \delta E_m \tag{8.11}$$

For a system of N-nodes including a slack node and having S nodes as a voltage controlled bus, there are ($2N$ S 2) linear equations to be solved. The equations are solved by Gaussian elimination and back substitution. Here in this method, the program operates upon and stores only non-zero elements. At the start of each iteration, the essential data are the initial estimate of latest iterate of node voltage, the nodal i.e. bus admittance matrix and the scheduled node quantities P_k and Q_k. From these the augmented Jacobian matrix is formed and triangularized in OPTIMUM ORDER (to be discussed later) either two rows or one row at a time, depending upon whether corresponding node is a load bus or voltage controlled bus, respectively. The matrix below (eqn. 8.12) shows the form of the upper triangular matrix resulting from Gaussian elimination of the augmented Jacobian matrix.

$$
\left[\begin{array}{cccccccccc|c}
1 & N'_{22} & H'_{23} & & & H'_{25} & N'_{25} & & & & \Delta P'_2 \\
 & 1 & J'_{23} & & & J'_{25} & L'_{25} & & & & \Delta Q'_2 \\
 & & 1 & H'_{34} & N'_{34} & H_{35} & N_{35} & H'_{36} & H'_{37} & N'_{37} & \Delta P'_3 \\
 & & & 1 & N'_{44} & H'_{45} & N'_{45} & H_{46} & H'_{47} & N'_{47} & \Delta P'_4 \\
 & & & & 1 & J'_{45} & L'_{45} & J_{46} & J'_{47} & L'_{47} & \Delta Q'_4 \\
 & & & & & 1 & N'_{55} & H'_{56} & H_{57} & N_{57} & \Delta P'_5 \\
 & & & & & & 1 & J'_{56} & J_{57} & L_{57} & \Delta Q'_5 \\
 & & & & & & & 1 & H_{67} & N_{67} & \Delta P'_6 \\
 & & & & & & & & 1 & N'_{77} & \Delta P'_7 \\
 & & & & & & & & & 1 & \Delta Q'_7
\end{array}\right] \tag{8.12}
$$

Here the Jacobian matrix is augmented on the right hand by the modified column of residual. The prime in the above matrix (eqn. 8.12) indicates the elements that have been altered from their original values in the elimination process, the unprimed elements are the new one which have been introduced in the elimination process. The scheme for storing this matrix in the computer memory is illustrated in the Tables 8.1 and 8.2. The primes are omitted in this display. The integers in the upper triangular table represent node number corresponding to columns of triangle and the alpha symbols represent elements of the triangularized Jacobian matrix. The negative sign on node identifies voltage controlled buses. Table 8.2 is an index by node numbers of the starting location in Table 8.1 of entries for each single or double row of the modified upper triangular matrix shown in equation 8.12. If the sign of the entry in Table 8.2 for node k is positive, the first three entries starting in the indicated location in Table 8.1 are ΔP_k, ΔQ_k and N_{kk}. These are followed by a signed column indicator m; if m is positive, it is followed by H_{km}, J_{km}, N_{km} and L_{km}; if m is negative, it is followed by H_{km} and J_{km}. These terms are followed by another column entry and the scheme repeates to the end of the row. Similarly, if the sign for the location of node k is negative, the first entry in Table 8.1 is ΔP_k followed by a column indicator m, if m is negative, it is followed by H_{km}, if positive by H_{km} and N_{km}. The end of row k is implied by the starting location of node $k+1$ in Table 8.2. There is no entry for $k = 1$, i.e. slack node. Tables 8.1 and 8.2 are rebuilt in every iteration.

Table 8.1

Loc.	Item	Loc.	Item	Loc.	Item
1	P_2	9	J_{25}	16	5
2	Q_2	10	N_{25}	17	H_{35}
3	N_{22}	11	L_{25}	18	N_{35}
4	−3	12	P_3	19	−6
5	H_{23}	13	4	20	H_{36}
6	J_{23}	14	H_{34}		
7	5	15	N_{34}		
8	H_{25}				

Table 8.2

Node	Location of row in upper triangular matrix
1	1
2	−12
3	24
4	40
5	−51 etc.

Concept of packed working row is shown below (Table 8.3).

Table 8.3

Row 3 before elimination of column 2	Loc.	1	2	3	4	5	6
	Column	2	−3	4	−6	7	
	H	H_{32}	H_{33}	H_{34}	H_{36}	H_{37}	
	J						
	N	N_{32}		N_{34}		N_{37}	
	L						
	Next Loc.	2	3	4	5	E	

Row 3 after elimination of column 2 and division by H'_{33}	Loc.	1	2	3	4	5	6
	Column	2	−3	4	−6	7	5
	H	X'	X	H'_{34}	H'_{36}	H'_{37}	H'_{35}
	J						
	N		X	N'_{34}		$'_{37}$	N'_{35}
	L						
	Next Loc.	2	3	6	5	E	4 etc.

Actually in addition to a compact storage scheme for a upper triangular, it is advantageous to use a compact working row as shown in Table 8.3. Here, because of optimal ordering which will be discussed later, the number of nonzero columns remains very small, thus gaining advantage of both memory length and speed. Because of exploitation of sparsity, i.e. optimal ordering (will be discussed later in this chapter), the load flow formulation as proposed by W.F. Tinney and his group has several advantages such as quick convergence, problem independent iterations, i.e. number of iterations required for convergence being independent of number of nodes, i.e. buses and faster solution and economy in memory requirement etc.

2. *Branch frame of ref.*

$$\bar{E}_{BR} = [Z_{BR}]\,\bar{I}_{BR}$$

or

$$\bar{I}_{BR} = [Y_{BR}]\,\bar{E}_{BR} \tag{8.13}$$

Here we obtain $b = n - 1$ linear indep. equations and Y_{BR} is sparse.

3. *Loop frame of ref.*

$$\bar{E}_{\text{loop}} = [Z_{\text{loop}}]\,\bar{I}_{\text{loop}}$$

or

$$\bar{I}_{\text{loop}} = [Y_{\text{loop}}]\,\bar{E}_{\text{loop}} \tag{8.14}$$

Here we obtain $= e - n + 1$ linear indep. equations. $[Z_{\text{loop}}]$ matrix is formed by inspection and is also sparse. Earlier approach to load flow problem was in the loop frame of ref. using $[Z_{\text{loop}}]$ but later on this method was discarded because of tedious data preparations etc.

8.5.2 *The Analysis of Large Structural Systems*

A structure is defined as an assembly of nodes or joints at which loads can be applied, connected by elements or members which behaves in a linear manner under load. The method on which structural engineering programs are based may be divided into two general categories. The first class may be termed as the method of nodal analysis. The rotations and displacement of a joint of a structure is taken as a basic variable, and all internal forces and moments are expressed in terms of them. The analysis is carried out by setting up and solving the joint equilibrium equations, which relate the joint displacements to the known applied loads.

The second class of methods may be called methods of mesh analysis. The basic unknowns are a set of redundant forces and moments and displacements of the structure are written in terms of them. The analysis is carried out by solving the equations of displacement compatibility.

The complete set of nodal equations can be written in the matrix form as,

$$\bar{F} - [Y]\,\bar{d} \tag{8.15}$$

where $\bar{F}$ is the column vector of the applied joint loads and $\bar{d}$ is the column vector of the joint displacements. The zero-nonzero configuration of the stiffness matrix $[Y]$ can be directly written from the line diagram of the structure. Here $Y_{PP} \neq 0$ and only those $Y_{pq} \neq 0$ where there is a link between node p and q. Thus in a majority of cases, Y is a sparse matrix.

8.5.3 The Genetic Theory

According to the present genetic theory, the fine structure of genes is obtained by alteration of some connected portion of this structure. By examining data obtained from suitable experiments, it can be determined whether or not the blemished portions of two mutant genes intersect or not, and thus intersection data for a large number of mutants can be represented as an undirected graph. Here also the graph and the associated matrix is sparse.

8.5.4 Sociology

Society is made up of groups and the use of matrix to represent a relationship between the members of groups is well known in sociometry. This relationship among the members of a group is usually binary in nature. Relationships such as communications, power and sociometric choice belong to this category and can be represented by a matrix in which all the elements are either zero or 1. The resultant matrix is a group matrix which is also sparse.

8.6 Optimal Elimination of Sparse Symmetric Systems

We will now formulate the sparsity problem based upon the mathematical concepts of dynamic programming (Appendix 1).

8.6.1 Problem Formulation

Consider a concrete physical situation such as a power system network. It consists of a set of busbars and a set of transmission lines connecting the busbars. A complete description of it must show the physical quantities attached to the various parts of the system (such as generation, load and line parameters) as well as the configuration of system itself. Mapping each busbar into a point and each transmission line into an edge, the physical system is mapped into a system of points and edges which is referred to as undirected symmetric finite graph. There may be occasions when a graph (i.e. collection of points connected by edges) may depict systems other than a power system as considered above. The points may stand for people, places, atoms, joints, etc. and edges may represent kinship relations, pipelines, bonds, trusses, etc. Diagrams like these are encountered under different names such as communication networks, circuit diagrams, family trees, organizational structures, simplex and sociograms.

Let a physical system be described by a set of n-linear algebraic equations of the form,

$$[A]\,\bar{x} = \bar{b} \tag{8.16}$$

The problem is to determine the solution vector $\bar{x}$ by the elimination process (say, Gaussian Elimination Method) so that the computational efforts and hence the computational time and cost are minimum.

Following Von Neumann, the number of multiplications required to obtain solution will be counted as a measure of computing time. In the restricted form, the numerical procedure is completely specified for a given system once the equation has been ordered. Further, if only the number of multiplications is to be counted, a reduced matrix M whose elements are defined to be:

$$\begin{aligned} m'_{ij} &= 1 \qquad \text{if } m_{ij} \neq 0 \\ &= 0 \qquad \text{if } m_{ij} = 0 \end{aligned} \tag{8.17}$$

contains all informations necessary for the problems under consideration.

Let for the system $Ax = b$ (equation 8.16), the total number of multiplications for the ith row is m_i and hence total for the system will be,

$$\varphi = \sum_{i=1}^{n} m_i \tag{8.18}$$

In general, the system (equation 8.16) will be given some arbitrary initial ordering 1, 2,. . ., n. Let ξ_i ($i = 1, 2,\ldots, n$) describe permutation for ordering. The problem is to find out ξ_i, for which ϕ (ξ_i) is a minimum. Since matrix [A] may be thought of as purely topological, this is a combinational problem concerned only with topology, i.e. connectivity of the system. Problem of this type fit into a very general classification of nonlinear programming problem.

8.6.2 *Graph Problem*

Optimal elimination is actually a topological problem which can be reformulated using notation from graph theory in order to facilitate the procedure. Some systems such as electrical networks and skeletal structures may be thought of as being their own graph. Thus a picture of one of these systems with slight modifications, could serve as its graph in spite of the facts that there may be more than one scalar quantities associated with each node. Other systems, such as those arise from difference equations, may have no natural graph. For these systems the following procedure may be used to construct a graph : with each equation in A, there is associated a node in the system graph and with each non-zero term a_{ij} $(i = 1, \ldots n;\ J = 1, \ldots n)$ there is associated an undirected branch between the ith and jth node. The system graph will be referred as G. During the elimination process it is modified just as rows of the system matrix are modified. Let G^i is a graph obtained by eliminating i-th node from G.

$G_i = [i = 1, \ldots, n]$ set consiting of i-th node and together with all nodes adjacent to ith node (two nodes i and j are adjacent if these are the end points of the same branch). Elimination of ith node from A corresponds to first forming for each node j in the graph:

$$G_j^{-i} = G_i \, UG_j \;; \quad \text{for } j \in G_i \in j = 1, 2, \ldots, n$$

for
$$= G_j \;; \qquad j \notin G_i \tag{8.19}$$

and then deleting ith node from each G_j^{-i} for which $i \in G_j^{-i}$, to form G_j^i ; successive elimination proceeds in a straight-forward manner i.e. elimination next of the k-th node would produce the set $G^{i},^{K}$. Introducing the norm $| G_j^{i, \ldots\ m} |$ which is defined to be number of nodes included in the sets $G_j^{i, \ldots, m}$ for an ordering $\xi_1, \xi_2, \ldots, \xi_n$, it is possible to identify $b_i = | G_i^{\xi_1 \xi_2 \ldots, m} |$ which in fact one plus the valence of the node ξ_i in the graph $G^{\xi_1, \xi_2, \ldots \xi_{i-1}}$.

Since the deletion of the vertex I alters the remaining graph only by an addition of branch JK if IJ and IK were branches. We have the following result :

Theorems

(i) Any graph which is obtained from another graph by deleting process is independent of the order in which nodes are eliminated.

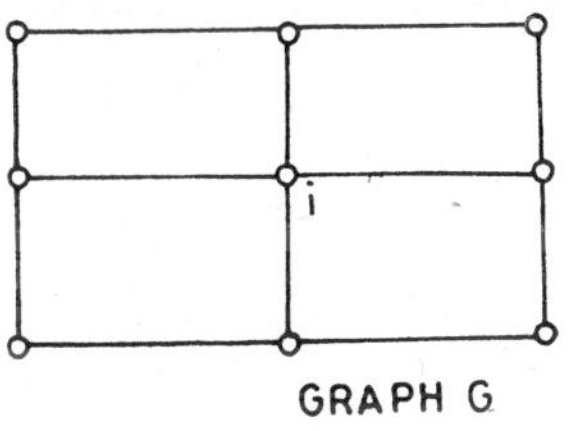

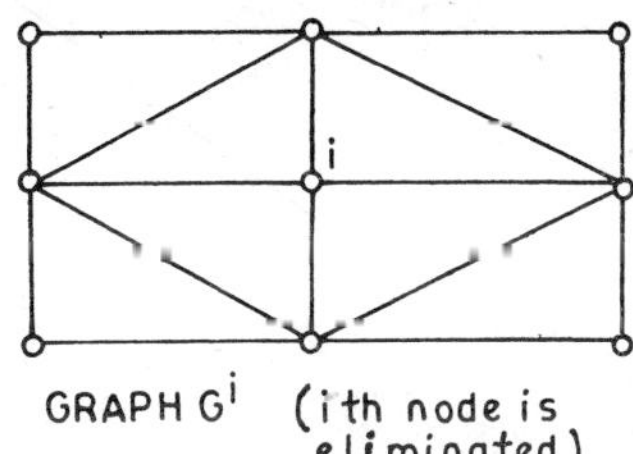

Fig. 8.2 Elimination of ith node from graph G.

(ii) When any two successive nodes in the optimal ordering are adjacent either in original graph or in the graph generated by the elimination process, the node with the smaller valence occurs first.

Example 8.1 : Take a physical system which can be described by the linear algebraic equations:

$$\begin{aligned} a_{11}x_1 + a_{12}x_2 &= b_1 \\ a_{21}x_1 + a_{22}x_2 + a_{23}x_3 &= b_2 \\ a_{32}x_2 + a_{33}x_3 + a_{34}x_4 + a_{35}x_5 &= b_3 \\ a_{43}x_3 + a_{44}x_4 + a_{45}x_5 &= b_4 \\ a_{53}x_3 + a_{54}x_4 + a_{55}x_5 &= b_5 \end{aligned}$$

which can be written in the matrix form as:

$$\begin{bmatrix} a_{11} & a_{12} & & & \\ a_{21} & a_{22} & a_{23} & & \\ & a_{32} & a_{33} & a_{34} & a_{35} \\ & & a_{43} & a_{44} & a_{45} \\ & & a_{53} & a_{54} & a_{55} \end{bmatrix} \begin{bmatrix} x_1 \\ x_2 \\ x_3 \\ x_4 \\ x_5 \end{bmatrix} = \begin{bmatrix} b_1 \\ b_2 \\ b_3 \\ b_4 \\ b_5 \end{bmatrix}$$

In a connected graph (Fig. 8.2) for the above equation nodes communicate between one another through the path between them. Thus if paths exist between the nodes adjacent to a node k such that if k is removed, flow in the graph is not interrupted, node k can be eliminated without affecting the remaining graph. In Fig. 8.3 elimination of node 1 will not affect the flow and hence will neither add any new branch in the graph nor add any new non-zero element in the off-diagonal location of the coefficient matrix A. However, if no path exists prior to the elimination, then new paths have to be created. For example, the elimination of node 3 in Fig. 8.3 affects the flow of path between nodes 2 and 4 only (and not between nodes 4 and 5) and hence a new link 2—4 will be added in the connected graph, and similarly, a new non-zero element a_{24} (which was zero earlier) will be added up in the coefficient matrix A.

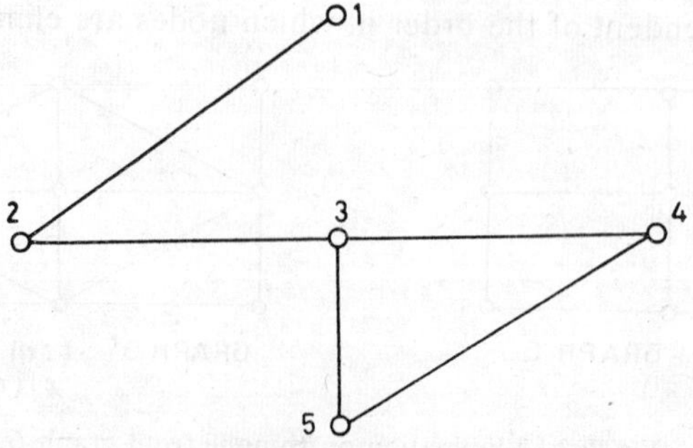

Fig. 8.3 Connected graph for equation.

To illustrate the point, consider the elimination of node k in the following cases (Fig. 8.4):

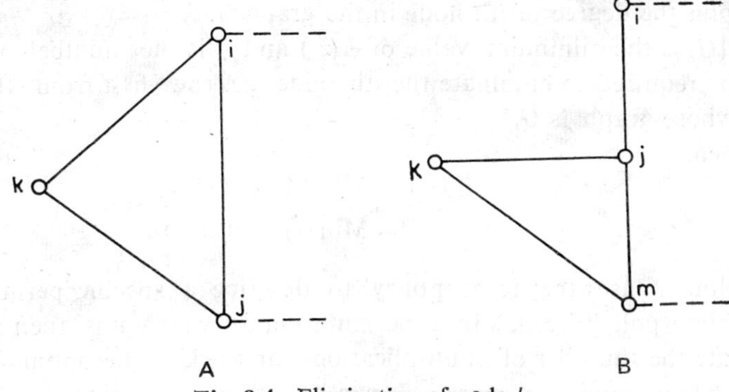

Fig. 8.4 Elimination of node k.

In case A, elimination of node k does not add any new edge as nodes i and j are connected by an edge (in the coefficient matrix the element a_{ij} will be non-zero). But in case B, elimination of node j results in the addition of an edge i—m and hence the element a_{im} which was zero earlier will appear now in the matrix A. This is represented in Fig. 8.5.

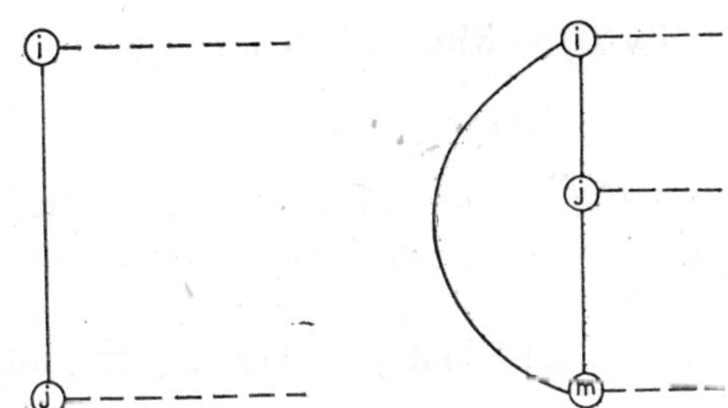

Fig. 8.5 Elimination of node k and edge i—m.

Thus, it is clear that for the coefficient matrix A of the linear system of algebraic equation $Ax = b$, there corresponds a graph $G(A)$ such that coefficient a_{ij} of the variable x_j in the ith equation of the system is the weight or traffic on the path between nodes i and j. Given a graph $G(A)$, the corresponding matrix can be constructed. There is associated with the matrix or graph $G(A)$ an incidence matrix M. It is sufficient to show that the elimination of node j from the graph $G(A)$ has the same effect on M as the elimination of variable x_j from the system equation $Ax = b$. When node j is eliminated, all the paths incident upon it are removed from the graph. The traffic going through node j must be redistributed as follows: for each node pair (k, i) which are neighbours of node j, the path between k and i is created or unaltered according as whether no path existed between k and i or a path existed between them. An unaltered path does not affect the matrix M but the created path adds new non-zero terms in the corresponding locations of M.

8.6.3 Functional Equations

Let $W(G)$ be defined to be the number of multiplications required to solve optimally the system (equation 8.16) whose graph is G, and let $d(i)$ be one plus the degree of ith node in the graph G.

$W(G)$ is the minimum value of $\phi(\xi_i)$ and e_i is the number of multiplications required to eliminate the ith node, i.e. row first from the given system whose graph is G.

Then,

$$W(G) = \text{Min}\,\phi(\xi_i)$$
$$= \underset{i}{\text{Min}}\,(e_i + W(G^i)) \tag{8.20}$$

Bellman uses the term 'policy' to describe a specific permutation, i.e., a certain policy results in a permutation for which it is then possible to evaluate the number of multiplications or work. The optimal policy corresponds to minimum work. It can be easily shown that the solution of eqn. 8.20 is unique while the optimal policy may not be.

Following Bellman, it is possible to choose any initial policy (method of ordering the nodes) and proceed iteratively to obtain the solution of the equation 8.20.

Let $W_0(G)$ be the number of multiplications required using this initial policy on the graph G. Then, we have from the eqn. 8.20,

$$W_N(G) = \underset{i}{\text{Min}}\,(e_i + W_{N-1}(G^i)) \tag{8.21}$$

$$N = 1, 2, \ldots, n$$

The sequence $(W_N(G))$ is clearly monotone decreasing sequence. The solution of eqn. 8.21 which is a dynamic programming problem, will yield the following result.

At each step in the elimination scheme, eliminate the node with smaller degree.

8.7 Direct Solution of Sparse Network Equations by Optimally Ordered Triangular Factorization

Usually, the objective in the matrix analysis is to obtain the inverse of the matrix of coefficients of a system of simultaneous linear network equations. However, for *sparse systems* which normally occur in the network problems, the use of inverse is very inefficient. This is because the matrix of equations formed from the given network condition is usually sparse while its inverse is full. Say, for example, the Y_{BUS} matrix of a given power system network is extremely sparse while its inverse (i.e. Z_{BUS} matrix) is full. By means of an appropriately ordered triangular factorization (i.e. decomposition), the inverse of a sparse matrix can be expressed as a product of sparse matrix factors, thereby gaining an advantage in speed, storage and reduction of round off error. The method consists of two parts:

(i) a scheme of recording the operations of triangular decomposition of a matrix such that repeated direct solution based on the matrix can be obtained without repeating the triangularization, and

(ii) a scheme of ordering the operations that tends to conserve this sparsity of original system.

The first part of the method is applicable to any matrix. Application of second part (i.e. ordering to conserve sparsity), is limited to sparse matrices in which the pattern of non-zero elements is symmetric and for which an arbitrary order of decomposition does not affect adversely the numerical accuracy, such matrices are normally characterized by strong diagonal. A large class of network problem fulfil this condition. Generally it is not worth considering the optimal ordering unless at least 80 per cent of the off-diagonal matrix elements are zero.

8.7.1 *Factored Direct Solution*

This scheme is applicable to any nonsingular matrix, real or complex, sparse or full, symmetric or nonsymmetric. The basic scheme is first presented for the most general case, i.e. a full nonsymmetric matrix. Symmmetry and sparsity are treated as special cases. Here it is shown how to derive an array of numbers from a nonsingular matrix, 'A' which can be used to obtain the effects of any or all of the following, i.e. A, A^{-1}, A^T, A^{T-1}. Following procedure is adopted to obtain this.

A. Triangular Decomposition: Usually the decomposition is accompalished by elimination of elements below the main diagonal in successive columns. From the point of view of computer programming for sparse network matrix, it is usually much more efficient to eliminate by successive rows.

Let us take an equation :

$$[A]\,\bar{x}=\bar{b}$$

Here A is a non-singular matrix, $\bar{x}$ is a column vector of unknowns, and b is a column vector of known constants. Let us take 4 equations in 4 variables :

$$
\begin{aligned}
a_{11}x_1+a_{12}x_2+a_{13}x_3+a_{14}x_4&=b_1\\
a_{21}x_1+a_{22}x_2+a_{23}x_3+a_{24}x_4&=b_2\\
a_{31}x_1+a_{32}x_2+a_{33}x_3+a_{34}x_4&=b_3\\
a_{41}x_1+a_{42}x_2+a_{43}x_3+a_{44}x_4&=b_4
\end{aligned}
\tag{8.22}
$$

$$
[A]=\begin{bmatrix} a_{11} & a_{12} & a_{13} & a_{14}\\ a_{21} & a_{22} & a_{23} & a_{24}\\ a_{31} & a_{32} & a_{33} & a_{34}\\ a_{41} & a_{42} & a_{43} & a_{44}\end{bmatrix},\ \bar{x}=\begin{bmatrix}x_1\\x_2\\x_3\\x_4\end{bmatrix}\ \text{and}\ \bar{b}=\begin{bmatrix}b_1\\b_2\\b_3\\b_4\end{bmatrix}
\tag{8.23}
$$

The augumented matrix $\bar{A}$ is obtained by adding to the fifth column, the column vector of known constants $\bar{b}$, i.e.

$$\bar{A}=\begin{bmatrix} a_{11} & a_{12} & a_{13} & a_{14} & b_1 \\ a_{21} & a_{22} & a_{23} & a_{24} & b_2 \\ a_{31} & a_{32} & a_{33} & a_{34} & b_3 \\ a_{41} & a_{42} & a_{43} & a_{44} & b_4 \end{bmatrix} \tag{8.24}$$

The first step is to divide the elements of first row by a_{11} as indicated below.

$$a_{1j}(1)=\frac{1}{a_{11}}\,a_{1j} \quad j=2,3,4$$

$$b_1(1)=\frac{1}{a_{11}}\,b_1 \tag{8.25}$$

The second step as indicated below, is to eliminate a_{21} from the second row by linear combination with the derived first row, and then to divide the remaining derived elements of the second row by its derived diagonal element.

$$a_{2j}^{(1)}=a_{2j}-a_{21}\,a_{1j}^{(1)} \quad j=2,3,4$$

$$b_2^{(1)}=b_2-a_{21}\,b_1^{(1)}$$

$$a_{2j}^{(2)}=\frac{1}{a_{22}^{(1)}}(a_{2j})^{(1)} \quad j=3,4$$

$$b_2^{(2)}=\frac{1}{a_{22}^{(1)}}\,b_2^{(1)} \tag{8.26}$$

Thus we obtain

$$\begin{matrix} 1 & a_{12}^{11} & a_{13}^{11} & a_{14}^{11} & b^1 \\ & 1 & a_{23}^{2} & a_{24}^{2} & b_2^2 \\ a_{31} & a_{32} & a_{33} & a_{34} & b_3 \\ a_{41} & a_{42} & a_{43} & a_{44} & b_4^1 \end{matrix} \tag{8.27}$$

The third step, as indicated below is to eliminate elements to the left of the diagonal of the third row and to divide the remaining derived elements of the row by the derived diagonal element.

$$a_{3j}^{(1)}=a_{3j}-a_{31}\,a_{1j}^{(1)}$$

$$b_3^{(1)}=b_3-a_{31}\,b_2^{(1)} \quad j=2,3,4$$

$$a_{3j}^{(2)}=a_{3j}^{(1)}-a_{32}^{(1)}\,a_{2j}^{(2)} \quad j=3,4$$

$$b_3^{(2)}=b_3^{(1)}-a_{32}^{(1)}\,b_2^{(2)}$$

$$a_{3j}^{(3)}=\frac{1}{a_{33}^{(2)}}\quad a_{3j}^{(2)}$$

$$b_3^{(3)} = \frac{1}{a_{33}^{(2)}} \quad b_3^{(2)} \tag{8.28}$$

Proceeding in the similar manner, all the elements to the left of the diagonal of the fourth row is eliminated and then remaining elements of the row is divided by the derived diagonal element as shown:

$$a_{4j}^{(1)} = a_{4j} - a_{41}\, a_{1j}^{(1)} \quad j = 2, 3, 4$$

$$b_4^{(1)} = b_4 - a_{41}\, b_1^{(1)}$$

$$a_{4j}^{(2)} = a_{4j}^{(1)} - a_{42}\, a_{2j}^{(2)}$$

$$b_4^{(2)} = b_4^{(1)} - a_{42}\, b_2^{(2)} \quad j = 3, 4$$

$$a_{4j}^{(3)} = b_{4j}^{(2)} - a_{43}\, a_{3j}^{(3)}$$

$$b_4^{(3)} = b_4^{(2)} - a_{43}\, b_3^{(3)} \quad j = 4$$

$$a_{4j}^{(4)} = \text{not to be calculated (as it is now reduced to one)}$$

$$b_4^{(4)} = \frac{1}{a_{44}^{(3)}} \quad b_4^{(3)} \tag{8.29}$$

Thus we obtain the following derived system:

$$\begin{matrix} 1 & a_{12}^{(1)} & a_{13}^{(1)} & a_{14}^{(1)} & b_1^{(1)} \\ & 1 & a_{23}^{(2)} & b_{24}^{(2)} & b_2^{(2)} \\ & & 1 & a_{34}^{(3)} & b_3^{(3)} \\ & & & 1 & b_4^{(4)} \end{matrix} \tag{8.30}$$

It should be noted that at the end of Kth, $K = 1, 2, 3, 4$, step, work on row 1 to k has been completed and row $K+1$ to n have not entered the process anyway. Moreover, for any row, the first element of the row (i.e. element in the first column) is eliminated by combining this row with the first row, the second element of the row (i.e. element in the second column) by combining this row with the second row and finally $k-1$ st. element is removed by combining this row with $k-1$ st. row. In the programming, the x_i's replaces b_i's, one by one as they are computed, starting with x_4 and working back to x_1. When 'A' is full and the order of the matrix 'n' is large, it can be shown that number of operation for triangular decomposition is approximately $n^2/3$ compared with n^3 for inverse

B. *Recording the Operations.* If the forward operation on b has been recorded so that they could be repeated, it is obvious that with this record and upper triangular matrix for back substitution, the equation $Ax = b$ can be solved for any vector $\bar{b}$ without repeating the triangularization. The recording of forward operation is, however, trivial. Each forward operation is completely defined by the row and column coordinates and value of a single element a_{ij}^{j-1} for $i > j$ that occurs in process.

The rules for recording the forward operations of triangularization are:

1. When the term $\frac{1}{a_{ii}}\, i-1$ is computed, store it in location ii.
2. Leave every derived term a_{ij}^{j-1} for $i > j$, in the lower triangular.

The final result of triangularizing A and recording the forward operations is given below:

$$\begin{matrix} d_{11} & u_{12} & u_{13} & u_{14} \\ l_{21} & d_{22} & u_{23} & u_{24} \\ l_{31} & l_{32} & d_{33} & u_{34} \\ l_{41} & l_{42} & l_{43} & d_{44} \end{matrix} \tag{8.31}$$

The above elements (equation 8.31) are defined in terms of derived system A, i.e.

$$\begin{aligned} d_{ii} &= 1/a_{ii} \quad (i-1) \\ u_{ij} &= a_{ij}^{(i)} \\ l_{ij} &= a_{ij}^{(j-1)} \end{aligned} \tag{8.32}$$

This matrix (eqns. 8.31 and 8.32) is known as the table of factors. It is convenient to define some special matrices in terms of elements of table of factors to represent the operations of direct solutions. Following are these non-singular matrices:

$$\begin{aligned} D_i&: \text{Row } i &&= 0, 0, \ldots, 0\ d_{ii}, \ldots, 0, 0 \\ D_i&: \text{Column } i &&= 0, 0, \ldots, 0\ l_{i+i,i}, \ldots, l_{ni} \\ L_i^*&: \text{Row } i &&= -l_{i,1}, \ldots, l_{ii-1}, 1, 0, \ldots, 0 \\ U_i&: \text{Row } i &&= 0, \ldots, 0, u_{i,i+1}, \ldots, u_{i,n} \\ u_i^*&: \text{Column } i &&= -u_{i,1}, \ldots, u_{i-1,i}, 1, 0, \ldots, 0 \end{aligned} \tag{8.33}$$

The inverse of those matrices (equation 8.33) are trivial. The inverse of identity matrix D involves only the reciprocal of element d_{ii}. The inverse of matrices L_i, L_i^*, U_i and U_i^* involve only the reversal of algebraic sign of the off-diagonal elements. The forward and back substitution operations on the column vector $\bar{b}$ that transforms it to $\bar{x}$ can be expressed as premultiplications by matrices D_i, L_i or L_i^* and U_i or U_i^* as shown below:

$$\begin{aligned} &U_1U_2 \cdots U_{n-1}D_nL_{n-1}L_{n-2} \cdots L_2D_2L_1D_1b = A^{-1}b = x \\ &U_1U_2 \cdots U_{n-1}D_nL_nD_{n-1}L_{n-1} \cdots L_3D_2L_2Db = A^{-1}b = x \text{ etc. etc.} \end{aligned} \tag{8.34}$$

The above equation (8.34) describes a sequence of operations on the vector $\bar{b}$ that is equivalent to premultiplication by A^{-1}. If 'A' is symmetric only d_{ii} and u_{ij} terms of table of factors are needed. Hence in this case the solution is obtained as follows:

$$U_1U_2 \cdots U_{n-1}D_nD_{n-1}U_{n-1}^T \cdots D_2U_2^TD_1U_{1b}^T = A^{-1}x = b \tag{8.35}$$

8.8 Sparsity and Optimal Ordering

When the matrix to be triangularized is sparse, the order in which rows are processed affects the number of nonzero terms in the resultant upper triangular matrix. If a programming scheme is used which processes and stores only nonzero terms, a very great savings in operations and computer memory can be achieved by keeping the table of factors as sparse as possible. The absolute optimal order of elimination would result into the least possible terms in the table of factors. An efficient algorithm for absolute optimal ordering has not been developed as yet and it appears to be impossible at the moment. However, several effective schemes which are near optimum have been developed.

Following are the descriptions of some scheme for optimal ordering:

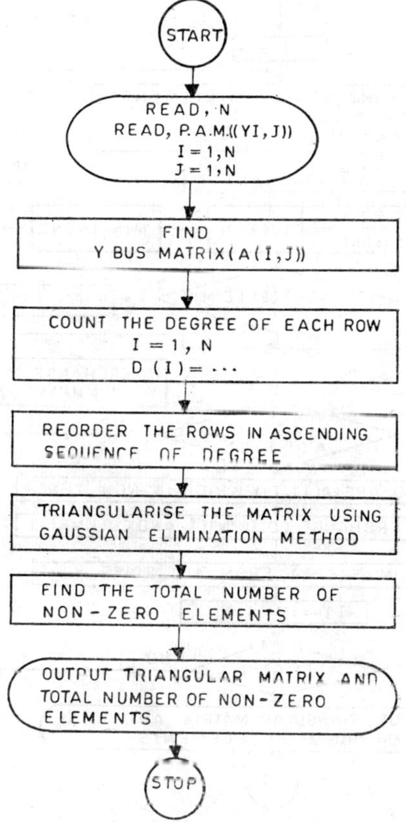

Fig. 8.6 Flow chart for scheme 1.

Scheme 1. Number the rows of the coefficient matrix A according to the number of nonzero off diagonal terms, before elimination. In this scheme the rows with only one off diagonal term are numbered first, those with two terms second etc. and those with most terms last. From the network point of view, the nodes are numbered, starting with that having the fewest connected branches (i.e. minimum degree) and ending with that having the most connected branches (i.e. maximum degree). This method does not take into account anything that happens during the elimination process but it is simple to program and fast to execute. The only information needed here is a list of the number of nonzero terms in each row of original matrix.

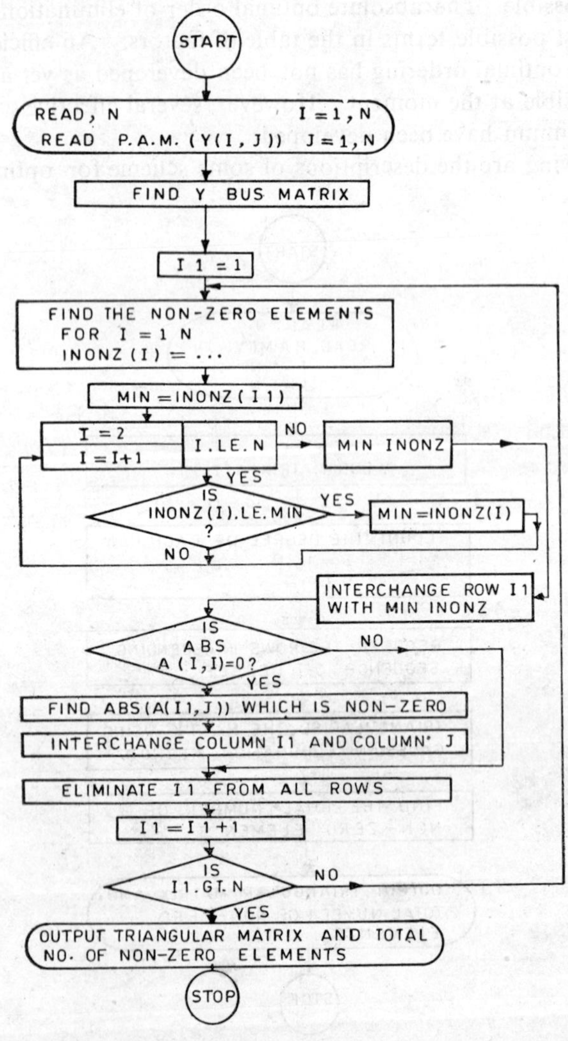

Fig. 8.7 Flow chart for scheme 2.

Scheme 2. Number the rows of the coefficient matrix A so that at each step of the process, the next row to be operated upon is the one with the fewest nonzero terms. If more than one row meets this criterion, select any one. From the network point of view, the nodes are numbered so that at each step of the elimination the next node to be eliminated is the one having the fewest connected branches (i.e. minimum degree). This method require simulation of elimination process to take into account the changes in the node branch connections effected at each step, i.e. this scheme requires a simulation of the effects on the accumulation of non-zero terms of the elimination process. Input information is a list by rows of the column numbers of the nonzero off diagonal terms, i.e. branches. This scheme, though takes longer time, is definitely better.

Scheme 3. Number the rows so that at each step of the elimination process the next row to be operated upon is the one that will introduce the fewest new nonzero terms. If more than one row meets this criterion

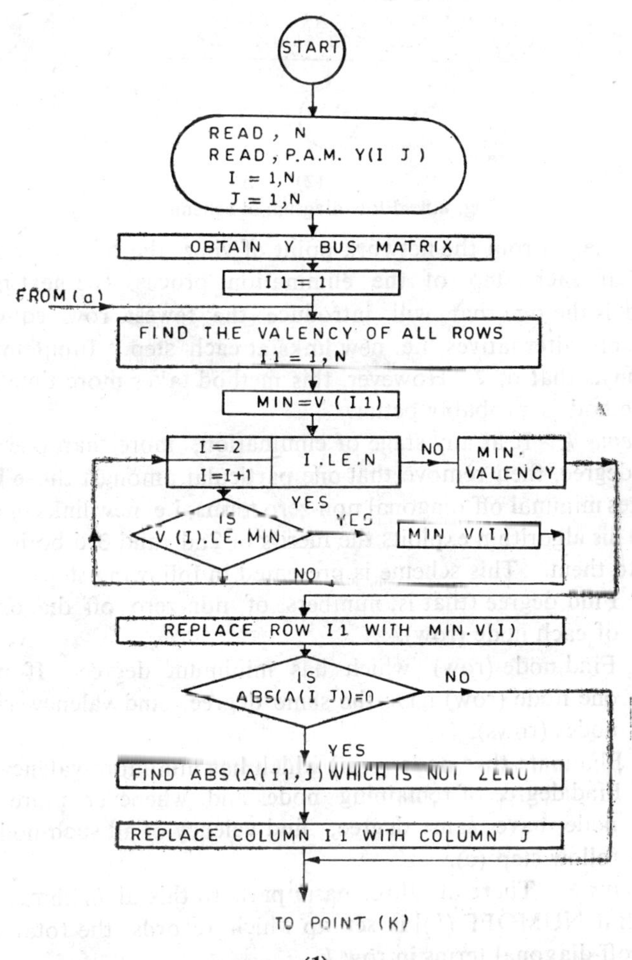

(1)

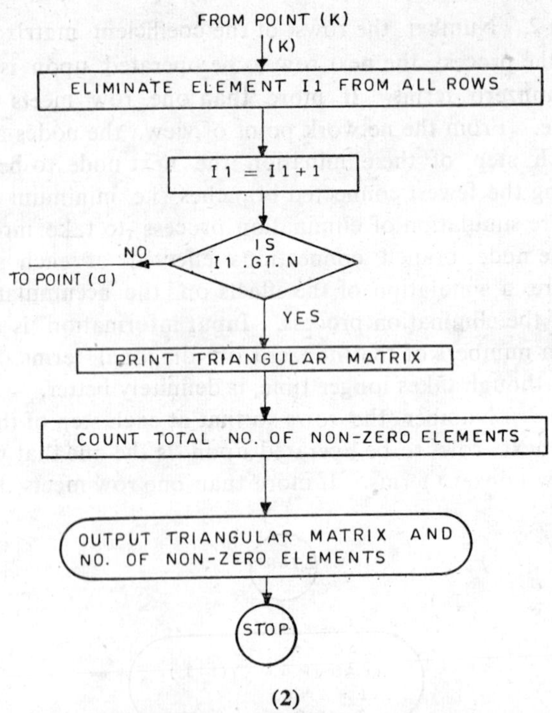

Fig. 8.8 Flow diagram of scheme 3.

select any one. From the network point of view the nodes are numbered such that at each step of the elimination process the next node to be eliminated is the one that will introduce the fewest row equivalent of every feasible alternatives, i.e. new links at each step. Input information is the same as that of 2. However, this method takes more time and hence second method is probably better.

Scheme 4. If at any stage of elimination, more than one node has the same degree, then remove that one particular amongst those first which also creates minimal off diagonal non-zero terms, i.e. new links in the system graph. This algorithm exploits the merits of 2nd and 3rd both and it is superior to them. This scheme is presented in following steps:

(a) Find degree (that is, numbers of non-zero off diagonal terms) of each node (row).

(b) Find node (row) which has minimum degree. If more than one node (row) have the same degree, find valency of all such nodes (rows).

(c) Eliminate that node (row) which has minimum valency.

(d) Find degree of remaining nodes and whenever more than one node have same degree, find valency of all such nodes. Then follow step (c).

Scheme 5. There are three basic parts to this algorithm. An array, [let us call it NUMOFF (k)] is set up which records the total number of non-zero off-diagonal terms in row k.

Part I of the algorithm searches this array once to see if there are any nodes with only one non-zero off-diagonal term. If one is found, it is numbered 1 and array NUMOFF is altered. There will be no additional non-zero terms created at this step of decomposition. The off-diagonal term of this new mode 1 will be located in some column *j*. Array NUMOFF is altered by reducing the recorded number of off-diagonal terms assciated with node *j* by one. If by this reduction of 1, the effective number of off-diagonal terms associated with node *j* is one or fewer, then the node *j* is numbered next and the process repeated. A single sweep through the array NUMOFF rapidly picks off every node that has only one or fewer effective non-zero off-diagonal terms.

Part II of the algorithm searches the remaining nodes for those which can be decomposed without increasing the number of non-zero terms. Suppose node *i* has associated with it two non-zero off-diagonal terms in

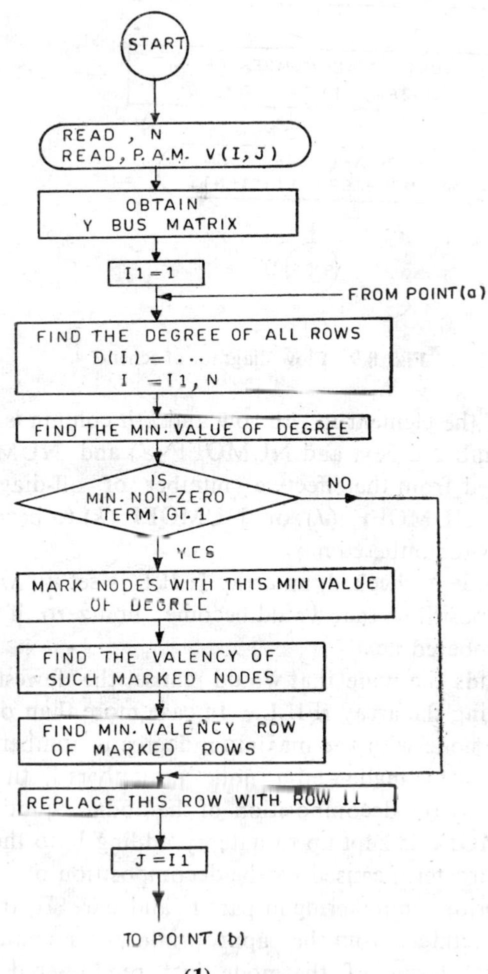

(1)

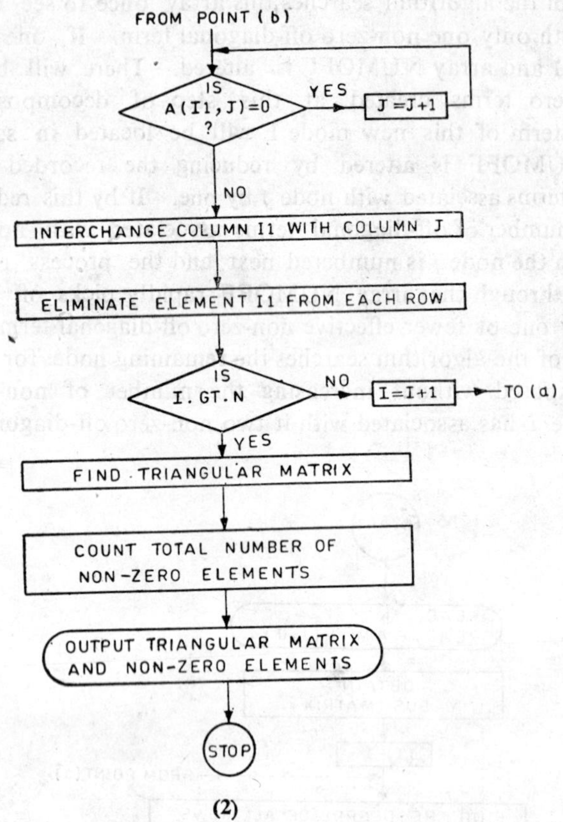

(2)

Fig. 8.9 Flow diagram of scheme 4.

row *j* and *k*. If the element in *j*th row and *k*th column is non-zero, then *i*th node is renumbered next and NUMOFF (*J*) and NUMOFF (*k*) would have one removed from the effective number of off-diagonal terms. If this causes the NUMOFF (*J*) or NUMOFF (*k*) to become 1, then that particular row is renumbered next.

As each node is checked, an array IFILL is set up which records the number of new positions that would become non-zero if that particular node were renumbered next.

Part III finds the node that would cause the fewest new non-zero terms by searching the array IFILL. In case more than one node satisfies this criteria, the node with the maximum degree is numbered next.

After a choice is made and a node renumbered, the new non-zero topology caused by the decomposition of the nodal equation is recorded. The array NUMOFF is kept up-to-date by adding 1 to the row in which each new non-zero term caused by the decomposition of that node appears. Also, as in all prior renumbering in part I and part II, the NUMOFF is altered by subtracting 1 from the appropriate rows containing the non-zero off-diagonal terms of the node just renumbered. If by this subtraction an effective number off 1 odd diagonal term appears in any of the non-renumbered rows, that row is immediately renumbered next.

After the book-keeping operations have been completed for renumbering of a row from part III, part II is entered at the begining. The search proceeds from this point as if it were the first entry into part II.

Before we discuss the other schemes of optimal ordering, we give the following definitions which are useful in understanding these schemes.

Degree. The degree of a node or vertex is the number of edges that are incident to the node. In the language of matrix, the degree of a node *i* is the number of non-zero terms at the off-diagonal locations in the row *i* of the corresponding graph.

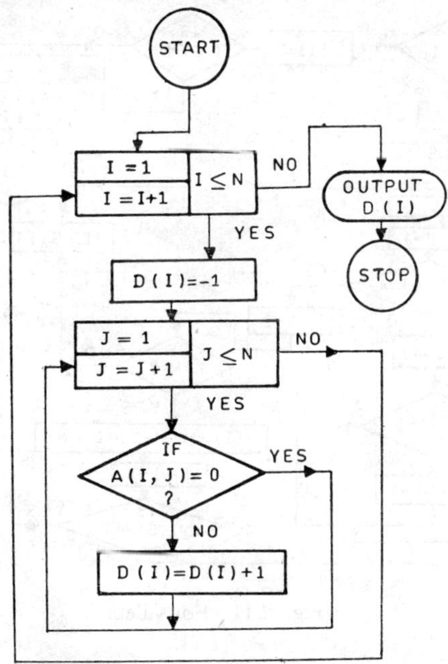

Fig. 8.10 Flow diagram for degree.

Valency. The valency of a node in a connected graph is the number of new paths (links) created or added amongst the remaining one as a result of elimination of the node. Hence the valency of the node *i* is defined as the number of new links added to the graph (that is, new non-zero elements generated in the corresponding coefficient matrix) because of the elimination of the node *i* (row *i*).

Scheme 6. The basis of this suggested algorithm (Fig. 8.12) is partial ordering. Every time a search is made for the single degree node and after eliminating it, the effect of its elimination is inspected on the remaining nodes. Then proceed sequentially to search the single or minimum degree node unless there is some node existing with degree less than the minimum. The algorithm can be presented in the following steps:

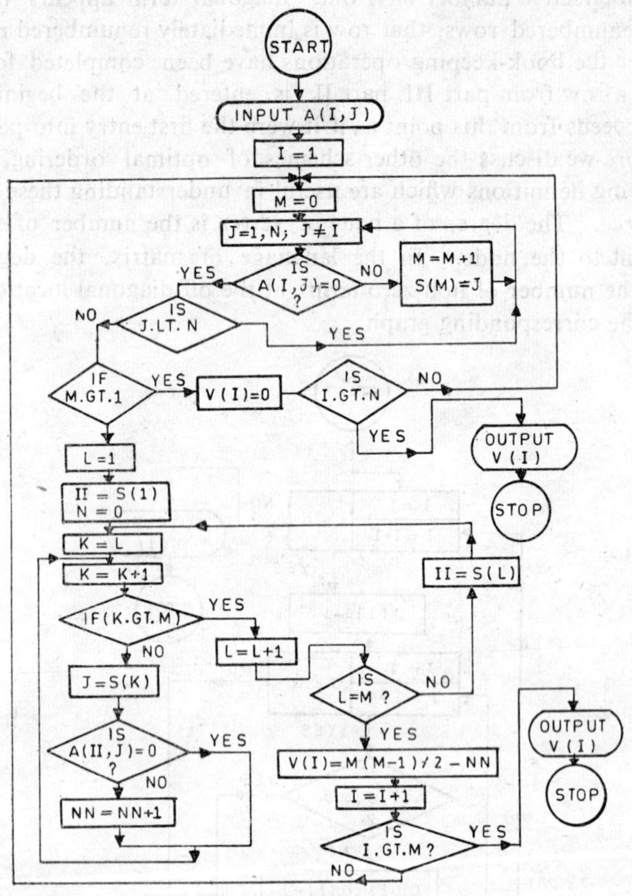

Fig. 8.11 For valency.

Step 1: Count degree of all nodes, that is number of non-zero terms in all rows.

Step 2: Search lists of nodes sequentially.

Step 3: Find node with minimum degree (MIN = 1). If there is such node, number it 1 and decrease degrees of associated nodes accordingly.

Step 4: Continue step 3 till there are nodes with degree 1.

Step 5: Increase MIN by 1. Find node with degree Min and number it. Adjust associated NUM array and Min = 1. If there is no such node repeat step 5.

Step 6: Repeat from Step 3 till all nodes are numbered.

Step 7: Triangularize numbered matrix using Gaussian elimination method.

Scheme 7. This algorithm (Fig. 8.13) is suggested to reduce the

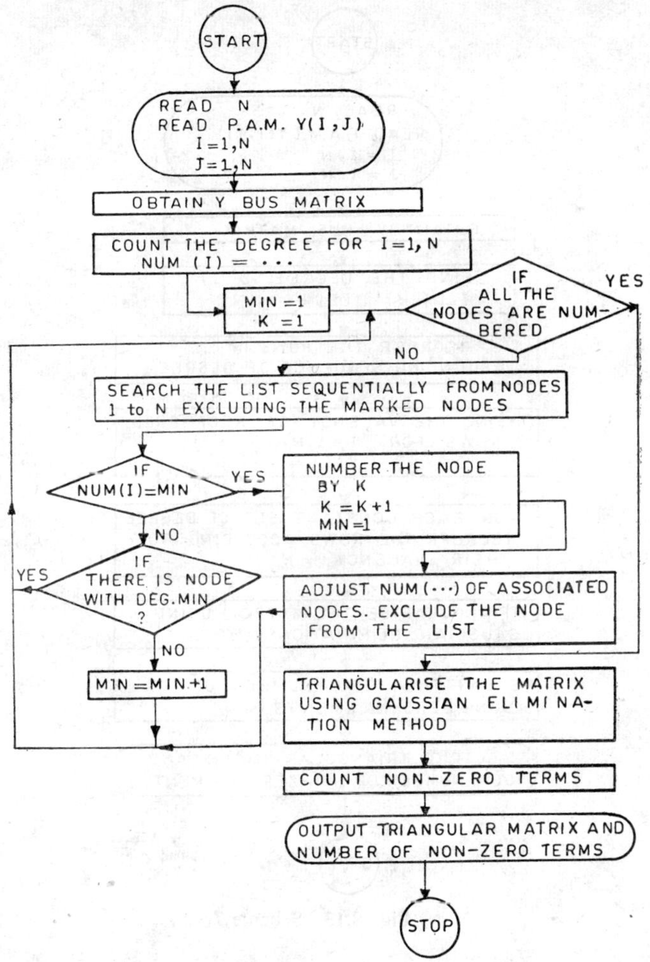

Fig. 8.12 Scheme 6.

(i) computational efforts and (ii) accumulation of new non-zero terms. Following are the steps of the algorithm.

Step 1: Find node or set of nodes whose degree is minimum. Once nodes of minimum degree is found, degree of other nodes is found.

Step 2: Arrange rows in ascending order of degree.

Step 3: Find valency of each node to calculate number of new non-zero terms added in other rows of matrix resulting from elimination of each node.

Step 4: For each set of degree, arrange rows in ascending order of valency.

Step 5: Obtain triangular matrix using Gaussian elimination method.

Scheme 8. In scheme 4, valency subroutine is called only if there is a tie between two nodes having equal number of edges but here at each stage

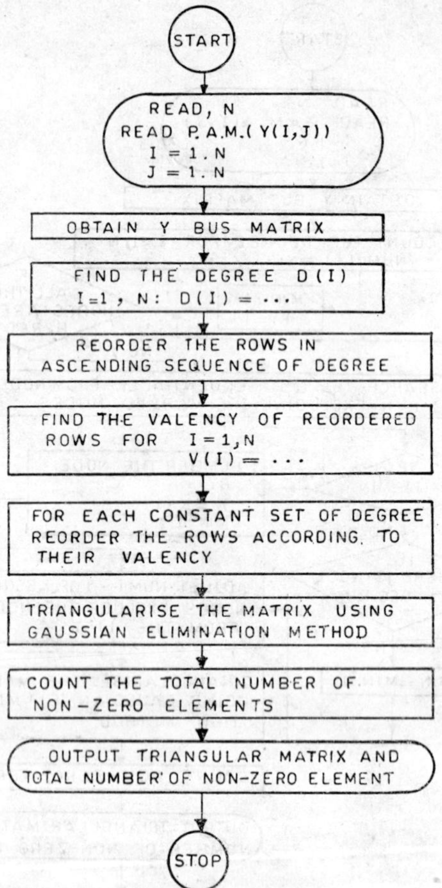

Fig. 8.13 Scheme 7.

of elimination a search for a minimum degree and minimum valency is made and that row is eliminated first. This algorithm (Fig. 8.14) can be explained in the following steps:

Step. 1: Find degree, that is, number of non-zero off diagonal terms of each row.
Step. 2: Arrange rows in ascending order of degree.
Step. 3: Find valency of all nodes.
Step. 4: Replace first row with minimum degree and minimum valency.
Step. 5: Eliminate first row.
Step. 6: Go to steps 1, 2, 3 and 4 for remaining set of nodes.
Step. 7: Eliminate next row.
Step. 8: Repeat steps until matrix is triangularized.

Scheme 9. Any node may be connected with one, two, three... different nodes. When a node is connected to *N* different nodes, it is

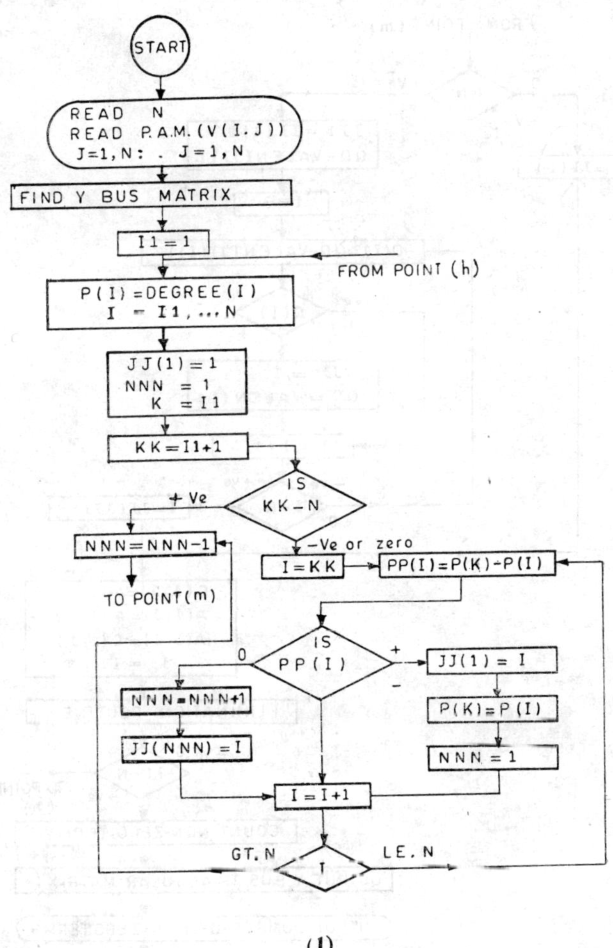
START
READ N
READ P.A.M. (V(I.J))
J=1, N: . J=1, N
FIND Y BUS MATRIX
I1=1
FROM POINT (h)
P(I)=DEGREE(I)
I = I1, ... N
JJ(1)=1
NNN = 1
K = I1
KK=I1+1
IS
KK−N
+ Ve
NNN=NNN−1
TO POINT(m)
−Ve or zero
I=KK
PP(I)=P(K)−P(I)
IS
PP(I)
0
+
−
NNN=NNN+1
JJ(NNN) = I
JJ(1) = I
P(K)=P(I)
NNN = 1
I = I+1
GT. N
I
LE. N

(1)

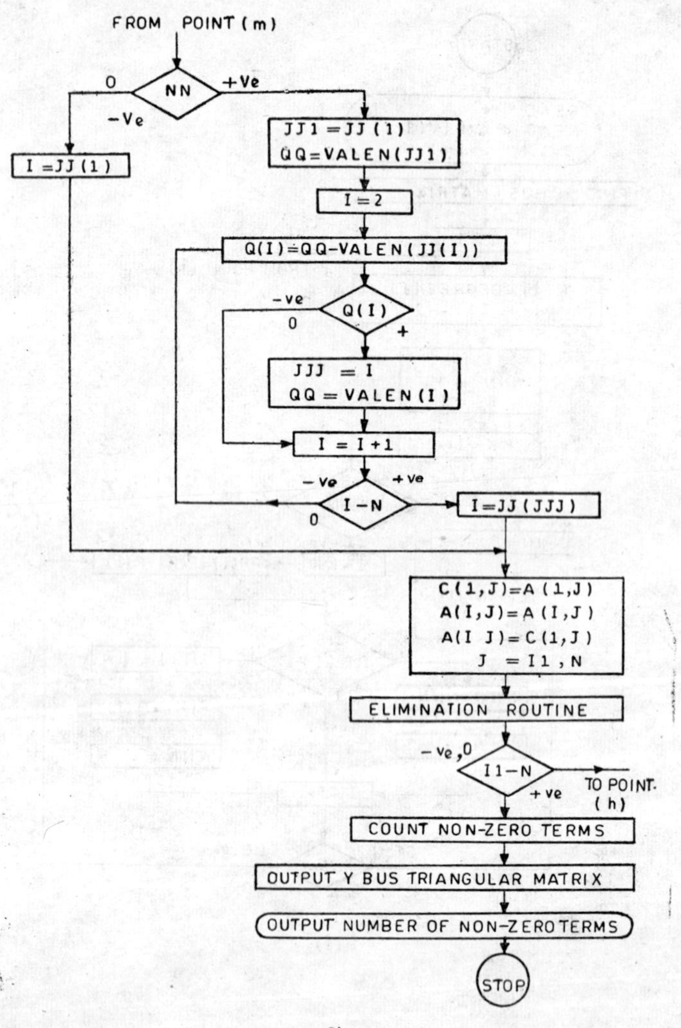

(2)

Fig. 8.14 Scheme 8.

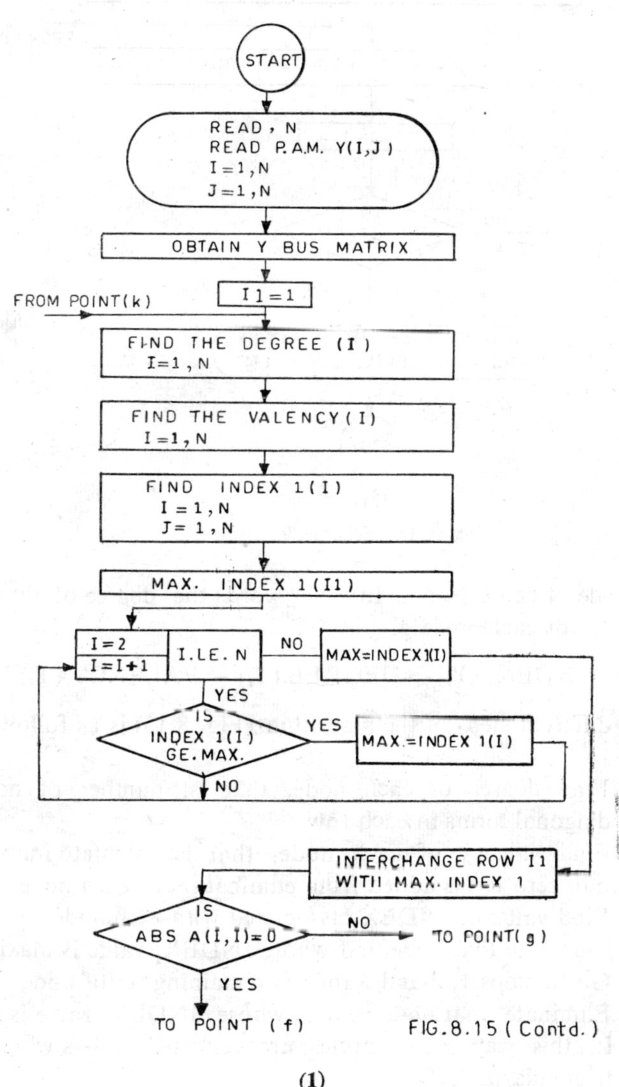

FIG.8.15 (Contd.)

(1)

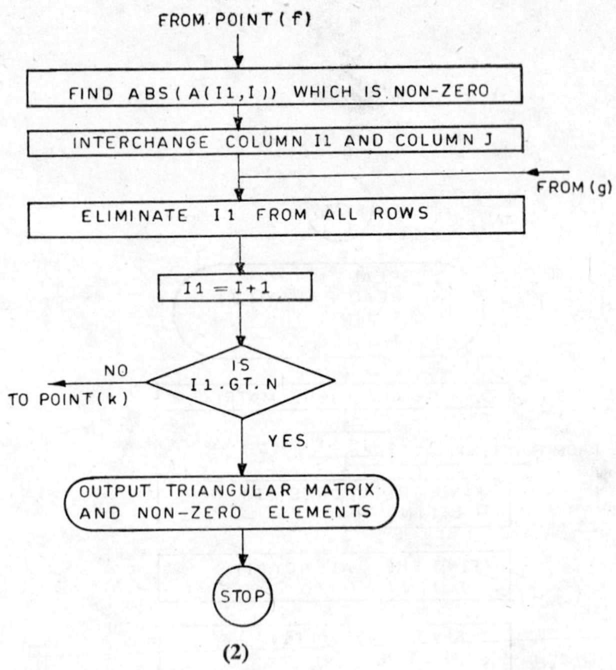

(2)

Fig. 8.15 Scheme 9.

called a node of category N. In other words the degree of that particular node is N. For each node Y_i.

$$\text{INDEX}_1\ (Y_i) = \text{DEGREE}\ (Y_i) - \text{VALENCY}\ (Y_i)$$

The different steps of the algorithm (Fig. 8.15) is as follows:

Step. 1: Find degree of each node, that is, number of non-zero off diagonal terms in each row.

Step. 2: Find valency of each node, that is, calculate number of new non-zero terms added from elimination of each node.

Step. 3: Find value of INDEX_1 associated with each node.

Step. 4: Eliminate that node first whose INDEX_1 value is maximum.

Step. 5: Go to steps 1, 2 and 3 for the remaining set of nodes.

Step. 6: Eliminate that node next to whose INDEX_1 value is maximum. In this way, these cycles are repeated unless whole matrix is triangularized.

Scheme 10. This algorithm is exactly the same except that here the INDEX_2 is defined instead of INDEX_1 in the following way:

$$\text{INDEX}_2(Y_i) = \text{DEGREE}(Y_i) - 2\text{X VALENCY}\ (Y_i)$$

Thus the procedure is same except that INDEX_2 is associated with the algorithm in place of INDEX_1.

Example 8.2. A comparative study of the merits and efficiencies of different algorithms have been made by testing these algorithms with the concrete example.

From the primitive admittance matrix (Fig. 8.16), the Y_{BUS} matrix formed for the network and triangularized by the different algorithms to

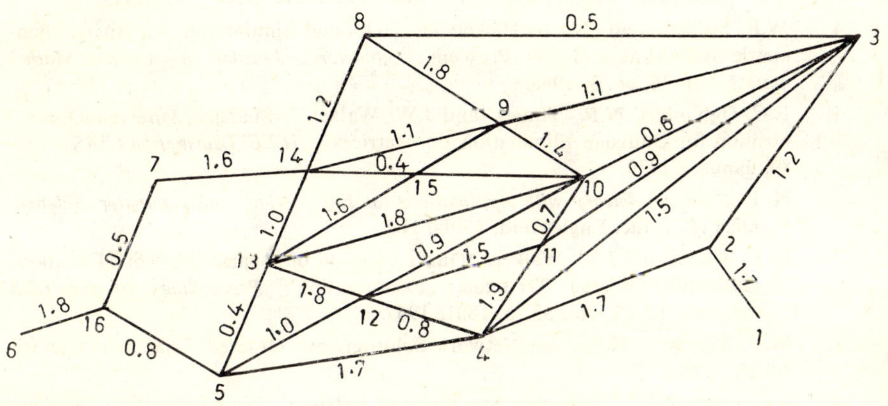

Fig. 8.16

study the relative merits. From the flow charts of these schemes, computer program have been made.

Result and Conclusion: These are given in Table 8.4 for all the schemes of near-optimal ordering as well as without using the technique of optimal ordering.

Table 8.4. Results

Title	*Total number of nonzero elements*	*Execution time (m sec)*
Without using optimal ordering	75	3016
Scheme 1	65	3950
Scheme 2	61	3983
Scheme 3	57	5000
Scheme 4	57	7166
Scheme 5	64	4233
Scheme 6	58	4616
Scheme 7	68	4000
Scheme 8	64	6133
Scheme 9	57	5650

References

1. W.P. Tinney and N. Sato. 'Techniques fo Exploiting the Sparsity of the Network Admittance Matrix', *IEEE Transac ns* (*PAS*), vol 82, p 944, December 1963.
2. L. Carpentier. 'Ordered Elimination'. Proceedings of the Power System Communication Conference, 1963.
3. R.P. Tewarson. *Sparse Matrices.* Academic Press, New York, 1973.
4. W.R. Spillers and Norris Hickerson, 'Optimal Elimination for Sparse Symmetric Systems as a Graph Problem'. *Quarterly Journal of Applied Mathematics*, vol. 26, no. 3, 1968.
5. E.C. Ogbuobiri, W.F. Tinney and J.W. Walker. 'Sparsity Directed Decomposition for Gaussian Elimination on Matrices'. *IEEE Transaction* (*PAS*), vol: 89, January 1970.
6. N. Deo. *Graph Theory with Applications to Engineering and Computer Science.* Prentice Hall, Inc. Englewood, Cliffs, N.J.
7. W.F. Tinney and J.W. Walker. 'Direct Solution of Sparse Network Equations by Optimally Ordered Triangular Factorization'. *Proceedings of the IEEE Transactions* (*PAS*) vol. 55, p. 1801, 1967.
8. W.F. Tinney. 'Notes on Network Solution by Ordered Triangular Factorization', 1968.
9. E.C. Ogbuobiri. 'Dynamic Storage and Retrieval in Sparsity Programming', *IEEE Transactions* (*PAS*), vol. 89, 1970.
10. R.D. Berry. 'An Optimal Ordering of Electronic Circuit Equations for a Sparse Matrix Solutions'. *IEEE Transactions* (*CAS*), vol. 18, p. 40, 1971.
11. F. Harry. 'Graph and Matrices'. *SIAM Review*, vol. 9, p. 83, 1967.
12. L. P. Singh and A. K. Goel. 'An Optimal Ordering for Sparse Systems'. *Journal of the Institution of Engineers* (*India*), vol. 57, pt EL 2, October 1976, p. 105.
13. L.P. Singh and H. C. Srivastava. 'Sparsity and Optimal Ordering'. *Journal of The Institute of Engineers* (*India*), vol. 57, pt BL 6, June 1977, p. 274.
14. K.K. Goyal and L.P. Singh. 'Optimal Elimination of Sparse Systems Using Dynamic Programming Technique'. Proceeding CS 9–81, March 1-4, New Delhi.

Chapter 9

Dynamic Analysis and Modelling of Machines

9.1 Introductory Remarks

We shall discuss in this chapter the dynamic analysis and modelling of synchronous machines and modelling of automatic voltage regulator and excitation systems, governing systems and prime mover and also modelling of induction machines. We shall start first, with the most simplified model of the synchronous machines and finally develop the detailed modelling of synchronous machines including excitation systems and prime mover and governing systems. These mathematical models are useful for the purpose of transient and dynamic stability studies.

9.2 Dynamic Analysis and Modelling of Synchronous Machines

We have discussed in Chapter 3, only an introduction to the synchronous machine modelling. We shall discuss in this chapter the dynamic analysis and detailed modelling of the synchronous machines including also the exciter and governer control. This is useful in the transient and dynamic stability studies of the power system network.

9.2.1 Simplest Model of the Synchronous Machine

The simplest model of a synchronous generator is a constant voltage source behind proper reactance. In this model, the voltage is assumed to be constant, only its phase angle changes. This model is very useful for the purpose of *transient analysis involving short period of study* say one second or less. In this model it is obvious that the changes in the flux linkages and saliency have been neglected. The machine equation in this case becomes,

$$E'_g = V + I_a r_a + jI_a X'_s \tag{9.1}$$

where E'_g is the transient voltage of the synchronous generator; V is the full load terminal voltage; I_a is the full load armature current and X'_s is the transient reactance.

The equivalent circuit and phasor diagram in this case will be as shown in Figs. 9.1 and 9.2 respectively.

In order to include the effect of saliency, we calculate a fictitious voltage "E_q" located at the quadrature axis of the synchronous generator. Direct axis is taken along the main pole axis while quadrature axis lags the direct axis by 90°. The expression for the voltage E_q in terms of full

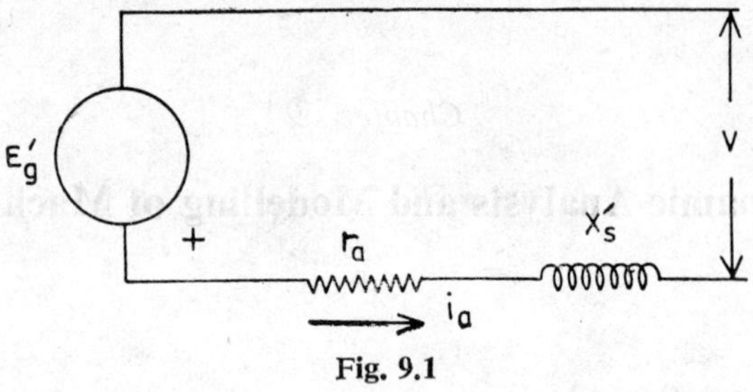

Fig. 9.1

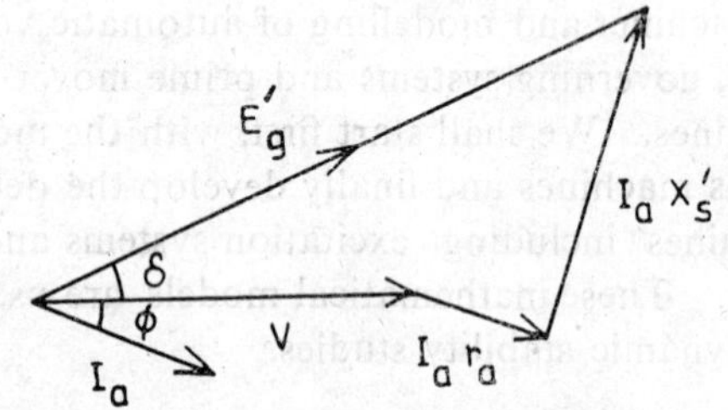

Fig. 9.2 δ is the torque angle which is assumed to change

load terminal voltage V and full load armature current I_a will be,

$$E_q = V + I_a r_a + jI_a X_q \tag{9.2}$$

Here X_q is the quadrature axis synchronous reactance.

The equivalent circuit and the phasor diagram for this case are given in Figs. 9.3 and 9.4 respectively. ω is rotor angular velocity which is equal to the synchronous speed ω_0 only during normal operation.

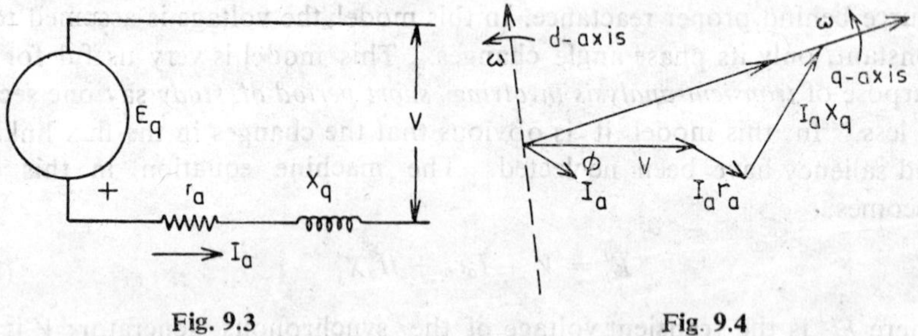

Fig. 9.3 Fig. 9.4

The voltage E_g (which is the excitation voltage i.e. the open circuit voltage) will be calculated as follows:

$$E_g = V + I_a r_a + jI_{ad} X_d + jI_{aq} X_q \tag{9.3}$$

where X_d $(= X_l + X_{ad})$ is the d-axis armature synchronous reactance and X_q $(= X_l + X_{aq})$ is the q-axis armature synchronous reactance. X_{ad} is the d-axis armature magnetizing reactance corresponding to the d-axis component of the armature reaction flux i.e. ϕ_{ad} and X_{aq} is the q-axis component of the armature magnetizing reactance corresponding to the q-axis component of the armature reaction flux i.e. ϕ_{aq}. Thus the phasor diagram including the effect of saliency will be as shown in Fig. 9.5.

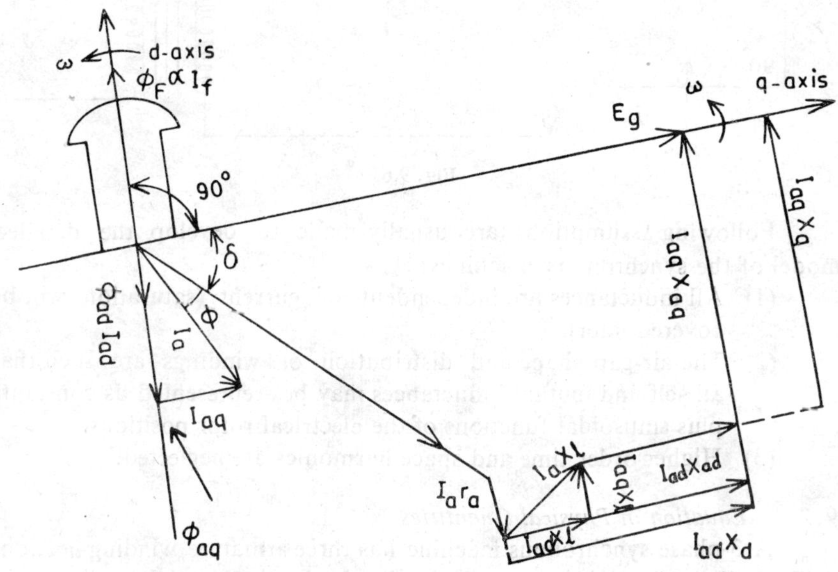

Fig. 9.5

Now we shall discuss the detailed modelling of the synchronous machine, initially only for the transient state and later on for the subs-transient state which includes the effects of damping.

Sign Convension. The flux linkage of a circuit or a portion thereof, produced by the current in the same circuit element, is considered to have the same sign as the current. Voltage induced by the flux linkage

$$V = -\frac{d\psi(t)}{dt} \tag{9.4}$$

Here $\psi(t)$ is the flux linkage and it is equal to Real $\psi_{max} e^{j\omega t}$. Then the voltage will be,

$$V(t) = Re(-j\omega\psi_{max} e^{j\omega t}) \tag{9.5}$$

i.e. the voltage phasor lags the flux phasor by 90°.

9.2.2 Circuit Equations

Each winding k in the machine (as shown in Fig. 9.6) can be described by an equation

$$V_k(t) = -R_k i_k(t) - \frac{d\psi_k}{dt} \tag{9.6}$$

where V_k is the voltage across the terminal of the winding. i_k is current out of rotor and stator i.e. armature winding (source convension has been used for both types of windings).

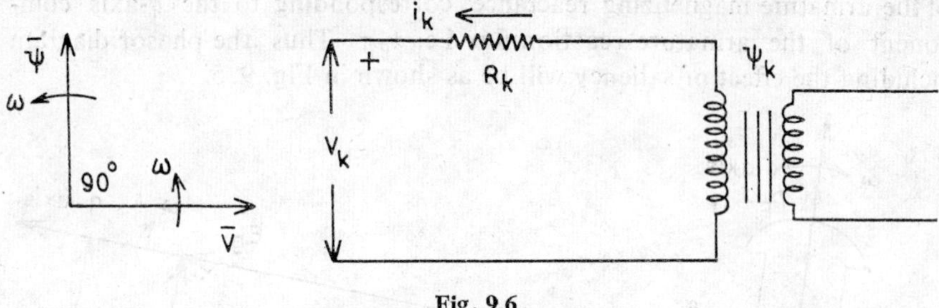

Fig. 9.6

Following assumptions are usually made to develop the detailed model of the synchronous machines:

(1) All inductances are independent of current (saturation will be covered later).

(2) The air-gap shape and distribution of windings are such that all self and mutual inductances may be represented as constants plus sinusoidal functions of the electrical rotor positions.

(3) Higher order time and space harmonics are neglected.

9.2.3 Equation in Physical Quantities

A 3-phase synchronous machine has three armature winding *abc*, one field winding '*f*' on the rotor with its flux in the direction of the direct axis and one fictitious winding '*g*' on the rotor with its flux in the quadrature axis (see Fig. 9.7). The fictitious winding '*g*' approximates the effect of eddy

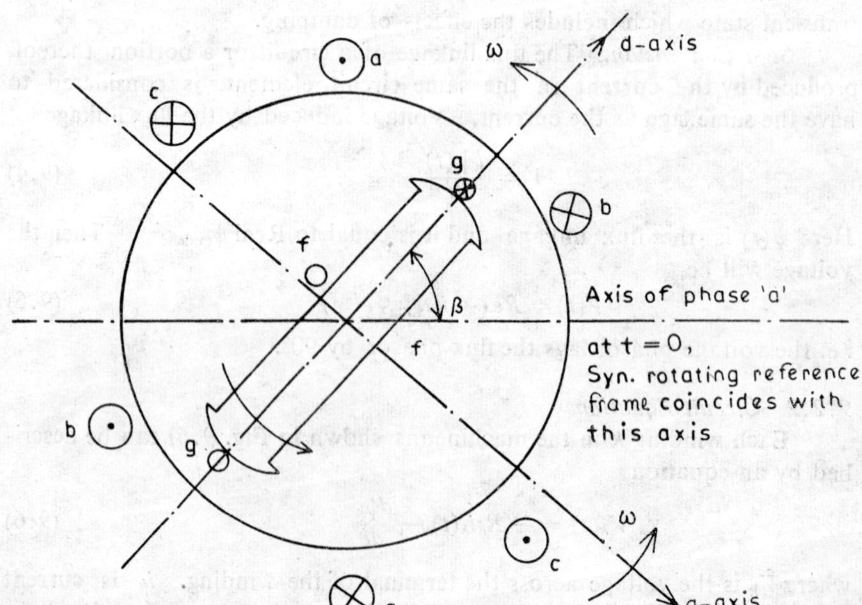

Fig. 9.7

currents circulating in the iron (rotor iron in round rotor machine and negligible in salient pole machine) and to some extent effect of damper windings. This fictitious winding is short circuited since it is not connected to any voltage source. Two additional windings could be added to the rotor (one along each axis) to represent damping and other effects more accurately. But considering all the assumptions made for stability studies and very short time constant of such windings (0.03 and 0.04 seconds), they are usually ignored. β_{mech} is the angle by which the rotor direct axis has turned beyond the axis of the armature phase '*a*'. The electrical rotor position

$$\beta = \beta_{elect} = \frac{P}{2}\,\beta_{mech}$$

where P is the number of poles.

With the rotation of the machine, angle β will go on changing with the constant angular velocity ω. We have, therefore

$$\beta = \omega t = \frac{P}{2}\,\beta_{mech} \tag{9.7}$$

or

$$\frac{d\beta}{dt} = \omega = \text{angular velocity of rotor} \tag{9.8}$$

We shall assume a synchronously rotating reference frame (i.e. axis) rotating with the synchronous speed ω_0 and which will be along the axis of phase '*a*' at $t = 0$. Then at any time t, it will be at $\omega_0 t$. Thus we have,

$$\beta = (\delta + \pi/2) + \omega_0 t \tag{9.9}$$

where δ is the displacement of the quadrature axis from the synchronously rotating reference axis and $(\delta + \pi/2)$ is the displacement of direct axis.

We have from eqn. 9.9,

$$\frac{d\beta}{dt} = \frac{d\delta}{dt} + \omega_0 \tag{9.10}$$

or

$$\frac{d\delta}{dt} = \frac{d\beta}{dt} - \omega_0 = \omega - \omega_0 \tag{9.11}$$

where

$$\omega_0 \text{ is the synchronous speed} = \frac{120f}{p}$$

$$\omega = \text{angular speed of rotor} = \frac{d\beta}{dt}$$

From equation (9.10), after taking second derivative, we get

$$d^2\beta/dt^2 = d^2\delta/dt^2 \tag{9.12}$$

δ is negative here, it will be positive if measured in the direction of ω_0.

The equation of the five windings can be written as one matrix (from eqn. (9.6)) as shown, i.e. $$[V(t)] = -[R][i(t)] - \frac{d}{dt}[\psi] \tag{9.13}$$

with $$[V] = \begin{bmatrix} V_a \\ V_b \\ V_c \\ V_f \\ 0 \end{bmatrix}; \quad i = \begin{bmatrix} i_a \\ i_b \\ i_c \\ i_f \\ i_g \end{bmatrix}$$

$$[R] = \begin{bmatrix} R_a & & & & \\ & R_b & & 0 & \\ & & R_c & & \\ & 0 & & R_f & \\ & & & & R_g \end{bmatrix} \text{ and } [\psi] = \begin{bmatrix} \psi_a \\ \psi_b \\ \psi_c \\ \psi_f \\ \psi_g \end{bmatrix} \tag{9.14}$$

The vectors of the flux linkages is proportional to the currents with matrix of self and mutual inductances as proportionality factor. Hence $\bar{\psi} = [L]\bar{i}$ i.e.

$$\begin{bmatrix} \psi_a \\ \psi_b \\ \psi_c \\ \psi_f \\ \psi_g \end{bmatrix} = \begin{bmatrix} L_{aa} & L_{ab} & L_{ac} & L_{af} & L_{ag} \\ L_{ba} & L_{bb} & L_{bc} & L_{bf} & L_{bg} \\ L_{ca} & L_{cb} & L_{cc} & L_{cf} & L_{cg} \\ L_{fa} & L_{fb} & L_{fc} & L_{ff} & L_{fg} \\ L_{ga} & L_{gb} & L_{gc} & D_{gf} & L_{gg} \end{bmatrix} \begin{bmatrix} I_a \\ I_b \\ I_c \\ I_f \\ I_g \end{bmatrix} \tag{9.15}$$

All the inductances except L_{ff} and L_{gg} are functions of β and thus they are time varying. The inductance L is symmetric, i.e.

$$L_{ba} = L_{ab} \text{ etc.}$$

Note: In all these analysis q-axis is taken lagging behind d-axis by 90°. This is in accordance with *IEEE* working group report 'Recommended phasor diagram for synchronous machine' paper No. 69 *TP* 143-*PWR*, presented at the *IEEE* Winter Power Meeting, New York, Jan. 26-31, 1969.

9.2.4 *Inductance of Synchronous Machine*

(*a*) *Rotor self inductances and stator to rotor mutual inductances :* Since stator, i.e. armature is a cylindrical structure, the self inductance of the field winding '*f*' and also for the fictitious winding '*g*' will not depend upon the rotor position. Hence

$$L_{ff} \text{ is constant}$$

and
$$L_{gg} \text{ is constant}$$

However, stator to rotor mutual inductances will vary periodically with β, i.e. with the rotation of the rotor. Between phase 'a' and the field winding 'f' on the rotor, for example, the mutual inductance will be maximum (refer to Fig. 9.7), M_f at $\beta = 0$ zero at $\beta = 90°$ and negative maximum M_f at $\beta = 180°$ and zero again at $\beta = 270°$.

Accordingly with space m.m.f. and flux distribution assumed sinusoidial, we have from the above physical reasoning, that

$$L_{af} = L_{fa} = M_f \cos \beta \tag{9.16}$$

We will also obtain similar expressions for phases b and c except that β is replaced by $\beta - 120$ and $\beta + 120$ respectively, i.e.

$$L_{bf} = L_{fb} = M_f \cos (\beta - 120)$$

and

$$L_{cf} = L_{fc} = M_f \cos (\beta + 120) \tag{9.17}$$

Similarly between phase 'a' and g winding on the rotor, the mutual inductance will be maximum 'M_g' for $\beta = -\pi/2$, zero at $\beta = 0$ and maximum $-M_g$ at $\beta = \pi/2$. Thus we obtain from the above,

$$L_{ga} = L_{ag} = -M_g \sin \beta$$

$$L_{bg} = L_{gb} = -M_g \sin (\beta - 120)$$

$$L_{cg} = L_{gc} = -M_g \sin (\beta + 120) \tag{9.18}$$

This is because for phases 'b' and 'c', angle 'β' will be replaced by '$\beta-120$' and '$\beta+120$' respectively.

Again because of stator symmetry, and also because of the magnetic axis of f-winding and g-winding are at right angle, we have

$$L_{gf} = L_{fg} = 0 \tag{9.19}$$

as there is no mutual coupling between f and g windings on the rotor.

(*b*) *Stator self inductances:* Self inductance of any stator phase will always have a positive value, but there will be a second harmonics variation because of different airgap geometry along d and q-axes. The self-inductance L_{aa} for example will be a maximum for $\beta = 0$, a minimum for $\beta = 90°$ and maximum, again for $\beta = 180°$, and so on.

Let us consider the flux linkage with phase 'a' only when it is excited. With the space harmonics ignored, the m.m.f. wave of phase 'a' is a cosine wave (space distribution) centred on the phase 'a' axis as shown in Fig. 9.8. In this case only phase a is excited. The peak amplitude is

$$F_a = N_a \, i_a \tag{9.20}$$

where N_a = effective turns/phase and i_a = phase 'a' current.

Now let this *MMF* is resolved into two components, one along d-axis (i.e. F_{da}) and the other along q-axis (i.e. F_{qa}). Their peak amplitudes are:

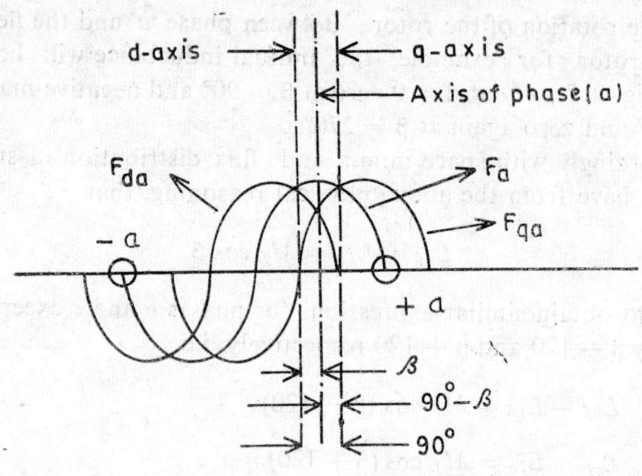

Fig. 9.8

$$F_{da} = F_a \cos \beta \tag{9.21}$$

$$F_{qa} = F_a \cos (90 - \beta) = F_a \sin \beta \tag{9.22}$$

The advantage of resolving m.m.f. is that the two components m.m.f. wave act on specific air gap geometry in their respective axes. Let

ρ_{gd} = permeance along d-axis

and ρ_{gq} = permeance along q-axis

These are known machine constants and their values can be found from flux plot for specific machine geometry. Hence the fundamental air gap flux per pole, along two axes are accordingly

$$\phi_{gda} = \text{comp. of flux along } d\text{-axis}$$

$$= F_{da}\, \rho_{gd} = F_a\, \rho_{gd} \cos \beta \tag{9 23}$$

$$\phi_{gqa} = \text{component flux along } q\text{-axis}$$

$$= F_{qa}\, \rho_{gq} = F_a\, \rho_{gq} \sin \beta \tag{9.24}$$

Let

ϕ_{gaa} = flux linking with the phase winding 'a'.

Hence the air-gap flux linking with the phase 'a' is

$$\phi_{gaa} = \phi_{gda} \cos \beta + \phi_{gqa} \sin \beta$$

$$= F_a\, (\rho_{gd} \cos^2 \beta + \rho_{gq} \sin^2 \beta)$$

$$= N_a i_a \left(\frac{\rho_{gd} + \rho_{gq}}{2} + \frac{\rho_{gd} - \rho_{gq}}{2} \cos 2\beta \right) \tag{9.25}$$

Since inductance is the proportionality factor relating flux linkages to current, the self inductance L_{gaa} of phase 'a' due to air gap flux when only phase a is excited, will be

$$L_{gaa} = \frac{N_a \phi_{gaa}}{i_a} \tag{9.26}$$

After substituting equation (9.25) in equation (2.26), we get,

$$L_{gaa} = N_a^2 \left(\frac{\rho_{gd} + \rho_{gq}}{2} + \frac{\rho_{gd} - \rho_{gq}}{2} \times \cos 2\beta \right)$$

$$= L_s' + L_m \cos 2\beta \tag{9.27}$$

where $\quad L_s' = N_a^2 \frac{\rho_{gd} + \rho_{gq}}{2}$

and $\quad L_m = N_a^2 \frac{\rho_{gd} - \rho_{gq}}{2}$

Here L_s' is constant terms and L_m is the amplitude of 2nd harmonics variation. To get the total self inductance L_{aa}, the leakage inductance L_{al}, representing the flux which does not cross the air gap to the rotor must be added i.e.

$$L_{aa} = L_{al} + L_{gaa} = L_{al} + L_s' + L_m \cos 2\beta$$

$$= L_s + L_m \cos 2\beta \tag{9.28}$$

where $L_s = L_{al} + L_s'$.

In the same way we can obtain the self inductance of phase b and c by replacing β by $\beta - 120°$ and $\beta + 120°$ respectively.

$$L_{bb} = L_s + L_m \cos 2(\beta - 120°)$$

$$= L_s + L_m \cos (2\beta - 240°)$$

$$= L_s + L_m \cos (2\beta + 120°) \tag{9.29}$$

and $\quad L_{cc} = L_s + L_m \cos 2(\beta + 120)$

$$= L_s + L_m \cos (2\beta + 240)$$

$$= L_s + L_m \cos (2\beta - 120) \tag{9.30}$$

(c) *Stator mutual inductances:* The mutual inductances between stator phases will also exhibit a second harmonics variation with β because of rotor shape. The mutual inductance between phase a and b, L_{ab} = 'L_{ba}' can be found by evaluating the air gap flux 'ϕ_{gba}' linking phase 'b' when only phase 'a' is excited. In the earlier equation of ϕ_{gaa} (equation (9.25)), if we replace 'β' by $\beta - 120$ as we now wish to evaluate flux-linking phase 'b', we obtain

ϕ_{gba} = flux linking with the phase winding 'b'

when only phase 'a' winding is excited

$$= \phi_{gda} \cos (\beta - 120) + \phi_{gqa} \sin (\beta - 120) \tag{9.31}$$

Now we have from the earlier equation (i.e. equations (9.23) and (9.24)),

$$\phi_{gda} = F_{da}\, \rho_{gd} = F_a\, \rho_{gd} \cos \beta$$

and $$\phi_{gqa} = F_{qa}\, \rho_{gq} = F_a\, \rho_{gq} \sin \beta \tag{9.32}$$

Therefore, after substituting the eqns. (9.32) in the eqn. (9.31), we get,

$$\phi_{gba} = F_a(\rho_{gd}\cos\beta\cos(\beta-120) + \rho_{gq}\sin\beta\sin(\beta-120))$$

$$= N_a i_a\left(-\frac{\rho_{gd}+\rho_{gq}}{4} + \frac{\rho_{gd}-\rho_{gq}}{2}\cos(2\beta-120)\right) \quad (9.33)$$

as $F_a = N_a i_a$

The mutual inductance between phase a and b due to air-gap flux is then

$$L_{gba} = \frac{N_a\phi_{gba}}{i_a} \text{ (as } N_b = N_a)$$

$$= N_a^2\left(-\frac{\rho_{gd}+\rho_{gq}}{4} + \frac{\rho_{gd}-\rho_{gq}}{2}\cos(2\beta-120)\right)$$

$$= -0.5L_s' + L_m\cos(2\beta-120) \quad (9.34)$$

where L_s' and L_m has the same meaning as used earlier in evaluating L_{aa} (i.e. equation (9.27)). There is also a very small amount of mutual flux which does not cross the air gap (i.e. flux around the ends of the windings). With this flux neglected, the mutual inductance is

$$L_{ab} = L_{ba} = -0.5L_s' + L_m\cos(2\beta-120)$$

$$= -M_s + L_m\cos(2\beta-120) \quad (9.35)$$

where $M_s = 0.5L_s'$ is a machine constant.

Similarly we can obtain

$$L_{bc} = L_{cb} = -M_s + L_m\cos 2\beta \quad (9.36)$$

and

$$L_{ca} = L_{ac} = -M_s + L_m\cos(2\beta+120) \quad (9.37)$$

Here L_s, L_m, M_s, M_g, M_f will now be regarded as known machine constants, which are determined either by tests or calculated by designer. We are giving below these machine reactances.

Machine Inductances

$$L_{ff} = \text{constant}$$

$$L_{gg} = \text{constant}$$

$$L_{fg} = 0 = L_{gf}$$

$$L_{af} = M_f\cos\beta = L_{fa}$$

$$L_{bf} = M_f\cos(\beta-120) = L_{fb}$$

$$L_{cf} = M_f\cos(\beta+120) = L_{fc}$$

$$L_{ag} = -M_g\sin\beta = L_{ga}$$

$$L_{bg} = -M_g\sin(\beta-120) = L_{gb}$$

$$L_{cg} = -M_g\sin(\beta+120) = L_{gc}$$

$$L_{aa} = L_s + L_m\cos 2\beta$$

$$L_{bb} = L_s + L_n\cos(2\beta+120)$$

$$L_{cc} = L_s + L_m \cos (2\beta - 120)$$

$$L_{ab} = - M_s + L_m \cos (2\beta - 120) = L_{ba}$$

$$L_{bc} = - M_s + L_m \cos 2\beta = L_{cb}$$

$$L_{ca} = - M_s + L_m \cos (2\beta + 120) = L_{ac} \quad (9.38)$$

9.2.5 *Park's Transformation to dqo Components*

The earlier equations (eqns. (9.13), (9.14) and (9.15)) are a set of differential equations describing the behaviour of machines. However, their solution is complicated by the fact that the inductances are function of rotor angle 'β', which is in turn is a function of time. And hence, these are the differential equations with variable coefficients which are the function of time with the result, their solution is quite complicated. This complication, however, can be avoided by transferring the physical quantities in the armature windings through a linear, time dependent and power invariant transformation called Parks transformation.

The Parks transformation matrix T_{park} transforms the field of phasors to the field of $d-q-o$ components. T_{park} which is a linear, time dependent matrix (i.e. an operator or system) is of the following form,

$$T^{d,q,o}_{\text{park}} = \frac{1}{\sqrt{3}} \begin{bmatrix} \sqrt{2} \cos \beta & \sqrt{2} \sin \beta & 1 \\ \sqrt{2} \cos (\beta - 120) & \sqrt{2} \sin (\beta - 120) & 1 \\ \sqrt{2} \cos (\beta + 120) & \sqrt{2} \sin (\beta + 120) & 1 \end{bmatrix} \quad (9.39)$$

This transformation matrix $T^{d,q,o}_{\text{park}}$ is orthogonal, i.e.

$$T^{-1}_{\text{park}} = T^{T}_{\text{park}}$$

and thus, it is power invariant transformation matrix. It is also clear from the above matrix T_{park}, that the entries in the column corresponding to '0' component is 1, sum of the entries of the remaining two columns are zero separately and all the columns are linearly independent. The phasor variables are transformed to a new set of variables with the matrix T_{park} as follows:

$$V^{abc}_{\text{phase}} = T_{\text{park}} V^{dqo}_{\text{park}}$$

$$i^{abc}_{\text{phase}} = T_{\text{park}} i^{dqo}_{\text{park}}$$

and $$\psi^{abc}_{\text{phase}} = T_{\text{park}} \psi^{dqo}_{\text{park}} \quad (9.40)$$

The inverse transformation is

$$V^{d,q,o}_{\text{park}} = T^{-1}_{\text{park}} V^{abc}_{\text{phase}} \quad \text{etc, etc.}$$

$$= T^{T}_{\text{park}} V^{abc}_{\text{phase}}$$

$$\text{i.e.} \begin{bmatrix} V^d \\ V_q \\ V_o \end{bmatrix} = \frac{1}{\sqrt{3}} \begin{bmatrix} \sqrt{2}\cos\beta & \sqrt{2}\cos(\beta-120) & \sqrt{2}\cos(\beta+120) \\ \sqrt{2}\sin\beta & \sqrt{2}\sin(\beta-120) & \sqrt{2}\sin(\beta+120) \\ 1 & 1 & 1 \end{bmatrix} \begin{bmatrix} V_a \\ V_b \\ V_c \end{bmatrix} \tag{9.41}$$

Hence after transformation, equations are

$$\begin{bmatrix} V_d \\ V_q \\ V_o \\ \hline V_f \\ 0 \end{bmatrix} = \left[\begin{array}{ccc|cc} \sqrt{\frac{2}{3}}\cos\beta & \sqrt{\frac{2}{3}}\cos(\beta-120) & \sqrt{\frac{2}{3}}\cos(\beta+120) & 0 & 0 \\ \sqrt{\frac{2}{3}}\sin\beta & \sqrt{\frac{2}{3}}\sin(\beta-120) & \sqrt{\frac{2}{3}}\sin(\beta+120) & 0 & 0 \\ \sqrt{\frac{1}{3}} & \sqrt{\frac{1}{3}} & \sqrt{\frac{1}{3}} & 0 & 0 \\ \hline 0 & 0 & 0 & 1 & 0 \\ 0 & 0 & 0 & 0 & 1 \end{array}\right] \begin{bmatrix} V_a \\ V_b \\ V_c \\ \hline V_f \\ 0 \end{bmatrix} \tag{9.42}$$

i.e. $$[V_{\text{park}}] = [T]^{-1}[V]_{\text{phase}} \tag{9.43}$$

This is identical for i and ψ and the inverse transformation is

$[V]_{\text{phase}} = [T][V_{\text{park}}]$ and is identical for i and ψ.

The Park's transformation matrix

$$[T] = \left[\begin{array}{ccc|cc} \sqrt{\frac{2}{3}}\cos\beta & \sqrt{\frac{2}{3}}\sin\beta & \sqrt{\frac{1}{3}} & 0 & 0 \\ \sqrt{\frac{2}{3}}\cos(\beta-120) & \sqrt{\frac{2}{3}}\sin(\beta-120) & \sqrt{\frac{1}{3}} & 0 & 0 \\ \sqrt{\frac{2}{3}}\cos(\beta+120) & \sqrt{\frac{2}{3}}\sin(\beta+120) & \sqrt{\frac{1}{3}} & 0 & 0 \\ \hline 0 & 0 & 0 & 1 & 0 \\ 0 & 0 & 0 & 0 & 1 \end{array}\right]$$

$$= \left[\begin{array}{c|c} T_{\text{park}} & 0 \\ \hline 0 & I \end{array}\right] \tag{9.44}$$

Here the newly recommended position of the quadrature axis lagging 90° behind direct axis is adopted. In Park's original equations (quadrature axis leading direct axis by 90°), the 2nd row in T^{-1} and the 2nd column T has negative sign, i.e.

$$T_{\text{park}}^{-1} = \begin{bmatrix} \sqrt{\frac{2}{3}}\cos\beta & \sqrt{\frac{2}{3}}\cos(\beta-120) & \sqrt{\frac{2}{3}}\cos(\beta+120) \\ -\sqrt{\frac{2}{3}}\sin\beta & -\sqrt{\frac{2}{3}}\sin(\beta-120) & -\sqrt{\frac{2}{3}}\sin(\beta+120) \\ \sqrt{1/3} & \sqrt{1/3} & \sqrt{1/3} \end{bmatrix}$$

and

$$T_{\text{park}} = \sqrt{1/3}\begin{bmatrix} \sqrt{2}\cos\beta & -\sqrt{2}\sin\beta & 1 \\ \sqrt{2}\cos(\beta-120) & -\sqrt{2}\sin(\beta-120) & 1 \\ \sqrt{2}\cos(\beta+120) & -\sqrt{2}\sin(\beta+120) & 1 \end{bmatrix}$$

Let us take the original equations (i.e. eqn. 9.13),

$$[V(t)] = -[R][i(t)] - \frac{d}{dt}[\psi]$$

i.e., after expansion,

$$\begin{bmatrix} V_a \\ V_b \\ V_c \\ V_f \\ 0 \end{bmatrix} = \begin{bmatrix} R_a & & & & \\ & R_b = R_a & 0 & & \\ & & R_c = R_a & & \\ & 0 & & R_f & \\ & & & & R_g \end{bmatrix} \begin{bmatrix} i_a \\ i_b \\ i_c \\ i_f \\ i_g \end{bmatrix} - \frac{d}{dt}\begin{bmatrix} \psi_a \\ \psi_b \\ \psi_c \\ \psi_f \\ \psi_g \end{bmatrix} \tag{9.45}$$

Hence for transformation, let us premultiply both sides of eqn. (9.45) by T^{-1}, we get

$$T^{-1}[v(t)] = -T^{-1}[R][i(t)] - T^{-1}\frac{d}{dt}[\psi]$$

$$[V_{\text{park}}] = -[R][I_{\text{park}}] - T^{-1}\frac{d}{dt}T[\psi_{\text{park}}] \tag{9.46}$$

because $T\psi_{\text{park}} = \psi_{\text{phase}} = \psi$.

Hence

$$[V_{\text{park}}] = -[R]\,[I_{\text{park}}] - T^{-1}\frac{d}{dt}T[\psi_{\text{park}}]$$

$$= -[R]\,[I_{\text{park}}] - T^{-1} * T\frac{d}{dt}[\psi_{\text{park}}] - T^{-1}\left[\frac{d}{dt}[T][\psi_{\text{park}}]\right]$$

$$= -[R][I_{\text{park}}] - \frac{d}{dt}[\psi_{\text{park}}] - T^{-1}\left[\frac{d}{dt}[T][\psi_{\text{park}}]\right] \tag{9.47}$$

It can be shown that $-T^{-1}\dfrac{d}{dt}[T]$ is a matrix (shown below) with zero entries except for $-\omega$ in the first row second column and $+\omega$ in the second row first column, i.e.

$$-T^{-1}\frac{d}{dt}T = \begin{bmatrix} \sqrt{\tfrac{2}{3}}\cos\beta & \sqrt{\tfrac{2}{3}}\cos(\beta-120) & \sqrt{\tfrac{2}{3}}\cos(\beta+120) \\ \sqrt{\tfrac{2}{3}}\sin\beta & \sqrt{\tfrac{2}{3}}\sin(\beta-120) & \sqrt{\tfrac{2}{4}}\sin(\beta+120) \\ \sqrt{\tfrac{1}{3}} & \sqrt{\tfrac{1}{3}} & \sqrt{\tfrac{1}{3}} \end{bmatrix}$$

$$\times \frac{d}{dt}\sqrt{\frac{1}{3}}\begin{bmatrix} \sqrt{2}\cos\beta & \sqrt{2}\sin\beta & 1 \\ \sqrt{2}\cos(\beta-120) & \sqrt{2}\sin\{\beta-120) & 1 \\ \sqrt{2}\cos(\beta+120) & \sqrt{2}\sin(\beta+120) & 1 \end{bmatrix}$$

$$= \begin{bmatrix} 0 & -\omega & 0 \\ \omega & 0 & 0 \\ 0 & 0 & 0 \end{bmatrix} \tag{9.48}$$

Therefore after the transformation, machine equations become as shown (from equation (9.47))

$$\begin{bmatrix} V_d \\ V_q \\ V_o \\ V_f \\ 0 \end{bmatrix} = -\begin{bmatrix} R_a & & & & 0 \\ & R_b=R_a & & & \\ & & R_c=R_a & & \\ & & & R_f & \\ & & & & R_g \end{bmatrix}\begin{bmatrix} i_d \\ i_q \\ i_o \\ i_f \\ i_g \end{bmatrix}$$

$$-\begin{bmatrix} \frac{d\psi_d}{dt} \\ \frac{d\psi_q}{dt} \\ \frac{d\psi_o}{dt} \\ \frac{d\psi_f}{dt} \\ \frac{d\psi_g}{dt} \end{bmatrix} + \begin{bmatrix} -\omega\psi_q \\ +\omega\psi_d \\ 0 \\ 0 \\ 0 \end{bmatrix} \tag{9.49}$$

where $\omega = \frac{d\beta}{dt}$ is an angular velocity of rotation.

Now $[\psi] = [L][i]$ i.e.

$$\begin{bmatrix} \psi_a \\ \psi_b \\ \psi_c \\ \psi_f \\ \psi_g \end{bmatrix} = \begin{bmatrix} L_{aa} & L_{ab} & \cdots & L_{ag} \\ L_{ba} & L_{bb} & \cdots & L_{bg} \\ \cdot & & & \\ \cdot & & & \\ L_{ga} & L_{gb} & \cdots & L_{gg} \end{bmatrix}\begin{bmatrix} i_a \\ i_b \\ \cdot \\ \cdot \\ i_g \end{bmatrix} \tag{9.50}$$

Then premultiplying eqn. (9.50) by T^{-1} and substituting $T[I_{\text{park}}]$ for i we get,

$$[T^{-1}][\psi] = T^{-1}[L]T[I_{\text{park}}]$$

i.e.

$$[\psi_{\text{park}}] = T^{-1}[L]T[I_{\text{park}}]$$

$$\begin{bmatrix} \psi_d \\ \psi_q \\ \psi_o \\ \psi_f \\ \psi_g \end{bmatrix} = \begin{bmatrix} L_d & 0 & 0 & \sqrt{\frac{3}{2}}M_f & 0 \\ 0 & L_q & 0 & 0 & -\sqrt{\frac{3}{2}}M_g \\ 0 & 0 & L_0 & 0 & 0 \\ \sqrt{\frac{3}{2}}M_f & 0 & 0 & L_{ff} & 0 \\ 0 & -\sqrt{\frac{3}{2}}M_g & 0 & 0 & L_{gg} \end{bmatrix} \begin{bmatrix} i_d \\ i_q \\ i_o \\ i_f \\ i_g \end{bmatrix} \quad (9.51)$$

where direct axis synchronous inductance $L_d = L_s + M_s + \frac{3}{2}L_m$

quadrature axis synchronous inductance $L_q = L_s + M_s - \frac{3}{2}L_m$

$$\text{Zero seq. inductance} = L_0 = L_s - 2M_s \quad (9.52)$$

Hence the machine performance equations in Park's components (instantaneous values) are (from equations (9.49) and (9.51)),

$$V_d = -R_a i_d - \frac{d\psi_d}{dt} - \omega\psi_q$$

$$V_q = -R_a i_q - \frac{d\psi_q}{dt} + \omega\psi_d$$

$$V_o = -R_a i_o - \frac{d\psi_o}{dt}$$

$$V_f = -R_f i_f - \frac{d\psi_f}{dt}$$

$$0 = -R_g i_g - \frac{d\psi_g}{dt} \quad (9.53)$$

and

$$\psi_d = L_d i_d + \sqrt{\tfrac{3}{2}}M_f i_f$$

$$\psi_q = L_q i_q - \sqrt{\tfrac{3}{2}}M_g i_g$$

$$\psi_o = L_o i_o$$

$$\psi_f = \sqrt{\tfrac{3}{2}}M_f i_d + L_{ff} i_f$$

$$\psi_g = -\sqrt{\tfrac{3}{2}}M_g i_q + L_{gg} i_g \quad (9.54)$$

Note: For sinusoidal steady state conditions, the flux phasor leads the voltage phasor by 90°. This means that V_q in the quadrature axis will be induced by the flux in the direct axis, i.e. $\omega\psi_d$ produce V_q. Similarly the flux in the q-axis will induce a voltage along d axis i.e. $-\omega\psi_q$ as shown in Fig. 9.10.

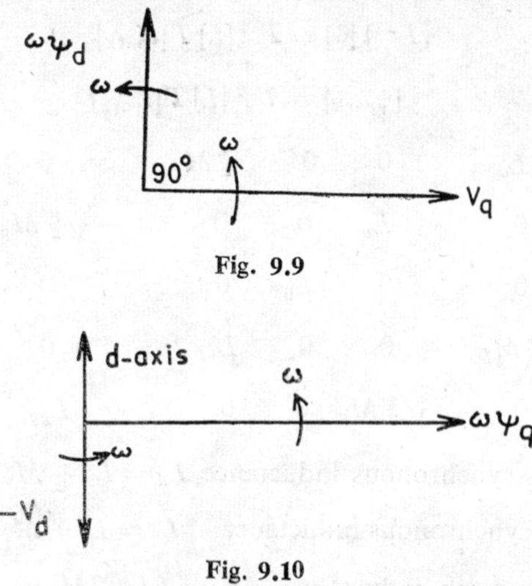

Fig. 9.9

Fig. 9.10

The flux linkages can be eliminated by inserting eqns. (9.54) into eqns. (9.53), which would give 5 equations in 9 unknowns viz. V_d, V_q, V_o, V_f, i_d, i_q, i_o, i_f and i_g. The missing four equations to solve for 9 unknowns are :

(i) One equation for the excitation voltage 'V_f'.

(ii) Three equations expressing the connections of the machine into the network. These will be discussed later.

Physical interpretation of equations (9.53) and (9.54): The terms $\omega\psi_d$ and $\omega\psi_q$ (which comes into existence because the transformation matrix 'T' is time dependent), are speed voltages (flux changes in space) and the terms $d\psi_d/dt$ and $d\psi_q/dt$ are the transformer voltages (flux changes in time). Usually transformer voltages are small compared with the speed voltages and may be neglected. It can be shown that neglecting transformer voltages corresponds to neglect of the harmonics and d.c. components in the transient solution for the stator voltages and currents. Neglecting of harmonics and d.c. components in the phase current is very common in machine analysis. The former is usually small and later dies away rapidly. Neither has a significant effect or influence on the average torque of the machine. However, if the transformer voltages are included, it should be realized that harmonics and d.c. terms are present in the electrical quantities, which make the solution of the network equations extremely difficult. It is preferable, therefore, to approximate the associated damping torques by additional terms in the swing equation.

Assumptions of balanced currents and voltages in the armature: Since the electro-magnetic transients in the network are much faster than the mechanical transients, the steady state phasor solutions on the network side is performed. Phasor diagram of synchronous machine (for over excited synchronous generator for lagging power factor) taking q-axis lagging d-axis by 90° will be as shown in Fig. 9.11.

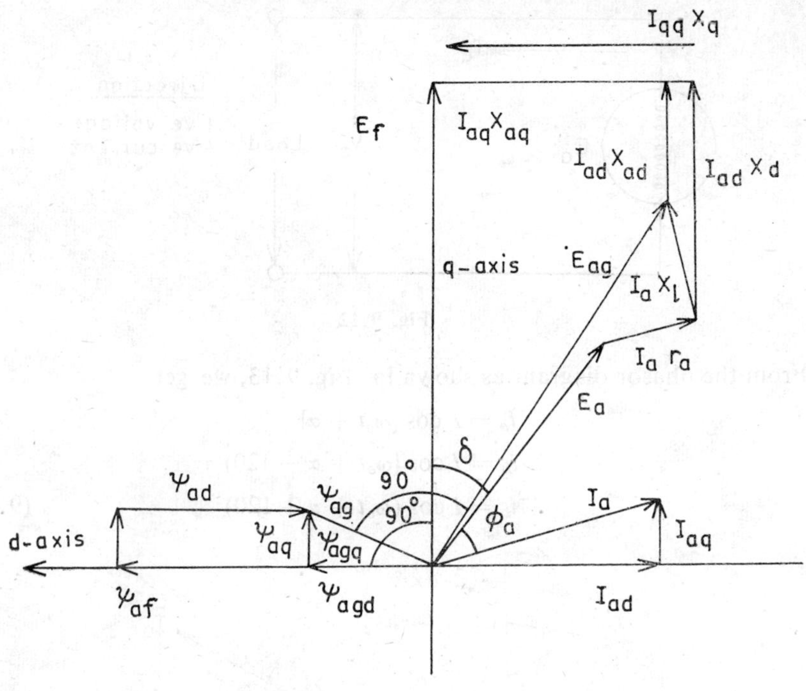

δ — torque or power angle

E_a — terminal voltage $= V_a$

ϕ_a — power factor angle

E_{ag} — voltage due to airgap flux i.e. voltage behind leakage reactance

E_{af} — voltage due to flux produced by main field current (i.e. open circuit voltage or excitation voltage)

$I_a r_a$ — voltage drop in the armature resistance

$I_a X_l$ — voltage drop in the leakage reactance

$I_{ad}X_{ad}$ — voltage drop across d-axis armature magnetizing reactance

$I_{aq}X_{aq}$ — voltage drop across q-axis armature magnetizing reactance

I_a — armature current

I_{ad} — direct axis component of armature current

I_{aq} — quadrature axis component of armature current

ψ_{ag} — flux linkages due to net air gap flux

ψ_{agd} — direct axis component of the flux linkages due to net air gap flux

ψ_{agq} — quadrature axis component of the flux linkages due to net air gap flux

ψ_{ad} — flux linkages due to direct axis component armature current

ψ_{aq} — flux linkages due to quadrature axis component armature current

ψ_{af} — flux linkages due to field winding current

X_d — d axis component of the sy. reactance $= X_l + X_{ad}$

X_q — q-axis component of the sy. reactance $= X_l + X_{aq}$

Fig. 9.11 Phasor diagram of salient pole synchronous generator.

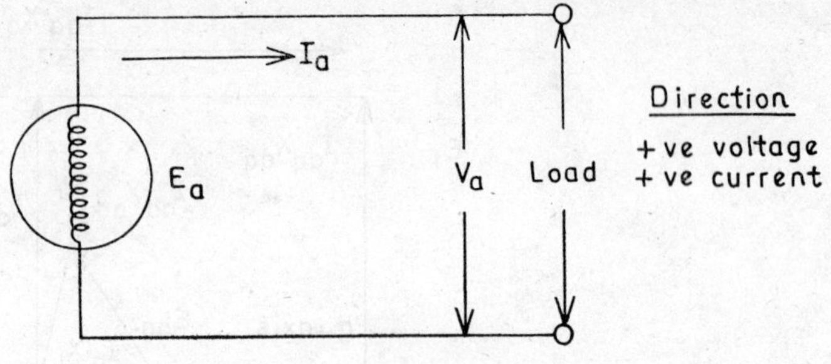

Fig. 9.12

From the phasor diagram as shown in Fig. 9.13, we get

$$i_a = I \cos [\omega_o t + \alpha]$$
$$i_b = I \cos [\omega_o t + \alpha - 120)$$
$$i_c = I \cos (\omega_o t + \alpha + 120) \qquad (9.55)$$

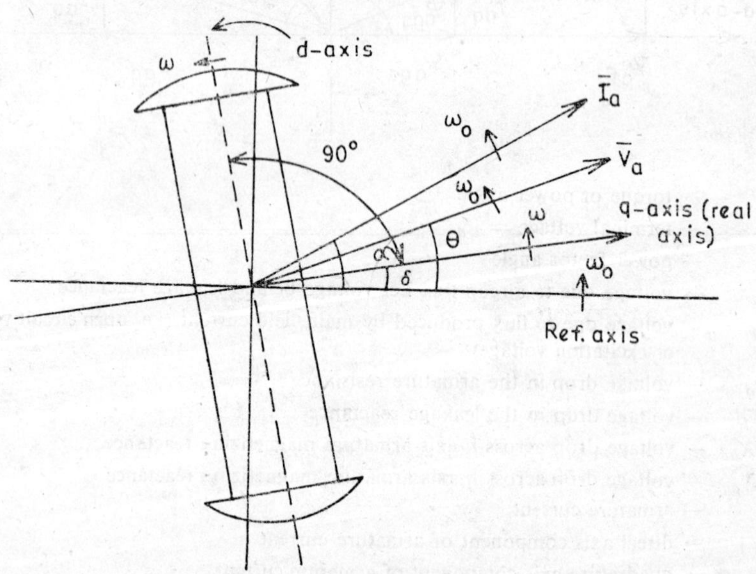

δ = Angle between sy. rotating ref. axis and q-axis
ω_0 = syn. speed
ω = Actual rotor speed

Fig. 9.13

Now by applying Park's transformation (i.e. $i_{\text{park}} = T^{-1}\, i_{\text{phase}}$), equation (9.55) becomes,

$$i_q = I \cos (\alpha - \delta) = I_q$$
$$i_d = I \sin (\alpha - \delta) = I_d$$
$$i_o = 0 = I_o \qquad (9.56)$$

As a phasor equation, equation (9.56) can be expressed as,

$$I_q + j\,I_d = I\,e^{j}\,(\alpha - \delta)$$

$$= \overline{I}\,e^{-j\delta} \quad \text{as} \quad \overline{I} = I\angle\alpha \tag{9.57}$$

Note : Since the voltage induced in the normal steady state operation lies in the quadrature axis, it is normal to adopt quadrature axis as the real axis and direct axis as the imaginary axis for the phasor representation.

Similarly, if the angle between phasor $\overline{V}_a$ (terminal voltage of phase '*a*') and reference axis is θ, we get (applying Park's transformation, i.e. $V_{\text{park}} = T^{-1}\,V_{ph}$),

$$V_q = V\cos(\theta - \delta) = V_q$$

$$V_d = V\sin(\theta - \delta) = V_d$$

$$V_o = 0$$

or with complex notation, we get

$$V_q + jV_d = V\,e^{j(\theta-\delta)}$$

$$= \overline{V}\,e^{-j\delta} \tag{9.59}$$

as $$\overline{V} = V\angle\theta$$

Assumptions:

(a) transformer voltage $d\psi_d/dt$ and $d\psi_q/dt$ being small, are neglected.

(b) balanced network currents and voltages are assumed.

The reasons of the above assumptions are that ψ_d and ψ_q changes slowly in time with the oscillations of angle δ and hence $d\psi_d/dt$ and $d\psi_q/dt$ are very small compared with $\omega\psi_d$ and $\omega\psi_q$. With these assumptions, we can rewrite the differential machine equations after dropping the terms for transformer voltages, zero sequence current, voltage and after substituting the terms of flux linkages ψ_d and ψ_q in terms of inductances. The variables i_g, i_f, ϕ_g, ψ_f and V_f are also replaced by new variables that are simple multiples of former (will be explained later) and equations (9.53) and (9.54) now becomes:

$$V_d = -R_a i_d - \omega\psi_q$$

$$V_q = -R_a i_q + \omega\psi_d$$

$$\psi_d = L_d i_d + \sqrt{\tfrac{3}{2}}M_f i_f$$

$$\psi_d = L_q i_q - \sqrt{\tfrac{3}{2}}M_g i_g \tag{9.60}$$

substituting flux linkages ψ_d and ψ_q in the equation of voltages V_d and V_q we get

$$V_q = -R_a i_q + \omega L_d i_d + \sqrt{\tfrac{3}{2}}\,\omega M_f i_f$$

$$V_d = -R_a i_d \quad \omega L_q i_q + \sqrt{\tfrac{3}{2}}\,\omega M_g i_g$$

$$\psi_f = \sqrt{\tfrac{3}{2}}M_f i_d + L_{ff} i_f$$

$$\psi_g = -\sqrt{\tfrac{3}{2}}M_g i_q + L_{gg} i_g$$

$$V_f = -R_f i_f - \frac{d\psi_f}{dt} \qquad (9.61)$$

Hence $\qquad d\psi_f/dt = -V_f - R_f i_f$

$$0 = -R_g i_g - \frac{d\psi_g}{dt} \qquad (9.62)$$

Hence $\qquad d\psi_f/dt = -R_g i_g$

$$I_q + j I_d = I e^{j(\alpha-\delta)}$$

and $\qquad V_q + j V_d = V e^{j(\theta-\delta)} \qquad (9.63)$

New variables here are I_f, I_g, ψ_f, ψ_g and V_f.

9.2.6 Phasor Diagram

Direct Axis: The direct axis is chosen to lead by 90° the phasor for the armature voltage generated by the flux produced by the field winding acting alone. The purpose of this choice is to establish the following reletions:

(i) The flux linkages produced by the field winding current have the same sence of direction as the positive direct axis. Positive field current produces positive direct axis magnetization.

(ii) The direct axis component of the armature current (I_{ad}) is in the negative direct axis. This is consistent with the physical concept that the magnetic effect of the armature tends to oppose that of the field current of the synchronous machine operating at a high leading or lagging power factor.

(iii) From a strictly mathematical point of view, the recommended reference positive direction for the direct axis is consistent with the use of a positive sign in the coefficient of Parks transformation to determine the direct axis component of current and voltage from the phasor quantities. Thus, the coefficient 'C_d' in the following transforming equations for a 3-phase machine is positive,

$$i_d = C_d \left(i_a \cos\beta + i_b \cos(\beta - 2\pi/3) + i_c \cos(\beta + 2\pi/3)\right) \qquad (9.64)$$

where 'β' is the angle measured in the electrical radians in the direction of synchronous angular velocity from the axis of phase 'A' to the axis of positive rotor flux, i.e. the direct axis. The value of C_d is $\sqrt{2/3}$.

Quadrature Axis: The positive direction of quadrature axis is taken to lag the direct axis by 90°. The reason for this choice is as follows:

(i) The direction of the voltage generated by the flux due to field current is in the positive quadrature axis.

(ii) In the mathematical analysis, the coefficient C_q in the following quadrature axis transformation is a positive number.

$$i_q = C_q \left(i_a \sin\beta + i_b \sin(\beta - 2\pi/3) + i_c \sin(\beta + 2\pi/3)\right) \qquad (9.65)$$

C_q here is equal to $\sqrt{2/3}$. Thus, both C_d and C_q are positive number if the choice of axes are made as above.

(iii) For a generator, the quadrature axis component of armature current 'i_{aq}' and the voltage due to field current E_{af} have positive values when referred to the reference direction of the positive axis.

(iv) It is common in system analysis to establish voltage as the phasor reference axis. Using this principle, the quadrature axis can be taken as real axis. The direct axis then becomes the imaginary axis, i.e. $E = E_q + jE_d$ and $I = I_q + jI_d$.

Direction or Sign of Flux Linkages: Flux linkage of a circuit or portion there is produced by the current in the same circuit element (self linkage) is considered to have the same sign as the current. Thus the self inductance of a circuit is always positive. Further the flux linkage in a circuit (1) due to current in another circuit (2) will be positive if the flux linking circuit (1) has the same sense as that produced by positive current in circuit (1). If the positive current in circuit (2) produces positive flux linkage in circuit (1) the mutual inductance between two circuits will be positive.

The scalar equation of the instantaneous voltage balance for the coil as shown in Fig. 9.14, can be written as,

$$e_1 = e_2 - iR$$

but $$e_2 = -\, d\psi/dt$$

Hence using source convention, we get

$$e_1 = -\frac{d\psi}{dt} - iR$$

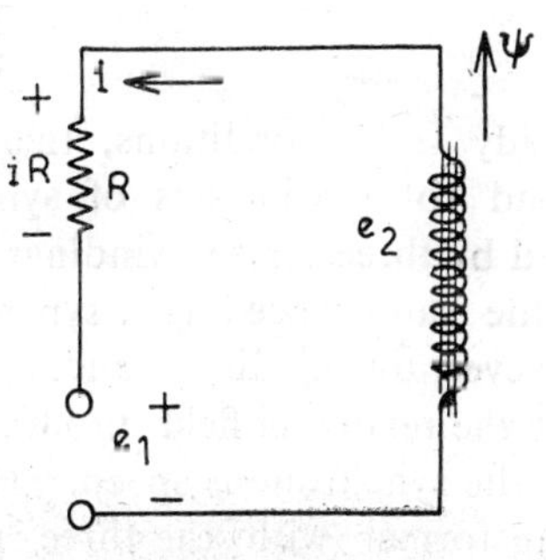

Fig. 9.14

9.2.7 Equivalent Circuit and Phasor Diagram

(i) *Nonsalient pole synchronous generator:* For this case $x_d = x_q$

$$E_{af} = E_a + i_a R_a + ji_a X_l + ji_a x_{ad}$$

$$= E_a + i_a R_a + ji_a X_d \qquad (9.66)$$

Thus, for this case, the machine is represented by a source (induced e.m.f.) E_{af} in series with the internal impedance $R_a + jX_d$.

(ii) *Salient pole synchronous generator:* Phasor diagram (Sub-transients effect).

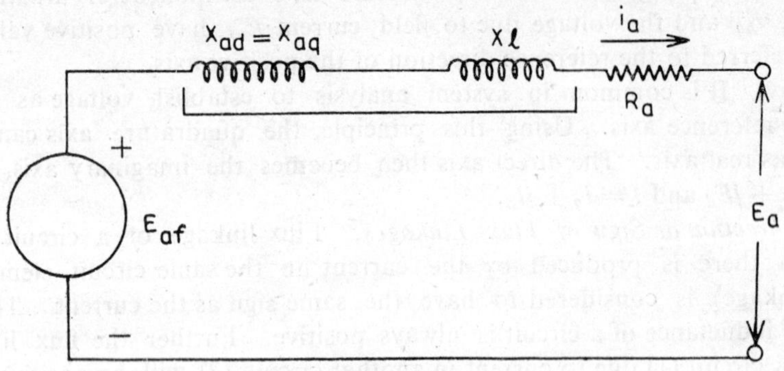

Fig. 9.15 Equivalent circuit of nonsalient pole synchronous generator.

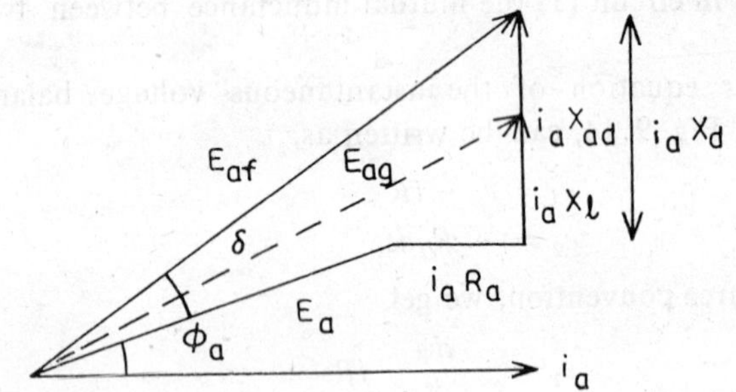

Fig. 9.16 Phasor diagram of nonsalient pole syn. generator.

9.2.8 *Reactances*

During normal steady state conditions, there is no transformer action between stator and rotor windings of synchronous machines as the resultant field produced by three phase windings on the stator and the rotor, both revolve with the same speed (i.e. synchronous speed) and in the same direction. However, during disturbances, the rotor speed is no longer the same as that of the revolving field produced by stator windings which always rotates with the synchronous speed. Hence the synchronous generator becomes a transformer with the three phase winding on the stator as the primary winding and the rotor windings as short circuited secondary winding of the transformer. The equivalent circuit of the synchronous generator excluding resistances will then be as shown in Fig. 9.18.

Now we can look these reactances, as armature leakage reactance (X_l) in series with armature magnetizing reactance (X_{ad}) which is in parallel with field reactance X_f and damper or fictitious 'g' winding reactance X_g. Initially all the reactances are in the circuit (just at the instant when the fault has occurred) and therefore initial or subtransient reactance are the lowest. However, after somtime, the g-winding (damping winding) is out of circuit as it has a very low time constant and hence we have only field

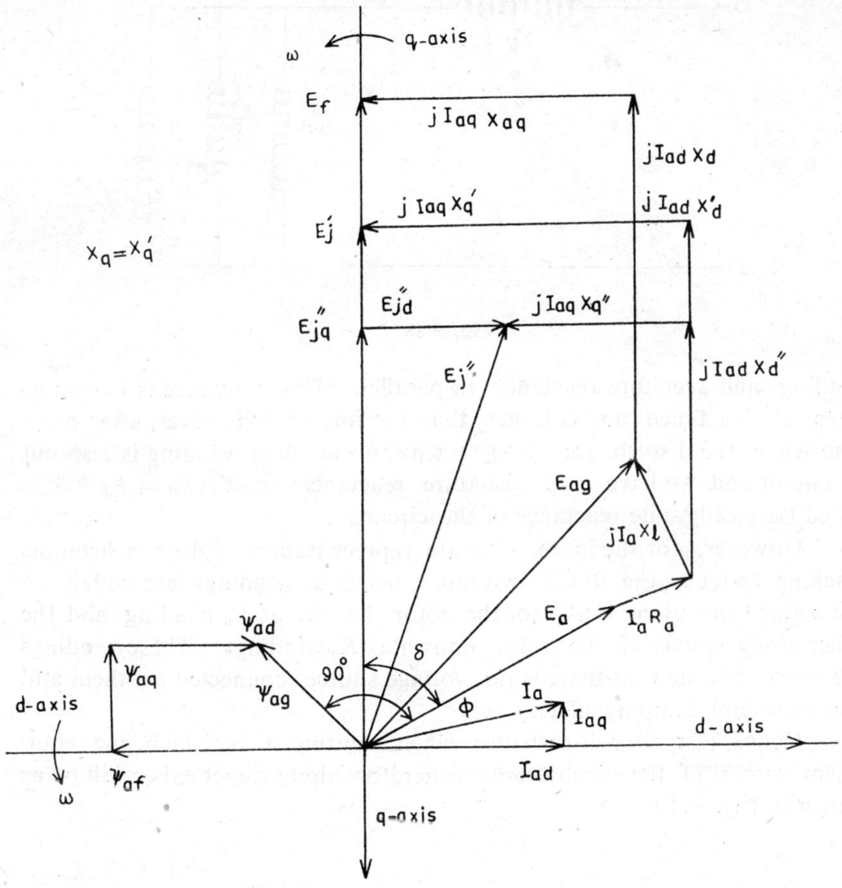

$I_{ad}X_{ad}$ = voltage across direct axis armature magnetizing reactance

$I_{aq}X_{aq}$ = voltage across q-axis armature magnetizing reactance

$I_{aq}X_q'$ = voltage across q-axis transient reactance

$I_{ad}X_d'$ = voltage across d-axis transient reactance

E_a = terminal voltage

E_{ag} = voltage due to air-gap flux

E_{af} = voltage due to main field (i.e. excitation voltage)

E_j' = voltage behind transient reactances

E_j'' = voltage behind sub-transient reactances

$I_{ad}X_d''$ = voltage across direct axis sub-transient reactance

$I_{aq}X_q''$ = voltage across quadrature axis sub-transient reactance

$I_{aq}X_q'$ = voltage across q-axis transient reactance

$$\left.\begin{aligned} X_d &= X_{ad} + X_l \\ X_q &= X_{aq} + X_l \end{aligned}\right\} \text{ and, } X_q' = X_q$$

Fig. 9.17 Phasor diagram for subtransient condition.

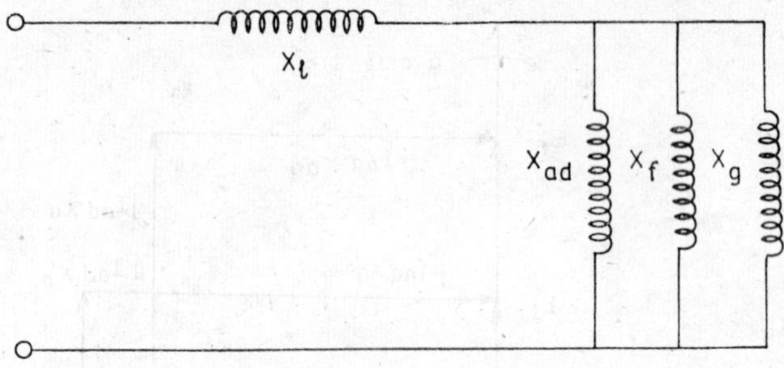

Fig. 9.18

winding and armature reactance in parallel. This reactance is known as transient reactance and is larger than the former. However, after some time when the disturbance altogether disappear, field winding is also out of circuit and we have only armature reactance i.e. X_d $(X_d = X_l + X_{ad})$ called the steady state reactance of the circuit.

However, for the more accurate repreeentation of the synchronous machine (refer to Fig. 9.17) two more fictitious windings are added on the rotor; one along d-axis of the rotor known as K_d winding and the other along q-axis of the rotor known as K_q windings. These windings are short circuited as there is no voltage source connected to them and they represent damping effect.

Hence, just after disturbance when hunting is very high, the equivalent circuit of the synchronous generator along direct axis, will be as shown in Fig. 9.19.

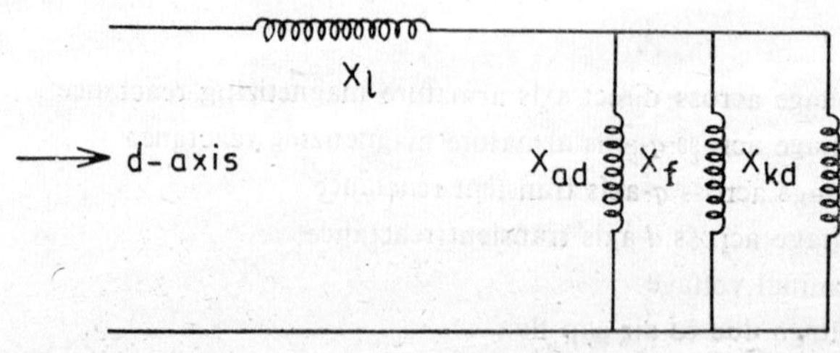

Fig. 9.19

The parallel combination of X_{ad}, X_f and X_{kd} is known as subtransient direct axis reactance X''_{ad}. Thus the subtransient direct axis synchronous reactance is

$$X''_d = X_l + X''_{ad}$$

This reactance is very small. After sometime, as the hunting becomes less, the winding K_d which has a very low time constant, is also out of circuit and thus the equivalent circuit becomes as shown in Fig. 9.20.

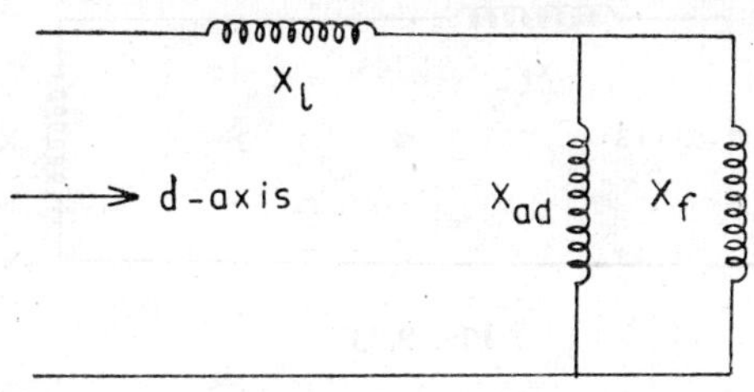

Fig. 9.20

The parallel combination of x_{ad} and x_f is known as X'_{ad} and hence $X'_d = X_l + X'_{ad}$ is called d-axis transient reactance. This will naturally be larger than X''_d. Finally, when the disturbance is altogether over, there will not be hunting of the rotor and hence there will not be any transformer action between the stator and rotor. At this time, we get the following circuit (Fig. 9.21).

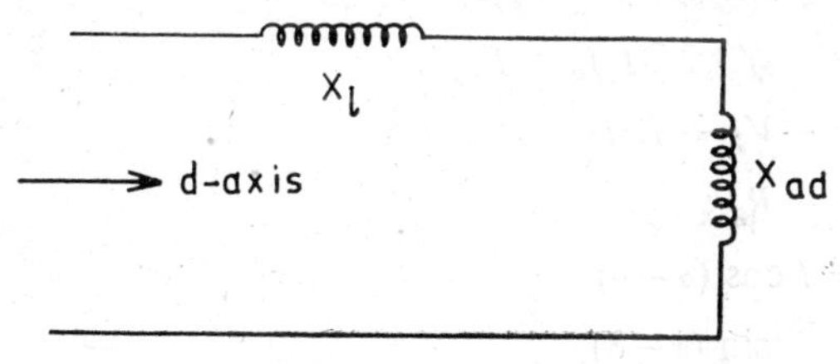

Fig. 9.21

Here, $X_d = X_l + X_{ad}$, is called direct axis synchronous reactance. It is very clear from the above that $X''_d < X'_d < X_d$.

Similarly along q-axis, the equivalent circuit, just after the disturbance will be as shown in Fig. 9.22.

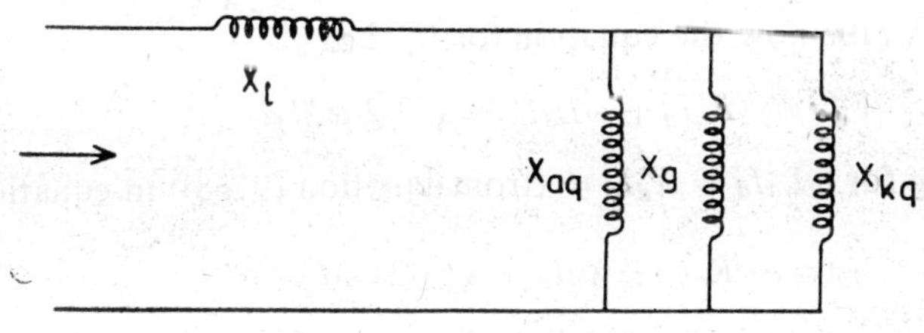

Fig. 9.22

Hence $X''_q = X_l$ plus parallel combination of X_{aq}, X_g and X_{kq}. This is a subtranslent q-axis reactance and is very small. After sometime, as hunting becomes less and less, both g-winding and K_q which have a very low time constant, will be out of circuit and hence the equivalent ciruit will be as shown in Fig. 9.23.

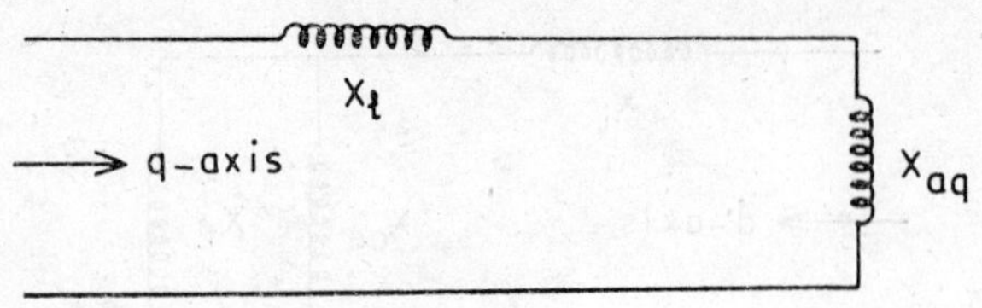

Fig. 9.23

And hence $X_q' = X_q = X_l + X_{aq}$ i.e. transient q-axis reactance is equal to q-axis synchronous reactance.

9.2.9 *Final Machine Dynamic Equations*

We have already shown that (refer to eqns.(9.61), (9.62) and (9.63)).

$$V_q = -R_a I_q + \omega L_d I_d + \sqrt{3/2}\ \omega M_f i_f$$
$$V_d = -R_a I_d - \omega L_q I_q + \sqrt{3/2}\ \omega M_g I_g$$
$$\psi_f = \sqrt{3/2}\ M_f I_d + L_{ff} I_f$$
$$\psi_g = -\sqrt{3/2}\ M_g I_q + L_{gg} I_g$$
$$d\psi_f/dt = -V_f - R_f I_f$$
$$d\psi_g/dt = -R_g I_g$$
$$i_q = I \cos(\alpha - \delta)$$
$$i_d = I \sin(\alpha - \delta)$$
$$V_q = V \cos(\theta - \delta)$$
$$V_d = V \sin(\theta - \delta) \qquad (9.67)$$

We define the following relationship,

$X_d = \omega L_d$ is a d-axis synchronous reactance

and

$X_q = \omega L_q$ is a quadrature axis synchronous reactance (9.68)

Let us take now the equation for V_q, i.e.

$$V_q = -R_a I_q + \omega L_d I_d + \sqrt{3/2}\ \omega M_f i_f \qquad (9.69)$$

Substituting for $\omega L_d I_d = X_d I_d =$ (from equation (9.68)) in equation (9.69), we get

$$V_q = -R_a I_q + X_d I_d + \sqrt{3/2}\ \omega M_f i_f \qquad (9.70)$$

Since $\psi_f = \sqrt{3/2}\ M_f I_d + L_{ff} I_f$ (from equation 9.67)), i.e.

$$I_f = \frac{1}{L_{ff}} \left(\psi_f - \sqrt{\frac{3}{2}}\ M_f I_d\right) \qquad (9.71)$$

Substituting this value of I_f from equation (9.71) in the expression for V_q in equation (9.70), we obtain,

$$V_q = -R_a I_q + X_d I_d + \sqrt{\tfrac{3}{2}}\ \omega M_f \left(1/L_{ff} \left(\psi_f - \sqrt{\tfrac{3}{2}}\ M_f I_d\right)\right)$$

$$= -R_a I_q + X_d I_d + \sqrt{\tfrac{3}{2}}\frac{\omega M_f \psi_f}{L_{ff}} - \tfrac{3}{2}\frac{\omega M_f^2 I_d}{L_{ff}} \tag{9.72}$$

Let us define $\sqrt{\tfrac{3}{2}}\dfrac{\omega M_f \psi_f}{L_{ff}} = E_q'$, as the subtransient voltage along q-axis since at this time both windings f and g are present along d-axis. And then substituting for E_q', in the expression for V_q (eqn. (9.72)), we get

$$V_q = -R_a I_q + X_d I_d + E_q' - \tfrac{3}{2}\frac{\omega M_f^2 I_d}{L_{ff}}$$

$$= -R_a I_q + \left(X_d - \tfrac{3}{2}\frac{\omega M_f^2}{L_{ff}}\right) I_d + E_q' \tag{9.73}$$

Let $X_d - \sqrt{\tfrac{3}{2}}\dfrac{\omega M_f^2}{L_{ff}} = X_d'$ is a transient d-axis reactance (note at this time, both d-axis component of the armature windings and f-winding are present).

Hence now, after substituting for X_d', we get,

$$V_q = -R_a I_q + X_d' I_d + E_q' \tag{9.74}$$

Similarly let us take the equation of V_d (from equation (9.67)), i.e.

$$V_d = -R_a I_d - \omega L_q I_q + \sqrt{\tfrac{3}{2}}\,\omega M_g I_g \tag{9.75}$$

But $\quad \omega L_q = X_q$ (from eqution (9.68))

Hence $\quad V_d = -R_a I_d - X_q I_q + \sqrt{\tfrac{3}{2}}\,\omega M_g I_g \quad$ (9.76)

But $\quad \psi_g = -\sqrt{\tfrac{3}{2}}\,M_g I_q + L_{gg} I_g$ (from equation (9.67)), i.e.

$$I_g = \frac{1}{L_{gg}}\left(\psi_g + \sqrt{\tfrac{3}{2}}\,M_g I_q\right) \tag{9.77}$$

Substituting equation (9.77) in equation (9.76), we get,

$$V_d = -R_a I_d - X_q I_q + \sqrt{\tfrac{3}{2}}\,\omega M_g\left(\frac{1}{L_{gg}}\left(\psi_g + \sqrt{\tfrac{3}{2}}\,M_g I_q\right)\right)$$

$$= -R_a I_d - X_q I_q + \sqrt{\frac{3}{2}}\cdot\frac{\omega M_g \psi_g}{L_{gg}} + \frac{3}{2}\frac{\omega M_g^2}{L_{gg}} I_q$$

$$= -R_a I_d - \left(X_q - \frac{3}{2}\frac{\omega M_g^2}{L_{gg}}\right) I_q + \sqrt{\tfrac{3}{2}}\frac{\omega M_g \psi_g}{L_{gg}} \tag{9.78}$$

Let us define,

$$\sqrt{\tfrac{3}{2}}\frac{\omega M_g \psi_g}{L_{gg}} = E_d' \tag{9.79}$$

as the transient voltage along d-axis as both q-axis component and g-windings are present to give transient state and also

$$X_q - \frac{3}{2}\frac{\omega M_g^2}{L_{gg}} = X_q' \tag{9.80}$$

i.e. q-axis transient reactance.

Substituting equations (9.79) and (9.80), in equation (9.78), we get,

$$V_d = -R_a I_d - X'_q I_q - E'_d \tag{9.81}$$

Thus from the above we have,

$$X'_d = X_d - \frac{3}{2}\frac{\omega M_f^2}{L_{ff}} \text{ as direct axis transient reactance}$$

$$X'_q = X_q - \frac{3}{2}\frac{\omega M_g^2}{L_{gg}} \text{ as } q\text{-axis transient reactance}$$

Let $T'_{do} = L_{ff}/R_f$ is a direct axis open circuit transient time constant

and $T_{qo} = L_{gg}/R_g$ is a quadrature axis open circuit transient time constant

and also,
$$E_d = \sqrt{\frac{3}{2}}\,\omega M_g I_g = \omega\psi_g$$

$$E_q = \sqrt{\frac{3}{2}}\,\omega M_f I_f = \omega\psi_f$$

i.e.
$$E = E_q + jE_d \tag{9.82}$$

After substituting eqn. (9.82) in equations (9.70) and (9.75) for V_q and V_d, we get,

$$V_q = -R_a I_q + \omega L_d I_d + \sqrt{\frac{3}{2}}\,\omega M_f I_f$$

$$= -R_a I_q + X_d I_d + E_q \tag{9.83}$$

and
$$V_d = -R_a I_d - \omega L_q I_q + \sqrt{\frac{3}{2}}\,\omega M_g I_g \tag{9.84}$$

$$= -R_a I_d - X_q I_q + E_d$$

But we have shown earlier (from eqns. (9.74) and (9.83)),

$$V_q = -R_a I_q + X'_d I_d + E'_q$$

Hence
$$-R_a I_q + X_d I_d + E_q = -R_a I_q + X'_d I_d + E'_q$$

i.e.
$$E'_q = E_q + (X_d - X'_d)\, I_d \tag{9.85}$$

Similarly, (from eqns. (9.81) and (9.84))

$$V_d = -R_a I_d - X'_q I_q + E'_d$$

$$= -R_a I_d - X_q I_q + E_d$$

i.e.
$$E'_d = -(X_q - X'_q)\, I_q + E_d \tag{9.86}$$

Let us take the phasor diagram (Fig. 9.24).

From the phasor diagram also, we get the same result, i.e.

$E'_q = E_q + I_d (X_d - X'_d)$ voltage behind d-axis transient reactance

$E'_d = -I_q (X_q - X'_q) + E_d$ voltage behind q-axis transient reactance

We get, from equation (9.86) after substituting eqns. (9.80), (9.82) and (9.67),

$$E'_d = -(X_q - X'_q) I_q + E_d$$

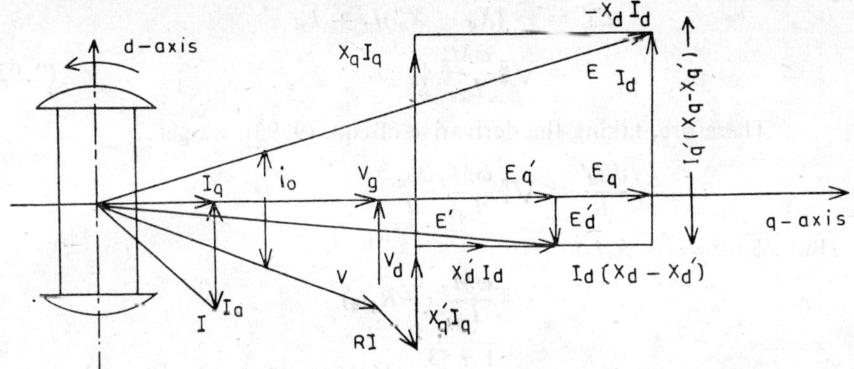

Fig. 9.24

$$= -\frac{3}{2}\frac{\omega M_g^2}{L_{gg}} I_q + \sqrt{\frac{3}{2}}\omega M_g I_g$$

$$= \sqrt{\frac{3}{2}}\frac{\omega M_g}{L_{gg}}\left(-\sqrt{\frac{3}{2}}M_g I_q + L_{gg} I_g\right)$$

$$= \sqrt{\frac{3}{2}}\frac{\omega M_g}{L_{gg}}\psi_g = E_d' \tag{9.87}$$

Similarly, after substituting eqns. (9.67) in eqn. (9.85), we get

$$E_q' = (X_d - X_d')\, I_d + E_q$$

$$= \frac{3}{2}\frac{\omega M_f^2}{L_{ff}} I_d + \sqrt{\frac{3}{2}}\,\omega M_f I_f$$

$$= \sqrt{\frac{3}{2}}\frac{\omega M_f}{L_{ff}}\left(\sqrt{\frac{3}{2}}M_f I_d + L_{ff} I_f\right)$$

$$= \sqrt{\frac{3}{2}}\frac{\omega M_f}{L_{ff}}\psi_f = E_q' \tag{9.88}$$

Hence we have from eqns. (9.87) and (9.88),

$$E_q' = \sqrt{\frac{3}{2}}\frac{\omega M_f}{L_{ff}}\psi_f \tag{9.89}$$

$$E_d' = \sqrt{\frac{3}{2}}\frac{\omega M_g}{L_{gg}}\psi_g \tag{9.90}$$

and

$$V_f = \sqrt{\frac{3}{2}}\frac{\omega M_f}{R_f} V_f \quad \text{open circuit voltage} \tag{9.91}$$

With these substitutions, we will get the following equations:

$$V_q = -R_a I_q + X_d I_d + E_q$$

$$V_d = -R_a I_d - X_q I_q + E_d$$

and also from the phasor diagram (and eqns. (9.85) and (9.86))

$$E_q' = (X_d - X_d') I_d + E_q = \sqrt{\frac{3}{2}}\frac{\omega M_f}{L_{ff}}\psi_f \tag{9.92}$$

$$E_d' = -(X_q - X_q')I_q + E_d$$

$$= \sqrt{\tfrac{3}{2}}\,\frac{\omega M_g}{L_{gg}}\,\psi_g \tag{9.93}$$

Therefore, taking the derivative of eqn. (9.93), we get,

$$\frac{dE_d'}{dt} = \sqrt{\tfrac{3}{2}}\,\frac{\omega M_g}{L_{gg}}\,\frac{d\psi_g}{dt}$$

(But $d\psi_g/dt = -R_g I_g$)

$$= \sqrt{\tfrac{3}{2}}\,\frac{\omega M_g}{L_{gg}}(-R_g I_g)$$

$$= -\frac{1}{T'_{qo}}\sqrt{\frac{3}{2}}\,\omega M_g I_g \text{ (as } T'_{qo} = L_{gg}/R_g \text{ and}$$

$$E_d = \sqrt{\tfrac{3}{2}}\,\omega M_g I_g)$$

$$= -\frac{1}{T'_{qo}}\,E_d \tag{9.94}$$

and from eqn. (9.92),

$$E_q' = (X_d - X_d')I_d + E_q$$

$$= \sqrt{\tfrac{3}{2}}\,\frac{\omega M_f}{L_{ff}}\,\psi_f$$

$$\frac{dE_q'}{dt} = \sqrt{\tfrac{3}{2}}\,\frac{\omega M_f}{L_{ff}}\,\frac{d\psi_f}{dt}$$

$$= \sqrt{\tfrac{3}{2}}\,\frac{\omega M_f}{L_{ff}}(-V_f - R_f I_f) \text{ as } d\psi_f/dt = -V_f - R_f I_f$$

$$= -\frac{1}{T'_{do}}\sqrt{\tfrac{3}{2}}\left(\frac{\omega M_f V_f}{R_f} + \omega M_f I_f\right) \text{ as } T'_{do} = L_{ff}/R_f$$

$$= -\frac{1}{T'_{do}}(V_f + E_q) \tag{9.95}$$

Thus we have,

$$dE'_q/dt = -1/T'_{do}(V_f + E_q)$$

and $$dE_d'/dt = -(1/T'_{qo})E_d$$

Therefore from earlier equations,

$$V_q = -R_a I_q + X_d I_d + E_q$$

$$= -R_a I_q + X_d I_d + E_q' - (X_d - X_d')I_d$$

$$= -R_a I_q + X_d' I_d + E_q' \tag{9.96}$$

and $$V_d = -R_a I_d - X_q I_q + E_d$$

$$= -R_a I_d - X_q I_q + E_d' + (X_q - X_q')I_q$$

$$= -R_a I_d - X_q' I_q + E'_d \tag{9.97}$$

$$V_q = -R_a I_q + X_d' I_d + E_q'$$

$$V_d = -R_a I_d - X_q' I_q + E_d'$$

i.e.
$$\begin{bmatrix} V_q \\ V_d \end{bmatrix} = \begin{bmatrix} -R_a & X_d' \\ -X_q' & -R_a \end{bmatrix} \begin{bmatrix} I_q \\ I_d \end{bmatrix} + \begin{bmatrix} E_q' \\ E_d' \end{bmatrix} \quad (9.98)$$

Thus we have the following 10 equations (8 algebraic and two differential) in 14 variables, viz. V_q, V_d, I_q, I_d, E_d, E_d', E_q, E_q', V_f, V, I, θ, α, δ):

$$V_q = -R_a I_q + X_d I_d + E_q$$

$$V_d = -R_a I_d - X_q I_q + E_d$$

$$E_q' = (X_d - X_d')I_d + E_q$$

$$E_d' = -(X_q - X_q')I_q + E_d$$

$$\frac{dE_q'}{dt} = -\frac{1}{T'_{do}}(V_f + E_q)$$

$$\frac{dE_d'}{dt} = -\frac{1}{T'_{qo}}E_d$$

$$V_q = V\cos(\theta - \delta)$$

$$V_d = V\sin(\theta - \delta)$$

$$I_q = I\cos(\alpha - \delta)$$

$$I_d = I\sin(\alpha - \delta) \quad (9.99)$$

The above equations (equations 9.99) can only be solved if 4 more equations are supplied. Those are:

(i) One for excitation system, i.e. V_f = constant.
(ii) 2 equations for the connection of the machine to the network
(iii) the 4th equation is the swing equation for the accelerating torque $f(\delta)$, expressing the connection of the machine to the primemover.

9.2.10 Extension of Machine Equations (inclusion of damper winding)

The 'f' and 'g' coils produce transient effect (i.e. X_d' and X_q') in the machine. The 'f' coil exist physically (field winding) whereas 'g' coil is hypothetical for representing the rotor eddy currents in the quadrature axis. However, it is quite difficult to calculate inductance for the 'g' coil.

A more accurate method of machine representation includes the subtransient effects of the damper windings (time constant shorter than those of transient effects). The damper windings can be approximated by two hypothetical coils both short circuited, namely the K_d—coil in the direct axis and K_g—coil in the quadrature axis.

The earlier machine equations can easily be modified to include the damper windings by substituting the scalar ('ψ_f' and 'ψ_f') by vectors,

$$\begin{bmatrix} \psi_f \\ \psi_{Kd} \end{bmatrix} \text{ and } \begin{bmatrix} \psi_g \\ \psi_{Kg} \end{bmatrix}$$

(same for currents and voltages).

Let us assume, that,

(a) Mutual inductances from stator coils 'a', 'b', 'c' (or its component 'd' and 'q' axes) to the 'Kd' coil is the same as 'f' coil, and to the K_g coil same as 'g' coil (fairly accurate).

(b) Mutual inductance between 'K_d' coil and 'f' coil (and K_g coil and 'g' coil) is the same as 'd'-component of stator coils and 'f'-coil (or 'q' component of stator to 'g' coil (less accurate)).

Then with these assumptions we obtain after Parks transformation, the following equations,

$$\begin{bmatrix} V_d \\ V_q \\ V_o \\ \begin{bmatrix} V_f \\ 0 \end{bmatrix} \\ \begin{bmatrix} 0 \\ 0 \end{bmatrix} \end{bmatrix} = - \begin{bmatrix} R_a & & & & & & 0 \\ & R_a & & & & & \\ & & R_a & & & & \\ & & & R_f & & & \\ & & & & R_{Kd} & & \\ & & & & & R_g & \\ & & & & & & R_{Kg} \end{bmatrix} \begin{bmatrix} i_d \\ i_q \\ i_o \\ i_f \\ i_{Kd} \\ i_g \\ i_{Kg} \end{bmatrix}$$

$$- \begin{bmatrix} d\psi_d/dt \\ d\psi_q/dt \\ d\psi_0/dt \\ d\psi_f/dt \\ d\psi_{Kd}/dt \\ d\psi_g/dt \\ d\psi_{Kg}/dt \end{bmatrix} + \begin{bmatrix} -\omega\psi\ q \\ \omega\psi\ d \\ 0 \\ 0 \\ \cdot \\ \cdot \\ 0 \end{bmatrix} \tag{9.100}$$

That is we get from the above:

$$V_d = -R_a I_d - \frac{d\psi_d}{dt} - \omega\psi_q$$

$$V_q = -R_a I_q - \frac{d\psi_q}{dt} + \omega\psi_d$$

$$V_o = -R_a i_0 - (d\psi_0/dt)$$

$$V_f = -R_f I_f - \frac{d\psi_f}{dt}$$

$$0 = -R_{Kd} I_{Kd} - (d\psi_{Kd}/dt$$

$$0 = -R_g I_g - \frac{d\psi_g}{dt}$$

$$0 = -R_{Kg} I_{Kg} - \frac{d\psi_{Kg}}{dt} \tag{9.101}$$

The earlier equations (eqns. 9.54) of flux linkages will also get modified, i.e.,

$$\begin{bmatrix} \psi_d \\ \psi_q \\ \psi_o \\ \begin{bmatrix} \psi_f \\ \psi_{Kd} \end{bmatrix} \\ \begin{bmatrix} \psi_g \\ \psi_{Kg} \end{bmatrix} \end{bmatrix} = \begin{bmatrix} L_d & 0 & 0 & [\sqrt{\tfrac{3}{2}}\,M_f & \sqrt{\tfrac{3}{2}}\,M_f] & 0 & 0 \\ 0 & L_q & 0 & 0 & 0 & [-\sqrt{\tfrac{3}{2}}\,M_g & -\sqrt{\tfrac{3}{2}}\,M_g] \\ 0 & 0 & L_0 & 0 & 0 & 0 & 0 \\ \begin{bmatrix} \sqrt{\tfrac{3}{2}}\,M_f \\ \sqrt{\tfrac{3}{2}}\,M_f \end{bmatrix} & \begin{matrix} 0 \\ 0 \end{matrix} & \begin{matrix} 0 \\ 0 \end{matrix} & \begin{bmatrix} L_{ff} & \sqrt{\tfrac{3}{3}}\,M_f \\ -\sqrt{\tfrac{3}{2}}\,M_f & L_{KdKd} \end{bmatrix} & & \begin{matrix} 0 \\ 0 \end{matrix} & \begin{matrix} 0 \\ 0 \end{matrix} \\ \begin{matrix} 0 \\ 0 \end{matrix} & \begin{bmatrix} -\sqrt{\tfrac{3}{2}}\,M_g \\ -\sqrt{\tfrac{3}{2}}\,M_g \end{bmatrix} & \begin{matrix} 0 \\ 0 \end{matrix} & \begin{matrix} 0 \\ 0 \end{matrix} & \begin{matrix} 0 \\ 0 \end{matrix} & \begin{bmatrix} L_{gg} & -\sqrt{\tfrac{3}{2}}\,M_g \\ -\sqrt{\tfrac{3}{2}}\,M_g & L_{KgKg} \end{bmatrix} & \end{bmatrix} \begin{bmatrix} i_d \\ i_q \\ i_o \\ \begin{bmatrix} i_f \\ i_{Kd} \end{bmatrix} \\ \begin{bmatrix} i_g \\ i_{Kg} \end{bmatrix} \end{bmatrix} \quad (9.102)$$

Thus from above, the flux linkages becomes :

$$\psi_d = L_d i_d + \sqrt{\tfrac{3}{2}}\, M_f I_f + \sqrt{\tfrac{3}{2}}\, M_f I_{Kd}$$

$$\psi_q = L_q i_q - \sqrt{\tfrac{3}{2}}\, M_g I_g - \sqrt{\tfrac{3}{2}}\, M_g I_{Kg}$$

$$\psi_o = L_o I_o$$

$$\psi_f = \sqrt{\tfrac{3}{2}} M_f I_d + L_{ff} I_f + \sqrt{\tfrac{3}{2}}\, M_f I_{Kd}$$

$$\psi_{Kd} = \sqrt{\tfrac{3}{2}}\, M_f I_d + \sqrt{\tfrac{3}{2}} M_f I_f + L_{KdKd} I_{Kd}$$

$$\psi_g = -\sqrt{\tfrac{3}{2}}\, M_g I_q + L_{gg} I_g - \sqrt{\tfrac{3}{2}} M_g I_{Kg}$$

$$\psi_{Kg} = -\sqrt{\tfrac{3}{2}} M_g I_q - \sqrt{\tfrac{3}{2}}\, M_g I_g + L_{KgKg} I_{Kg} \quad (9.103)$$

9.2.11 Swing Equation

The link between electrical and mechanical side of the synchronous machine is provided by the Dynamic Equation for the acceleration of the combined turbine i.e. prime mover and synchronous generator rotor, which is usually called swing equation.

Let T_S is a net torque supplied to the synchronous generator. T_S is actually the torque supplied by the prime mover minus torque due to no-load rotational losses in the synchronous machine which include frictional losses in the bearing, windage loss and coreloss. Let T_e is the electromagnetic torque developed in synchronous generator. Then any difference between T_S and T_e will cause acceleration or retardation depending upon whether T_S is greater or less than T_e. If T_S exceeds T_e, then there will be an accelerating torque T_a.

Then we have

$$T_S - T_e = T_a \quad (9.104)$$

Similarly if P_S is the net shaft power and P_e is the electromagnetic power developed in the synchronous generator, then the power causing acceleration will be,

$$P_S - P_e = P_a \quad (9.105)$$

Both values are positive for the generator action. However, for motor, both T_S and T_e (i.e. P_S and P_e) are negative indicating for acceleration, T_e should exceed T_S. If ω is the actual rotor speed, then we have,

$$P_a = T_a\, \omega = P_S - P_e \quad (9.106)$$

But $T_a = I\alpha$ where I is the moment of inertia and α is the angular acceleration in rad/sec^2. Hence we have

$$P_a = I\alpha\omega = I\omega\alpha \quad (9.107)$$

but $M = I\omega$, is the angular momentum in Joule-second/rad. Thus we have

$$P_a = M\alpha = P_S - P_e \quad (9.108)$$

If β is the space angle i.e. the angle by which rotor d-axis is ahead of the axis of a phase, then

$$\beta = \omega t$$

where ω is the actual rotor speed. δ is the angle between synchronously rotating reference frame and q-axis. Assuming ω_o to be synchronous speed, we have (see equation 9.9)

$$\beta = \omega_o t + \delta + \pi/2 \tag{9.109}$$

$$d\beta/dt = \omega_o + \frac{d\delta}{dt} \tag{9.110}$$

i.e. $$d\beta/dt - \omega_o = d\delta/dt \tag{9.111}$$

or $$\omega - \omega_o = d\delta/dt \tag{9.112}$$

and also

$$d^2\beta/dt^2 = d^2\delta/dt^2 = \frac{d\omega}{dt} = \alpha \tag{9.113}$$

Hence the equation,

$$P_a = M\alpha = P_S - P_e$$

$$= \frac{Md^2\beta}{dt^2} = P_S - P_e \tag{9.114}$$

i.e. $$Md^2\beta/dt^2 = Md^2\delta/dt^2 = P_S - P_e \tag{9.115}$$

Equation (9.115) is known as swing equation. In this equation, the angular momentum M ($= I\omega$) is not constant because ω changes during the disturbance. Thus the above swing equation is a second order differential equation with variable coefficient with the result its solution is very difficult. Equation (9.115) for a single synchronous generator connected to the infinite bus will be as given below:

$$Md^2\delta/dt^2 = P_S - P_e = P_S - \frac{E_g E_m}{X}\sin\delta$$

$$= P_S - P_{\max}\sin\delta \tag{9.116}$$

where $P_{\max} = \frac{E_g E_m}{X}$.

Here E_m is the internal voltage of the synchronous machine, E_g is the voltage of the infinite bus, X is the transient reactance of the machine plus the reactance of the connecting wire and δ is the torque angle. We obtain one such equation for each machine and therefore for multi-machine system, we need number of such swing equations. The solution of the equation will be the values of δ for different time t. The curve between δ and t is known as swing curve and if the curve indicates that δ after reaching its maximum value start decreasing or δ oscillates about equilibrium position, the synchronous machine has not lost stability.

Now angular momentum $M = I\omega = \gamma\omega$ where $I = Y$ is the moment of inertia in Kgram-meter² and ω is the angular velocity in rad/sec. Then we have the swing equation

$$Md^2\beta/dt^2 = P_S - P_e \tag{9.117}$$

Substituting $M = Y\omega = I\omega$, we get

$$Y\omega\, d^2\beta/dt^2 = P_S - P_e$$

$$= T_S\omega - T_e\omega_0 \qquad (9.118)$$

Dividing equation (9.118) by the rotar angular velocity ω, we get

$$Yd^2\beta/dt^2 = T_S - T_e\frac{\omega_0}{\omega} = T_S - T_e \qquad (9.119)$$

Since $\omega_0/\omega = 1$ as the rotor speed does not differ much from the synchronous speed ω_0.

In equation (9.119), the effect of damping which opposes the acceleration or retardation of the rotor has been neglected. The effect of damping can be included by including two terms, one constant term T_d and the other proportional to the change of speed. Thus by including damping, the above torque equations (equation 9.119) will be as follows:

$$\frac{Yd^2\beta}{dt^2} = T_S - T_e - T_d - \frac{K}{P/2}\left(\frac{d\beta}{dt} - \omega_0\right) \qquad (9.120$$

$$\frac{Yd^2\beta}{dt^2} + \frac{K}{P/2}\left(\frac{d\beta}{dt} - \omega_0\right) = T_S - T_e - T_d \qquad (9.121)$$

where T_d is the damping torque, K is an absolute synchronous damping constant $= MR^2/g$ and $P/2$ is pole pairs.

Now we have, from equation (9.109),

$$\beta = \omega_0 t + \delta + \pi/2 \qquad (9.122)$$

or $$d\beta/dt - \omega_0 = d\delta/dt \qquad (9.123)$$

or $$d^2\beta/dt = d^2\delta/dt^2 = \alpha \qquad (9.124)$$

With the substitution of equation (9.123) in the swing equation (torque eqn. (9.121)), we get

$$\frac{Yd^2\delta}{dt^2} + \frac{K}{P/2}\frac{d\delta}{dt} = T_S - T_e - T_d \qquad (9.125)$$

Inertia constant H is defined as the Mega joules of stored energy of a machine at synchronous speed per mega-volt ampere of the machine rating. Hence,

$$H = \text{Inertia constant}$$

$$= \frac{\text{Stored energy in Mega joules}}{\text{Machine rating in Mega volt-amps}}$$

Let G = Machine rating in Mega-volt-amps

Then stored energy in Mega-joules

$$GH = \tfrac{1}{2}I\omega_0^2 = \tfrac{1}{2}M\omega_0 = \tfrac{1}{2}y\omega_0^2 \qquad (9.126)$$

i.e. $$Y = I = 2GH/\omega_0^2 \qquad (9.127)$$

Substituting equation (9.127) in equation (9.125), we get

$$\frac{2GH}{\omega_0^2}\frac{d^2\delta}{dt^2}+\frac{K}{P/2}\frac{d\delta}{dt}=T_S-T_e-T_d \tag{9.128}$$

Multiplying equation (9.128) by the synchronous speed ω_0. we obtain,

$$\frac{2GH}{\omega_0}\frac{d^2\delta}{dt^2}+\frac{\omega_0 K}{P/2}\frac{d\delta}{dt}=T_S\omega_0-T_e\omega_0-T_d\omega_0$$

$$=\frac{P_S\omega_0}{\omega}-\frac{P_e\omega_0}{\omega_0}-\frac{P_d\omega_0}{\omega}$$

$$=P_S-P_e-P_d \tag{9.129}$$

Since $\omega/\omega_0=1$.

Dividing equation (9.129) by the machine rating G, we obtain the swing equation in p.u. power, i.e.

$$\frac{2H}{\omega_0}\frac{d^2\delta}{dt^2}+K_d\frac{d\delta}{dt}=P'_S-P'_e-P'_d \tag{9.130}$$

Here K_d is a damping torque coefficient and P'_S, P'_e and P'_d are per unit shaft power, electromagnetic power and damping power respectively.

Equation (9.130) is a 2nd order linear differential equation (i.e. swing equation) with constant coefficient because inertia constant H is a true constant and its value also remains nearly the same for a range of ratings.

Equation (9.130) is an additional equation for δ. However, in actual solution, two first order differential equations are written rather than one second order differential equation (i.e. eqn. (9.130)), i.e.

$$\frac{2H}{\omega_0}\frac{d\omega}{dt}+K_d(\omega-\omega_0)=P_{\text{Mech}}-P_{\text{elec}}-P_{\text{damp}}$$

$$=P'_S-P'_e-P'_d \tag{9.131}$$

and

$$\omega-\omega_0=\frac{d\delta}{dt} \tag{9.132}$$

9.3 Excitation System

The excitation systems of synchronous generators and motors have an extreme effect on system stability and when evaluated on the basis of increased power carrying capacity per increase in system cost, they are by far the most economical source of increased stability limits.

With this realization, there has been considerable effort extended in recent years to achieve the maximum benefits from the excitation systems resulting in more complex means for controlling the excitation of synchronous plant. Furthermore the size of modern steam generating sets have reached a level during this period where conventional D.C. rotating exciters could not provide the required excitation power and it became necessary to develop a new generation of excitation systems using an A.C. power source.

As a result, there is a considerable range of excitation system types as well as variations between manufacturers of a given type. There are readily available in the technical literature many publications of models of excitation systems with various degrees of sophistication of representation and we will consider here only two aspects of excitation systems.

(1) How accurate a representation is necessary for stability analysis,

(2) Which parameters are dominant and need be included in most studies.

With the wide range of excitation system in service and the profusion of people performing analysis of them, there has been considerable variation in definitions of basic quantity and in the nomenclature used. Responsible bodies through out the world have been working to unify this field and in recent years the IEEE (7, 8) has been quite active in this direction with two useful committee reports.

Figure 9.25 provides a summary of the basic elements of modern excitation systems. To represent an excitation system in digital computer dynamic stability analysis, it is necessary to derive transfer functions for each of the elements in the form of the first order differential equations, i.e. in the form

$$e_2 = \frac{Ke_1}{1+PT} \tag{9.133}$$

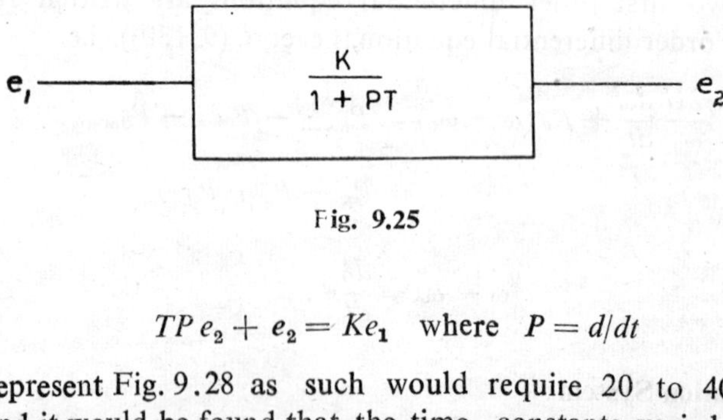

e_1 —— $\frac{K}{1 + PT}$ —— e_2

Fig. 9.25

Therefore,

$$TP\,e_2 + e_2 = Ke_1 \quad \text{where} \quad P = d/dt$$

To represent Fig. 9.28 as such would require 20 to 40 of these elements and it would be found that the time constants varied from 0.5 to 0.0005 seconds. Such a model would not be practical to multi-machine stability studies and some means of reducing it to managable proportion is required.

In practice, it is possibe to reduce the problem by eliminating some of the loops in the block diagram of the model, combining elements in series and discarding the smallest time constants in each group. In this manner the typical models given in Fig. 9.26 can be obtained as good approximations. These models are detailed enough for most system stability studies and result in quite acceptable accuracy. The largest errors with use of models of this simplicity usually result from having incorrect data.

Some people have explored the possibility of reducing such control systems by determining the eignvalues of the systems discarding those lightly damped and rebuilding a system with reduced number of eigenvalues but there seems little chance of this technique being practical at this stage.

It is important when reducing the excitation system if the elements of the result can readily be identified with actual physical components (or component groups) since the influence on system stability of individual components is important.

Figure 9.26 is an excellent example of this type of regular model. The first block (T_R) represents the voltage setting circuit, the second block (K_A) the automatic voltage regulator, the third (K_E) with attendent saturation function represents the self-excited main exciter and finally the the derivative stabilizing loop (K_F).

(i) In most instances, the time constant T_R is small enough to be neglected.

(ii) The automatic voltage regulator usually comprises several control loops and a simple reduction is necessary to the form $K_a/(1+sT_a)$. Voltage regulator gain (K_a) has an important effect on power system performance while the time constant T_a has much smaller influence owing to the larger T_e in series. It is important that input and output limits be included with this component, especially input limits when the system has additional special signal inputs such as frequency and its derivatives.

The output limits may be derived by dividing the exciter ceiling voltage (positive and negative) by exciter transfer function.

If nothing is known about the input limits, we try

$$\text{Input limit} = \frac{\text{Output limit} \times 1.25}{K_a \text{ main}}$$

(iii) The time constant of the exciter (T_e) is dominant time constant in the excitation system. If it is not possible to obtain data for the main exciter saturation function then a useful approximation is to increase (T_e) by 20 percent and decrease the exciter ceiling voltage by 20 percent.

(iv) The derivative stabilizing loop $SK_f/(1+ST_f)$ is extremely important and if omitted the exciter system and main generator will be unstable for most practical values of K_a. It can only be omitted when there are additional input signals to the excitation system such as frequency derivatives etc.

A useful value of K_f is 0.1 to 0.15 and T_f varies in the range of 0.5 to 2.0 seconds.

Stability studies where generators experience operation for some time at speeds greatly different from nominal synchronous speed, generally require modified excitation system parameters as they are usually steady state values and applicable only over a certain frequency (and terminal voltage) range, say ± 15 percent.

Because of high gain of an excitation system (100—400) errors in forward path gain K_a are more important than errors in most other parameters (including generator and network).

The excitation systems often contains other features such as voltage drop compensations to compensate for the voltage drop in some impedance between the generator and the rest of the network, which can be allowed for by adding vectorially to the terminal voltage a signal I_c

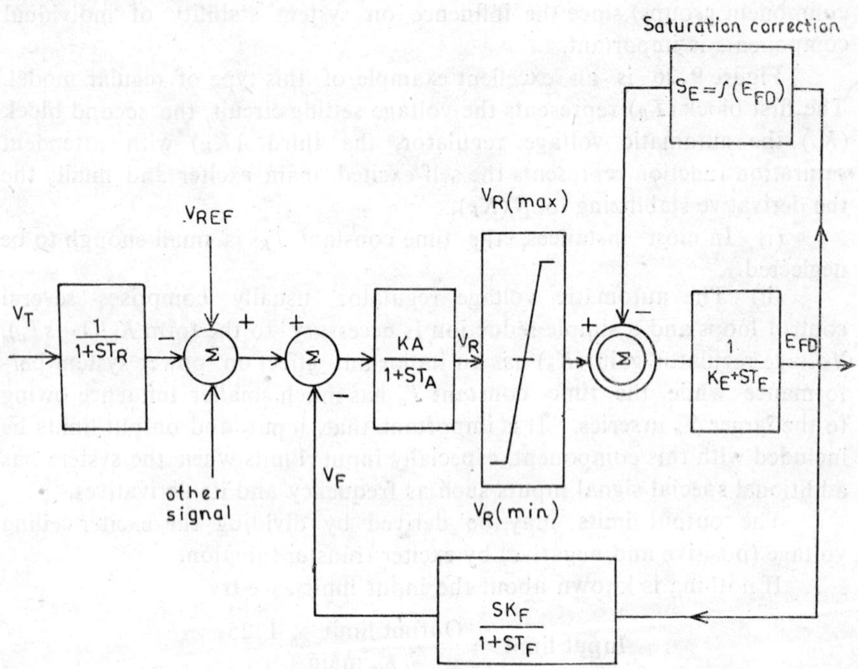

E_{FD} = Exciter output voltage (applied to generator field)

I_{FD} = Generator field current

I_T = Generator field terminal current

K_A = Regulator gain

K_E = Exciter constant related to self excited field

K_F = Regulator stabilizing circuit gain

S_E = Exciter saturation function

T_A = Regulator amplified time constant

T_E = Exciter time constant

T_F = Regulated stabilizing circuit time constant T_{F1} and T_{F2}

T_R = Regulated input filter time constant

V_R = Regulator output voltage

V_T = Terminal voltage of generator applied to the regulator input.

Fig. 9.26

$$I_c = (I_d + I_q)(r_c + jX_c)K_c \tag{9.134}$$

where I_d, I_q = components of stator currents

r_c = resistances compensated for

X_c = reactances compensated for

K_c = gain of compensation network

(i.e. percentage of impedence compensated for).

The resultant voltage is then used as an input to the standard excitation model instead of terminal voltage.

Remember, parameters given can be quite different from those actually of the excitation system because of change in electrical circuits with time and inaccurate calculation of control settings. In addition, they apply only for input signals with a certain period of oscillations i.e. considering a sinusoidal input to an excitation (when 60 cps A.C. terminal voltage is interpreted by the envelope and is D.C.) the parameters are quoted as being typical on a certain frequency range (0-20 cycles per minute).

The first transfer function in Fig. 9.26 is a simple time constant T_R representing regulator input filtering. For most system, T_R is very small and may be considered to be zero. The first summing point compares the regulator reference with the output of the input filter to determine the voltage error input to the regulator amplifier. Most computer program do not require V_{REF} but rather internally calculate the proper value by assuming V_T at $t = 0$. The second summing point combines the voltage error input with the excitation damping loop signal. The main regulator transfer function is represented as a gain K_A and a time constant T_A. Following this, the maximum and minimum limits of the regulator are imposed so that the large input error signals may not produce a negative output which exceeds the practical limit.

The next summing point substracts a signal which represents the saturation function, $S_E = f(E_{FD})$ of the exciter. That is, the exciter output voltage (or generator field voltage E_{FD}) is multiplied by a nonlinear saturation function and subtracted from the regulator output signal. The resultant is applied to the exciter transfer function $1/(K_E + ST_E)$.

Major loop damping is provided by the feedback transfer function.

$SK_F/(1 + ST_F)$, from the exciter output E_{FD} to the first summing point.

V_{REF} = regulator reference voltage setting

V_{RH} = field rheostat setting

V_T = generator terminal voltage

ΔV_T = generator terminal voltage error

Note : It should be emphasized that there is an inter-relation between exciter ceiling $E_{FD\,MAX}$, regulator ceiling $V_{R\,MAX}$, exciter saturation, S_E and K_E. The following expression must be satisfied under steady state conditions

$$V_R - (K_E + S_E)\, E_{FD} = 0 \tag{9.135}$$

$$E_{FD\ \mathrm{MIN}} \leqslant E_{FD} \leqslant E_{FD\ \mathrm{MAX}} \tag{9.136}$$

At the ceiling, or $E_{ED} \propto E_{FD\ \mathrm{MAX}}$

$$V_{R\ \mathrm{MAX}} - (K_E + S_{E\ \mathrm{MAX}})\, E_{FD\ \mathrm{MAX}} = 0 \tag{9.137}$$

K_E is always specified either as input data or program logic, to permit automatic calculation. In addition, for the remaining three constants $V_{R\ \mathrm{MAX}}$, $S_{E\ \mathrm{MAX}}$ and $E_{FD\ \mathrm{MAX}}$, the specification of any two establishes the third.

The exciter saturation function is defined as multiplier of the exciter output E_{FD} to represent the increase in exciter excitations requirement because of saturation.

9.3.1 *Other Input Signals*

In last several years, increase emphasis on the system design to timprove dynamic stability have resulted in the use of other regulator inpu signals in addition to the terminal voltage. These signals are chosen to provide positive damping of the power systems oscillations to improve generator stability and damping tie line oscillations.

Some of these signals are : accelerating power, speed, frequency and rate of change of terminal voltage. When used, they are added at the voltage reference summing point to the terminal voltage error as indicated in Fig. 9.26. Work is still going on this area.

Thus we conclude that the proposed excitation system (definition) or synchronous machine will be as shown in Fig. 9.27.

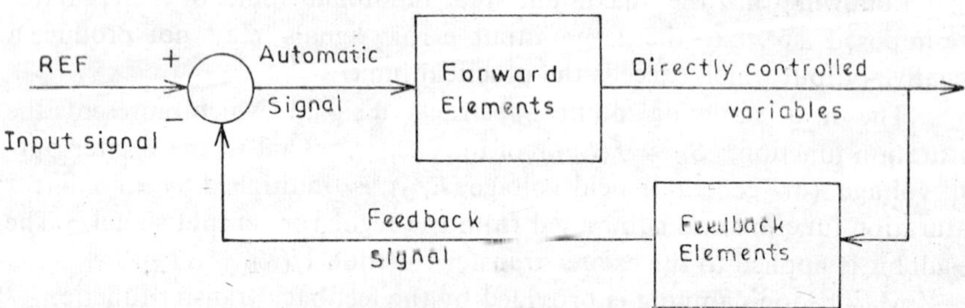

Fig. 9.27 Essential element of an automatic feed back control system.

Exciter System : Source of field current for the excitation of synchronous machine and includes exciter, regulator and manual control.

A synchronous machine excitation control system is a feedback control system as shown in Fig. 9.28.

9.4 The Prime Mover and Governing System

The effects on the system performance of prime movers and governing system is not usually apparant until 0.5 to 1.0 seconds after a disturbance. Consequently, where stability is being lost on the first

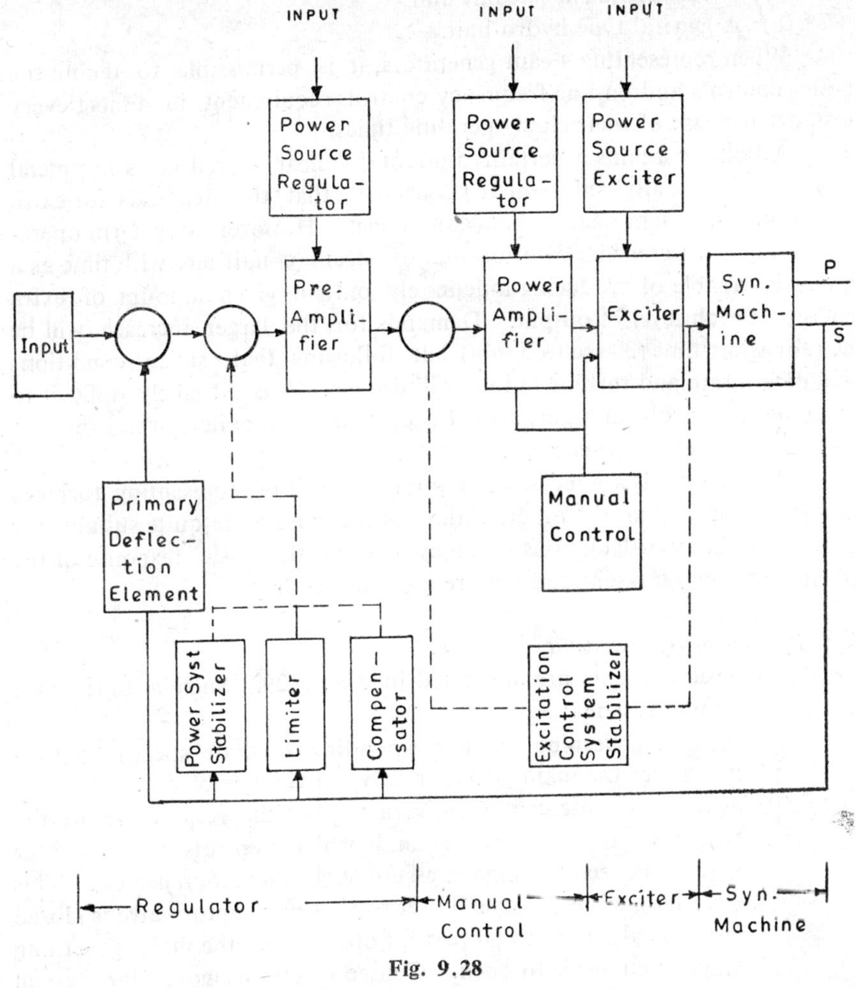

Fig. 9.28

swing (i.e. within 0.5 sec.) modelling can usually be shortened by omitting prime movers and governors.

However, when the system is marginally stable or when the stability is decided on the second and subsequent swings, then governer and prime movers play an important part.

It is regrettable, then, that the prime movers and governors are so difficult to represent and that the representations used are so inaccurate. They are distressingly non-linear, contains dead bands and delays, the parameters change with generator output and with age, and it is usually inconvenient to obtain the parameters for model from the generating units themselves so that much of the data used consists of typical values only.

There are numerous publication in the IEEE transactions of steam and hydrounit governor and turbine models. We shall consider only some aspects of two different generating unit, viz.,

(i) A reheat system (steam) unit

(ii) A general type hydro unit.

When representing steam generators, it is permissible to ignore the boiler controls and on line frequency controls equipment in almost every instance because of their slow operating times.

A boiler contains a certain amount of heat stored in its hot metal and this is usually sufficient to guarantee that the demands for extra steam during system disturbances can be met. However, long term operation study must consider deterioration of steam conditions with time as a boiler is capable of producing indefinitely only a given amount of extra energy at each level of output. Demands for the larger increase will be met for short time (30 sec to 5 min) but following that, steam conditions will deteriorate and turbine output will decline. It is extremely difficult to examine this problem rigorously at present because boiler-turbine models are not comprehensive enough.

Reference 9 is a good starting point to learn of representing turbines and governors. And the models, that is proposed, are quite suitable for general stability calculations as they exhibit the basic response of the plant that they represent and require a minimum data.

9.4.1 *Reheat System Unit*

The basic elements are indicated in Fig. 9.29. Such a unit has a several governing systems:

(i) Primary governing system (responding to shaft speed) controlling either the main governer valve or throttle blades.

(ii) Secondary (interceptor) governing system responding to the frequency of the turbines and which controls the interceptor valves between the high pressure stage and the reheater. This governing system is usually set so that interceptor valve is closed and it is about 25 to 50 percent open before the main governing valves commence to open. Consequently this governing system can usually be ignored.

(iii) Anticipatory governing system responding to the accelerating power of the unit. This system is usually set not to operate when either,

(a) the generator output is greater than a certain value (i.e. 25 percent of the maximum output), or

(b) the turbine mechanical power output is less than a certain value (i.e. 80 percent of the maximum capability).

Should both of these conditions be violated, then the governer is activated.

This governing system is activated only when the unit suffers loss of a large percentage of its load and on sensing this condition the emergency stop valves (located adjacent to the main governing valves) are closed rapidly to prevent dangerous overspeed. A reset time delay is included so that when electrical and mechanical powers revert to within settings, the emergency stop valves will open after a certain time.

This governing system is usually applied only on some modern large steam units.

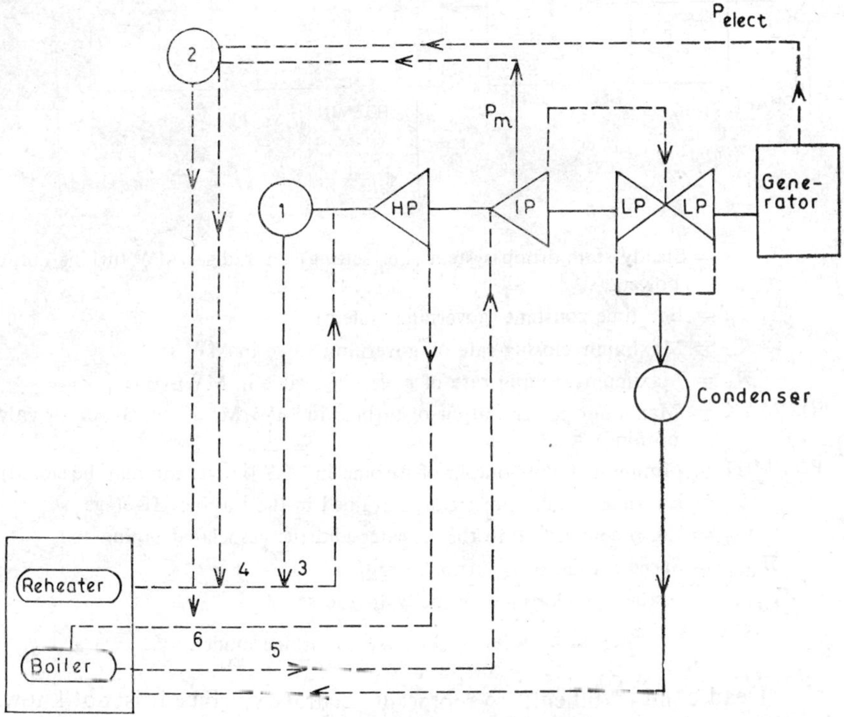

1—Primary governing system; 2—Anticipatory governor system; 3—Main governing valve on throttle blade; 4—Combined stop and emergency; 5—Interceptor governor valve; 6—Combined stop and emergency valve.

Fig. 9.29 Basic Component of Reheat System

(iv) Emergency overspeed trip. Should the shaft velocity exceeds a preset value then this governer will close the combined stop and emergency valves and shut the set down. Starting up is a lengthy process and usually this set would not figure any further in the stability calculations.

A model capable of representing such a set with sufficient accuracy for general stability studies is shown in Fig. 9.30. This is the reduction of the elements representing the system in Fig. 9.29.

Delays and deadbands are associated with:

(a) The speed sensing mechanism, friction and blacklash.
(b) Overlapping of oil ports in the servo-system as well as friction.
(c) Friction in the main governing valve.

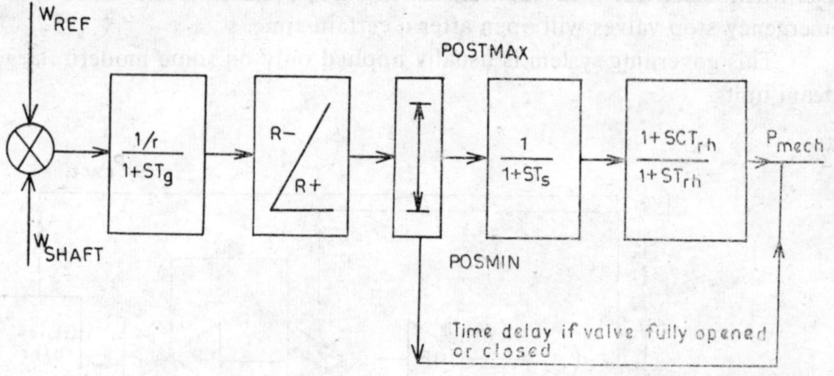

r = Steady state droop system (i.e. setting) in rad/sec/MW turbine output power

T_g = Eq. time constant (governing system)

R_- = Maximum closing rate of governing valve in MW/sec

R_+ = Maximum opening rate of governing valve in MW/sec

POS MAX = Maximum power output of turbine in MW (Maximum Governor valve opening)

POS MIN = Minimum power output of turbine in MW (Governor may be closed)

T_s = Eq. time constant of steam entrained in the turbine HP stage

T_{rh} = Eq. time constant in the reheater and the associated piping

W_{REF} = Speed setting of governor in rad/sec

W_{SHAFT} = Actual aogulor rotor velocity in rad/sec

Fig. 9.30 Simple reheat steam turbine model.

Dead band is difficult to represent accurately, since it is not known just where in the dead band the system is located, at the commencement of the study.

The representation of the governing system is the most inaccurate section of the model in Fig. 9.30, especially from neglecting the non-linearities and assuming constant steady state and transient droop.

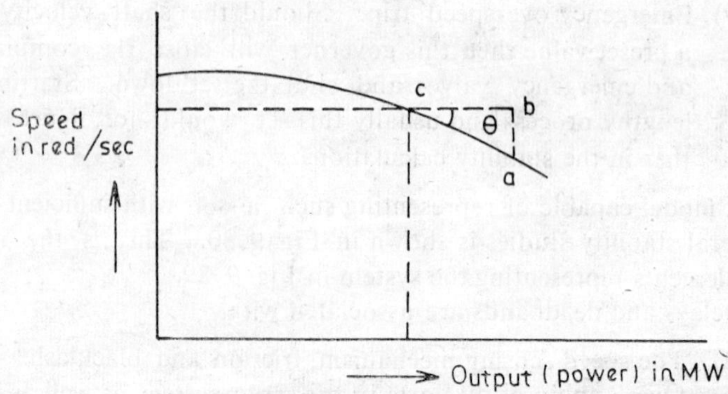

Steady state performance characteristics of a typical turbine (steam) generator.

Steady state droop is defined as follows: Let us take power-speed characteristics of a typical turbo-generator.

Steady state droop is the slope of the power versus speed curve of a turbo generator and hence it is equal to

$$\tan \theta = ab/bc$$

$$= \frac{\text{Speed in rad/sec}}{\text{Power in MW}} \qquad (9.138)$$

i.e. rad/sec/MW.

The turbine representation is a reduction of the following representation.

Let GP_1 = power developed in the turbine H.P. stage

Th = Time constant associated with entrained steam in the H.P. stage.

GP_2 = Power developed in the subsequent turbine stage.

T_r = Time constants associated with entrained steam in the reheater and connected pipe work.

$T1$ = Time constant associated with entrained steam in IP and L.P. stages of the turbine.

Then the expression for the total turbine shaft power P_t as a function of governer valve opening G is

$$P_t = \left(\frac{P_1}{(1+ST_h)} + \frac{P_2}{(1+ST_h)(1+ST_R)(1+ST_1)}\right) G$$

$$= \left(\frac{P_1\,(1 + S(T_r + T_1) + S^2 T_r T_1) + P_2}{(1+ST_h)(1+ST_r)(1+ST_1)}\right) G$$

But as reheater time constant is lower

$$P_t = \frac{(1+SCT_r)}{(1+ST_h)(1+ST_r)} (P_1 + P_2)\, G \qquad (9.139)$$

where C = Fraction of power developed in the HP stage.

In the p.u. system (i.e. 1 p.u. valve opening occurs at 1 p.u. turbine output power) expression (9.139) becomes,

$$P_t = \frac{(1+SCT_r)\, G}{(1+ST_h)(1+ST_r)} \qquad (9.140)$$

To represent non-reheat sets we simply replace T_r by T_1 in expression (9.140).

9.4.2 *Hydro-Unit*

The representation of a hydro unit is highly dependent on the type of prime mover (i.e. whether Pelton wheel, Francis or Kaplan turbine) as the method of controlling the speed of the unit is quite different for each of these.

There are three types of turbine with respect to head conditions such as:

(i) Low head—upto 100′ high, Sp. speed (90–180 rpm) and speed (100–400 rpm). These are propeller turbine (reaction turbine).

(ii) Medium head—50′ to 1000′, Sp. speed (90–200 rpm) and speed (100–400 rpm). These are Francis turbine (reaction turbine), and

(iii) High head: From 800′ and above; Sp. speed (7–3 rpm) and speed (120–720 rpm). These are impulse turbine (Pelton wheel) η Vs output of these turbines are shown in the following figure.

Note: Adjustable blade propeller turbine is known as Kaplan turbine.

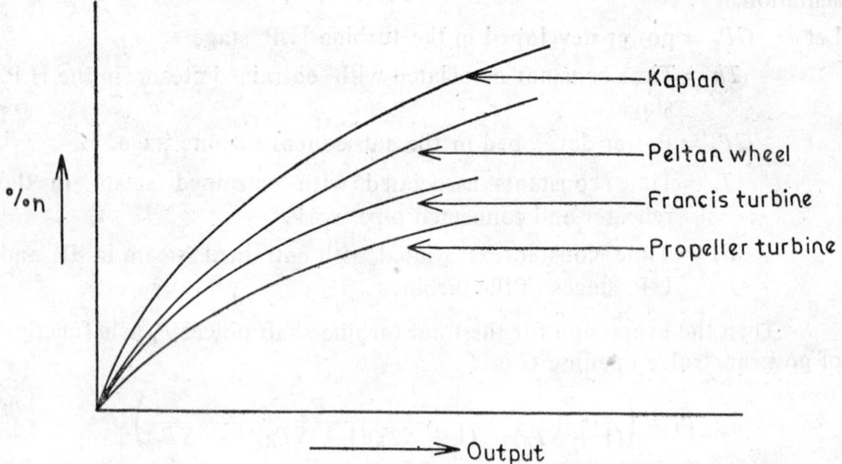

Pelton wheel turbine operates at high heads and long penstocks (i e. high pressure channel) so that the effect of inertia of water column is appreciable and must be represented. Shaft speed is decreased by the fast insertion of diffuser blades i.e. needles in the water jet followed by a slower constriction of the jet orifice. Shaft speed is increased by opening the jet orifice which is a very slow process compared to diffuser, i.e. needle insertion.

Francis turbines are controlled by varying the openings in a circular wicket gate around the turbine and although the opening rate is usually slower than closing rate, the difference is not as marked as for Pelton wheel. Penstocks may or may not require representation depending upon the head of the set. On Kaplan unit, governor also changes the blade angle of the propeller to correspond with the gate opening (position). On the impulse turbine, the governor controls the position of the needles and jet deflectors.

A general model suitable for representing hydro-set with first order of accuracy in most stability studies is shown in Fig. 9.31 which is based on that developed by Kirchmayer (see ref. 9). Here the assumptions are:

(i) Neglecting deadband, delays and nonlinear performance in the governing system.

(ii) Neglecting the variation in head of the set with daily (or seasonal) use.

(iii) Neglecting the charge in the output of turbine with gate opening i.e. dP/dG variation as G (i.e. gate opening) alters.

(iv) Assuming a constant equivalent water starting time constant.

The transient droop setting r and dash pot (i e. damping) recovery time constant (T) are quite important in most stability studies, as the study time ususally is too short for the effective droop to reduce to the steady state value.

Representation of the water column inertia is important as there is an initial tendency for the turbine torque to change in the opposite direction to that finally produced when there is a change in the wicket gate in the case of reaction turbine or office opening in the case of impulse turbine.

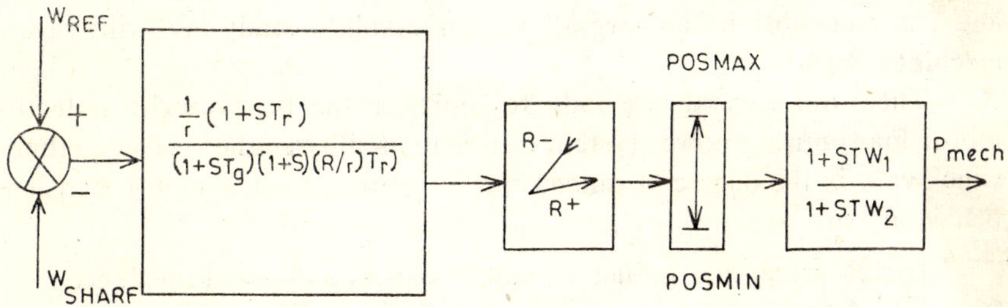

r = Steady state droop setting in rad/sec/MW turbine output power

R = Transient droop setting in rad/sec/MW turbine output power

T_r = Recovery time constant of the temperature droop dash pot

T_g = Eq. governor system time constant

R_- = Maximum closing rate of the governor valve in MW/sec

R_+ = Maximum opening rate of the governor valve in MW/sec

POS MAX = Maximum power output of turbine in MW (maximum governor valve opening)

POS MIN = Minimum -do- (usually governor valve is fully closed)

Fig. 9.31 Simple general hydro-turbine model.

9.5 Induction Machine Modelling

In most power system transient stability studies, the system loads are represented by shunt impedances. Commonly the shunt impedances are treated as static impedances. However, all loads do not behave as shunt impedances and, methods have been developed for taking into account more correctly the proper variation of real and reactive power with system voltage and frequency changes.

In the case of induction machines, it seems necessary to take into account the load inertia so that mechanical transient is properly considered. In some cases, these transients are of equal importance to those arising from generator swing. In the case of synchronous motor loads, it is evident that they can be represented in practically the same way as the generators. In the case of induction motor loads, several representations may be used depending upon the degree of refinement desired in the representation. One representation that can be made is to use the induction motor steady state equivalent circuit. This permit the calculation of electrical torque at each point in a step-by-step swing curve calculation, taking into account mechanical transients, but assuming that all electrical time constants are negligible. Because of the relative small effective rotor time constant of many induction motors, this representation is, in many cases, perfectly valid and reasonable. In some cases, however, where the induction motor rotor resistance is small or the reactance external to the motor is especially large, so that the rotor effective time constant is too large to neglect it has seemed desirable to include also the rotor electrical transients. Certain special studies have been made including stator electrical transients also. However, this has been shown to be unnecessary and impracticable in an overall system stability study involving several machines.

There are several methods to represent the behaviour of induction motor loads during power system transient stability studies. The methods, which vary in the degree of refinements in representing a motor transients behaviour, are

(i) Representing induction motor load as a simple impedance.
(ii) Approximate representation of an induction motor's steady state behaviour.
(iii) Representation of the induction motors behaviour including the motor's mechanical transients only.
(iv) Representation of an induction motors behaviour including both mechanical and rotor electrical transients.
(v) Representation of an induction motor's behaviour including the motor's mechanical, rotor electrical and stator electrical transients.

We will discuss these methods along with the applications.

9.5.1 Representation of the Induction Motor Load as a Simple Impedance

For the usual transient stability study, made by hand or upon the A.C. network analyzer, the system loads other than critical synchronous motors are commonly represented by shunt impedances. Various methods have been developed for taking into account of the variation in the real and reactive power of the loads with changes in the system voltages and frequency. The two most common representations are:

(a) Constant impedance (real and reactive power varies directly as the square of the voltage and independent of system frequency).
(b) Constant prefault current (real and reactive power varies directly as the voltage and is independent of system frequency).

The representation of composite or induction motor loads is most commonly made by assuming that the load impedance remains constant. This is the simplest representation.

9.5.2 *Approximate Representation of an Induction Motor's Steady State Behaviour*

When the system load at any point is primarily composed of induction motors, an approximate equivalent circuit of the induction motor as shown in Fig. 9.32 may be used. Prior to the system disturbance X_0, and R are varied to produce the desired motor terminal conditions. During the transient disturbance, the resistance R is varied in order to

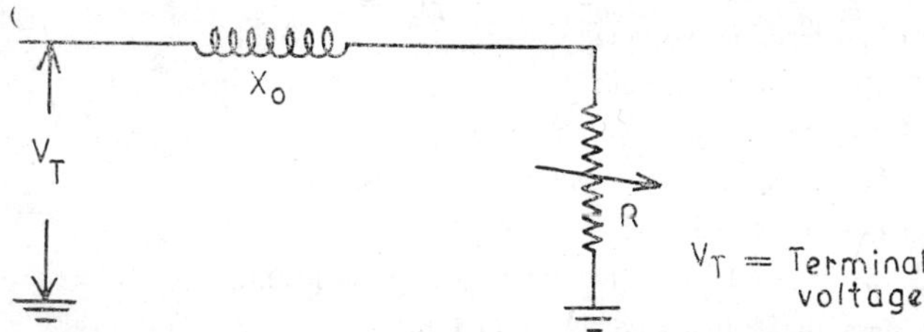

Fig. 9.32

maintain constant power. The reactance X_0 remains constant. If desired, the familiar induction motor steady state equivalent circuit, as shown in Fig. 9.33, may be used instead of the approximate equivalent circuit (neglecting core loss component), and the secondary resistance r_r/s is varied so as to produce constant pre-fault power.

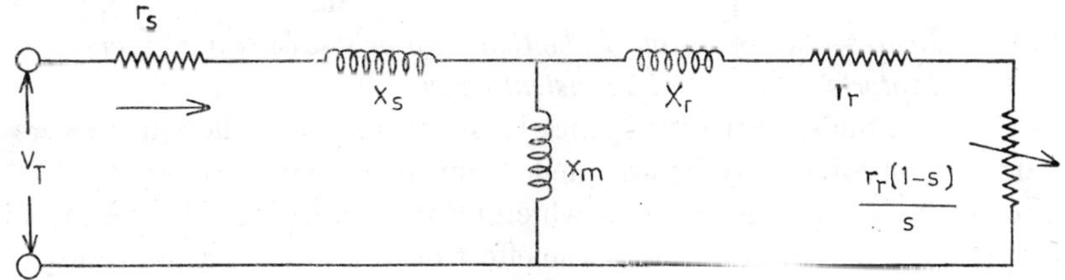

Fig. 9.33 Steady state equivalent circuit of induction motor (excluding core loss component).

Since $r_r + r_r\left(\frac{1-S}{S}\right) = r_r/s$, the above equivalent circuit becomes as shown in Fig. 9.34.

$$S \text{ is the rotor slip} = \frac{N_1 - N_2}{N_1} \tag{9.141}$$

where N_1 is the synchronous speed $= \frac{120f}{P}$ and N_2 is the actual rotor speed.

These representations are usually satisfactory when it is desired to improve the representation of the system loads but again where the load's

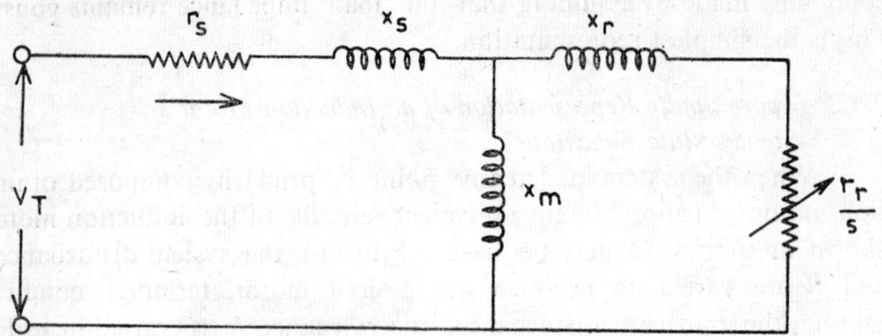

V_t—terminal voltage

I_t—terminal current

r_s—stator resistance/phase in p.u.

X_s—stator leakage reactance/phase in p.u.

r_r—rotor resistance per phase in p.u.

x_r—rotor leakage reactance per phase in p.u.

x_m—magnetizing reactance in p.u.

Fig. 9.34

exact behaviour is relatively unimportant in its effects on the overall system stability. However, where the concentration of induction motors are such that their transient behaviour has a significant effect upon system stability limits, or where it is necessary to evaluate the transient of motor themselves, then a more elaborate representation is required. These are discussed below.

9.5.3 *Representation of an Induction Motor's Behaviour Including the Motor's Mechanical Transient Only*

When the system disturbance is such that the induction motors are partially or completely interrupted from their power source by a short circuit, the major time constant which must be considered is the machines effective rotor time constant. The effect of machine's stator transients is to introduce a d.c. component of current in the armature phase current and pulsating electrical torques. The system study is impracticable except during a 3-phase terminal fault.

The value of effective rotor time constant of a motor is a function of external system parameters as well as of a motor constants. In many power systems, the effective rotor time constant is small. Where this is true, in analysing the transient behaviour due to short circuits, it is frequently possible to neglect both stator and rotor time constants. For this case, the behaviour of the induction motor is analysed using the familiar steady state equations with appropriate values of reactance, resistance and slip. The steady state equivalent circuit for representing the transient behaviour of an induction motor on the a.c. network analyser is given in Fig. 9.35 (core loss component has been neglected).

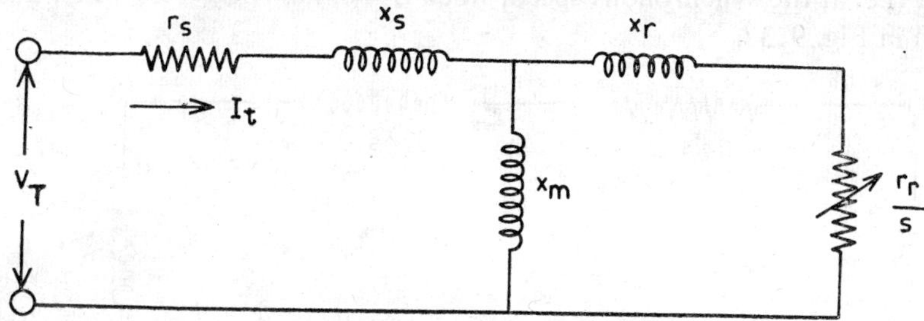

Fig. 9.35 Approximate equivalent circuit of induction motor.

9.5.4 Representation of an Induction Motor's Behaviour Including Both Mechanical and Rotor Electrical Transients

Where the induction motor is completely interrupted from a synchronous power source during a disturbance due to an open circuit condition, or where the motor and the system parameters are such that the effective rotor time constant should be considered, then the induction motor transient behaviour can be computed using transient equations. The effects of stator electrical transients on system response usually can be neglected. The equivalent circuit given with Fig. 9.36 is used to represent the transient behaviour of an induction motor including the effects of mechanical transients and rotor electrical transients with a single time constant.

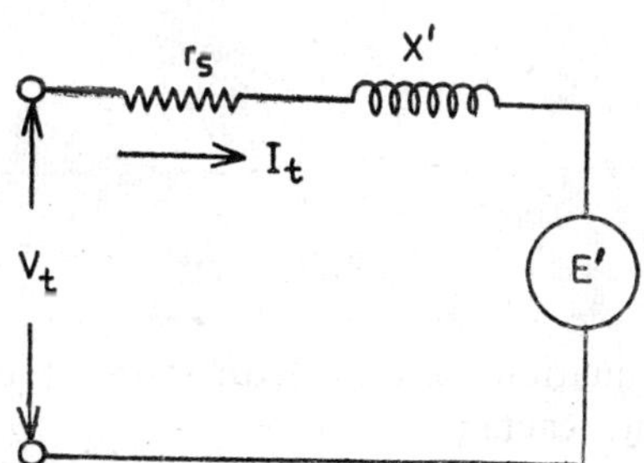

X'—transient reactance; E'—voltage behind transient reactance

Fig. 9.36 Simplified representation of induction motor for transient analysis.

Since the rotor resistance r_r is small compared to reactances, it can be neglected in calculating X and X'. From the steady state equivalent circuit (Fig. 9.35), the open circuit reactance X is,

$$X = X_s + X_m \tag{9.142}$$

(This corresponds to the synchronous reactance for synchronous machines.)

Since the steady state equivalent circuit (Fig. 9.35) during open circuit (i.e. at the synchronous speed, when $S=0$, i.e. $r_2/S=\infty$) will be as shown in Fig. 9.37.

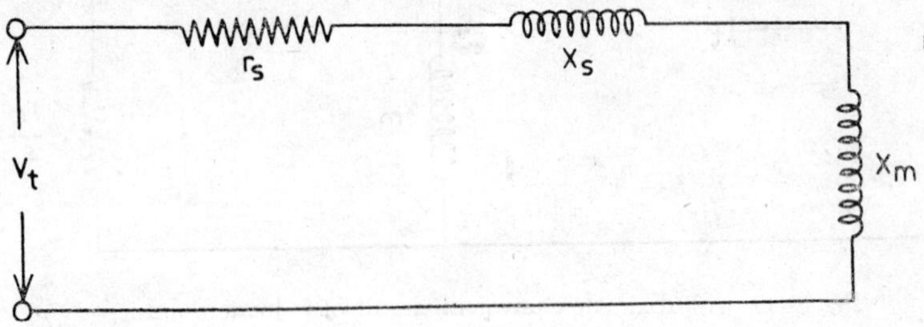

Fig. 9.37

Similarly the blocked rotor reactance X' = transient reactance (compare the transient reactance of synchronous machines)

$$= X_s + \frac{X_m X_r}{X_m + X_r} \tag{9.143}$$

as the equivalent circuit of the induction motor now becomes (as r_r is neglected) as shown in Fig. 9.38.

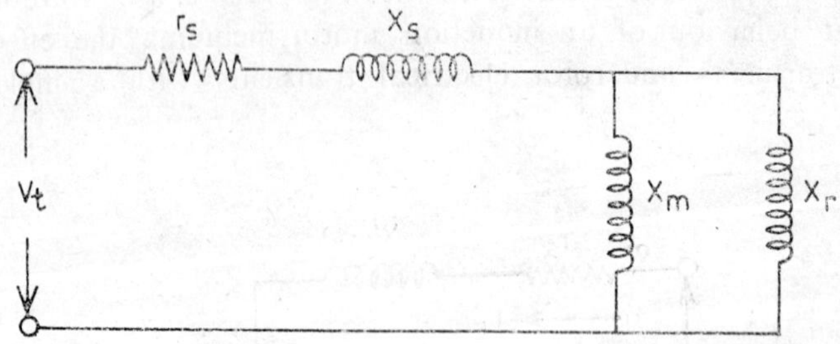

Fig. 9.38

The differential equation describing the rate of change of voltage (i.e. E') behind transient reactance X' is

$$dE'/dt = -j2\pi f SE' - \frac{1}{T^o}(E' - j(X - X')\, I_t) \tag{9.144}$$

where the rotor open circuit time constant T^o in seconds is

$$T^o = \frac{X_r + X_m}{2\pi f\, r_r} = \frac{L_r + L_m}{r_r} \tag{9.145}$$

and the terminal current is (from Fig. 9.36)

$$I_t = (V_t - E')\frac{1}{r_s + jX'} \tag{9.146}$$

where X and X' can be obtained from the steady state equivalent circuit of the induction motor as discussed earlier.

The differential equation (9.144) along with the equation of mechanical motion of the rotor and the equation for the motor electrical torque is solved in order to represent the behaviour of the induction motor where the effective rotor time constant is to be included in the analysis.

9.5.5 Representation of an Induction Motor's Behaviour Including Motor's Mechanical Transients, Rotor Electrical Transients and the Stator Electrical Transients

The effect of the stator transient is to produce d.c. offset in the stator currents and fluxes when sudden disturbances take place in the stator circuit. However, normally stator transients are neglected because due to this neither the rotor fluxes nor the average electrical torque are appreciably changed and the pulsating torque characteristics is also eliminated. This representation is quite impracticable.

References

1. Brereton, D.S., Lewis, D.G., Young, C.C., "Representation of induction motor loads during power system stability studies", *Transaction AIEE* Vol. 76, PAS, Part III, Aug. 1957, pp. 451-461.
2. Gabbard, J.L. Jr. and Rose, J.E., "Digital computation of induction motor transient stability", *Trans. AIEE*, Vol. 76, PAS, Part III, Dec. 1957, pp. 970–977.
3. Stagg, G.W. Elabiad, A.H., *Computer Methods in Power System Analysis*, McGraw-Hill, New York.
4. Fitzgerald, A.E., Kingsley, C., and Kusko, A., *Electric Machinery*, III edition, McGraw-Hill, New York.
5. Stevenson, W.D., *Elements of Power System Analysis*, McGraw-Hill, New York.
6. Kirchmayer, L.K., *Economic Operation of Power Systems*, Wiley, New York, 1958.
7. Proposed excitation system definitions for synchronous machines, *IEEE*, Vol. PAS 88, No. 8, Aug. 1969, pp. 1248–1258.
8. Computer representation of exciter systems, *IEEE*, Vol. PAS–87, June, 1968, p. 1460.
9. Kirchmayer, L.K., "*Economic Control of Interconnected Systems*", Wiley, N.Y. 1959.
10. Deo, N. "*Graph Theory with Applications to Engineering and Computer Science*, Prentice Hall, Inc. Englewood, Cliffs, N.J.
11. Bellman, R., *Dynamic Programming*, Princeton University Press, Princeton, N.J., 1957.
12. Recommended phasor diagram of synchronous machine, IEEE Committee Report, *IEEE PAS*, Vol. PAS–88, No. 11, Nov. 1969 pp 1593–1610.

Chapter 10

Stability Studies

10.1 Introductory Remarks

In Chapter 9, we discussed, in detail, the dynamic analysis and modelling of synchronous machines and also of induction machines. Such studies as well as mathematical models are very useful in stability studies which we are going to discuss in this chapter.

The problem of stability was not very important during early days when we had what is known as urban systems. But now, for the present day Grid system, which is the interconnections of different generating stations located at different places and complex nature of load systems, the stability studies have become very important. The basic purpose for conducting stability studies is to see whether the proposed or the existing system will remain in stability i.e. in equilibrium during or after any conceable disturbances. In other words, whether the different elements of the power system will remain in synchronism during or after such disturbances. The datas obtained from the stability studies are usually voltages, internal machine angle (i.e. load or torque angle), currents, powers, speeds and torques of the machines as well as voltages at the buses and power flows in the lines of the power system network during and immediately after the disturbances. The major disturbances which cause stability problems are the loss of generations, excitations, loss of transmission facilities, switching operations, momentary changes in loads and faults etc.

The datas obtained from stability studies help us to design adequate protection schemes of the power system network. We have discussed in this chapter the stability, its definition and classifications and the factors affecting stability. We have also discussed mathematical formulation and solution technique of stability problem for a large scale power system network involving multi-machine system.

10.2 Stability and Stability Limit

Power system network can be divided into three major subnetworks such as generation subnetwork, transmission subnetwork and distribution subnetwork. Generation subnetwork consists mostly of 3-phase synchronous generators, transmission subnetwork consists mostly of 3-phase transmission network and distribution subnetwork consists of distribution systems including loads. Loads can be either static, dynamic such as synchronous motors and induction motors etc. or composite loads.

During normal operations i.e. during steady state conditions, the different components of the power system remain in equilibrium with respect to each other. However, during disturbances which may be as simple as very gradual change in load or as large or complex as sudden change in loads, loss of generations, transmission facilities etc, the different components of the power system may not remain in equilibrium with respect to each other. In other words, the different components of the power system may not be in synchronism. The ability of the different elements of the power system to remain in synchronism or to remain in equilibrium is known as stability. The precise definitions of stability and stability limit as published by American Institution of Electrical Engineers are as follows:

Stability when used with reference to the power system, is that attribute of the system, or part of the system, which enables it to develop restoring forces between the elements there of, equal to or greater than the disturbing forces so as to restore a state of equillibrium between the element.

A stability limit is the maximum power flow through some particular point in the system or the part of the system to which stability limit refers to is operating with stability.

Stability and stability limit can be classified into different catagories according to the magnitude of disturbances i.e. whether it is a small disturbance such as gradual change in load or a large disturbance such as loss of generations, excitations, transmission facilities and switching operations and sudden change in loads etc. and also according to the duration of disturbances. Following are the classifications of stability according to the magnitude and nature of disturbances.

Steady state stability limit If the magnitude of disturbance is small such as very gradual change in load, the dynamics of rotating machines will not be effected and hence dynamical equations of rotating machines such as synchronous machines etc. will not appear in the mathematical formulation of the power system for the purpose for stability studies. This is the simplest case of stability and is referred to as steady state stability of the system.

Steady state stability limit is the maximum flow of power possible through any particular point of the power system without the loss of stability i.e. equilibrium operation when the load is increased very gradually.

However very gradual change in load is only a theoretical concept because in actual practice even the increase in load is gradual, the dynamics of rotating machines will be effected hence dynamical equations of rotating machines may appear in the mathematical formulation of the power system network for the purposes of stability studies. This gradual changes in loads etc. are known as small disturbances. Such small disturbances which also effect the dynamics of rotating machines i.e. where the dynamical equations of synchronous machines including the effects of regulator and automatic voltage regulator are included, is referred to as Dynamic stability.

Dynamic stability is actually an extension of steady state stability where we also include the effects of synchronous machines and automatic voltage regulator by their dynamical (i.e. differential) equations. The Dynamic stability is concerned with small disturbances lasting for a long time and therefore in Dynamic stability studies we always neglect non-linearities i.e. we assume the system to be linear.

Transient stability limit: If the magnitude of disturbances is very large, the dynamics of the rotating machines will be effected and hence the dynamical equations of the rotating machines including that of the automatic voltage regulator and exciter, prime mover and governer etc. will appear in the mathematical formulation for the purpose of stability studies. Examples of large disturbances as mentioned earlier are sudden change in loads, loss of generations, excitations, transmission facilities, and switching operations and faults etc. The effects of large disturbances of course, lasting for short durations, cause transient stability. Because of nature of disturbances, the non-linearities are always included in the transient stability studies. The transient stability limit is defind as follows:

Transient stability limit is the maximum flow of power possible through any particular point of the power system network without the loss of stability i.e. equillibrium operations, when sudden disturbance occurs.

10.3 Steady State Stability Limit

Since the steady state stability is caused by small disturbances such as very gradual change in load, the dynamics of rotating machines will not be involved and hence steady state stability limit can be determined by the following power system equation,

$$P = \frac{E_g E_m}{X} \sin \delta \tag{10.1}$$

Here P is the power transfer from the sending end to the receiving end, E_g and E_m are the voltages of sending end and receiving end of the power system network respectively, δ is the internal (i.e. power) angle and X is steady reactance of the connecting network.

For a two machine system (see Fig. 10.2), E_g may be the internal voltage of the synchronous generator, E_m the internal voltage of the synchronous motor, δ is the angle between the internal voltages of generator and motor, X is sum of the synchronous reactances of the synchronous generator and synchronous motor and also reactances of the connecting network and P is the power supplied by the synchronous generator to the synchronous motor.

Maximum power will be transferred when $\delta = 90°$, hence substituting $\delta = 90°$ in equation 10.1, we get

$$P_{max} = \frac{E_g E_m}{X} \tag{10.2}$$

By substituting equation (10.2) in (10.1), we get,

$$P = \frac{E_g \cdot E_m}{X} \sin \delta$$

$$= P_{maxm} \cdot \sin \delta \qquad (10.3)$$

Equation (10.3) represents the steady state stability limit of the power system. It is clear from equation (10.3) that the steady state stability limit can be increased by:

1. Increasing system voltages E_g or E_m.
2. Decreasing system reactance X.

In fact, increasing steady state stability limit, by decreasing reactance X is economical and also effective. System reactances can be decreased as follows:

1. Using duplicate lines. By using duplicate lines i.e. double circuit, reactances are automatically decreased. In addition, the duplicate circuit also improves the dependability as well as flexibility of the system.
2. By using series capacitor. By using series capacitor, automatically line reactances are decreased. Moreover, there are other advantages of using series capacitors such as to improve the voltage regulation of the system and also to improve the system power factor.

10.3.1 Transient State Stability Limit

Before we discuss the transient stability problem of a large scale power system network, we will consider initially the simplified model of the power system network such as a single synchronous machine connected to infinite bus or a two machine system where one machine can be a synchronous generator while the other synchronous motor for the purpose of transient stability studies. In such situations, it is not necessary to solve the swing equation and to plot and inspect swing curve in order to determine whether the internal machine angle (i.e. torque angle δ) is increasing infinitely or oscillating about equilibrium position. The principle by which such phenomena can be accessed, in other words transient state stability limit is determined without solving the swing equation is called *Equal area criterion for stability.* This phenomena, though applicable only to the simplified power system network as discussed earlier, helps us to understand how certain physical factors affect the transient state stability limit and thus provides conceptual understanding about the transient stability problem.

Equal Area Criterion for Stability: In order to understand the principal of equal area criterion for stability, let us take the swing equation (9.115),

$$P_s - P_e = M \frac{d^2\delta}{dt^2}$$

i.e.

$$\frac{d^2\delta}{dt^2} = \frac{1}{M}\left[P_s - P_e\right] \tag{10.4}$$

Let

$$\frac{d\delta}{dt} = x\ ;\quad \left[\frac{d\delta}{dt}\right]^2 = x^2 \tag{10.5}$$

Now

$$\frac{d}{dt}x^2 = 2x\frac{dx}{dt} \tag{10.6}$$

Substituting equation (10.5) into (10.6), we get

$$\frac{d}{dt}\left[\frac{d\delta}{dt}\right]^2 = 2\frac{d\delta}{dt}\frac{d^2\delta}{dt^2} \tag{10.7}$$

Substituting $\frac{d^2\delta}{dt^2}$ from equation (10.4) into equation (10.7) and rearranging we get

$$\frac{d}{dt}\left[\frac{d\delta}{dt}\right]^2 = 2\frac{d\delta}{dt}\frac{P_s - P_e}{M} \tag{10.8}$$

Cancelling dt from both sides of equation (10.8) and integrating, we get

$$\left[\frac{d\delta}{dt}\right]^2 = \int_{\delta_o}^{\delta}\frac{2[P_s - P_e]}{M}\,d\delta \tag{10.9}$$

i.e.

$$\frac{d\delta}{dt} = \sqrt{\int_{\delta_o}^{\delta}\frac{2\,(P_s - P_e)}{M}\,d\delta} \tag{10.10}$$

Initially the system is stable i.e. the synchronous machine rotates with the synchronous speed [$\omega = \omega_o$] and $d\delta/dt = 0$. During the transient disturbances, $\omega \neq \omega_o$ and $d\delta/dt \neq 0$. However if after the disturbance, $d\delta/dt = 0$, i.e. the synchronous machine again rotates with the synchronous speed [$\omega = \omega_o$], it is employed that the stability limit has not been exceeded and hence the system is stable.

From the above we conclude that the system is stable after the transient disturbance if

$$\frac{d\delta}{dt} = 0$$

Therefore, from equation (10.10), we have for the stability criterion,

$$\frac{d\delta}{dt} = \int_{\delta_o}^{\delta}\frac{2\,(P_s - P_e)}{M}\,d\delta = 0$$

i.e.

$$\int_{\delta_o}^{\delta} [P_s - P_e]\, d\delta = 0 \tag{10.11}$$

Since angular momentum M and 2 cannot be zero.

Let us take an example of a synchronous motor connected to the infinite bus. The motor is operating with torque angle δ_o and the corresponding power out put is P_o. The power input to the motor at the torque angle δ_o is

$$\frac{|E_g|\,|E_m|}{X} \sin \delta_o = P_{max} \sin \delta_o$$

where E_g is the voltage of infinite bus, E_m is the internal voltage of the synchronous motor and X is the sum of transient reactance of the synchronous motor as well as the reactance of connecting line between the motor and the infinite bus.

Now suppose power output of the motor is suddenly increased to P_s from its initial value of P_o. Since power input is the same as original, the motor slows down with the result the torque angle δ_o increases and part of the energy stored as a kinetic energy in the rotating part is released and delivered as output so as to fill up the gap between input and output.

As shown in Fig. 10.1, with the increase in δ, the power input increases and finally when $\delta = \delta_s$, the power input matches with the new power output P_s. However, due to inertia of the rotating part, rotor may go past the equilibrium angle δ_s and will oscillate between δ_o and δ_m and eventually will rotate with angle δ_s, if transient stability limit is not exceeded.

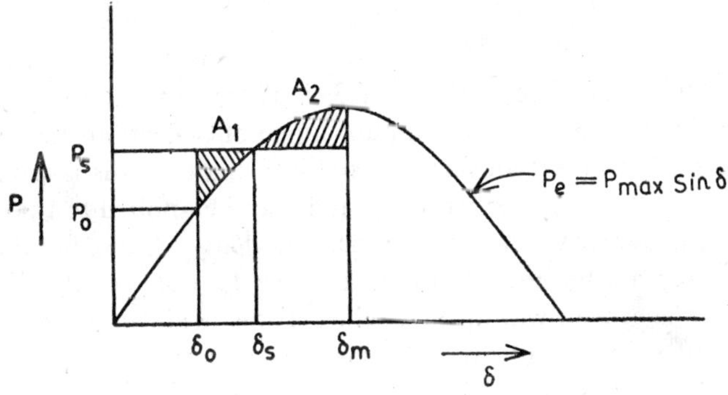

Fig. 10.1 (a)

From Fig. 10.1a, the shaded area

$$A_1 = \int_{\delta_o}^{\delta_s} [P_S - P_e]\, d\delta$$

and area

$$A_2 = \int_{\delta s}^{\delta m} (P_e - P_s)\, d\delta$$

i.e.,

$$A_1 - A_2 = \int_{\delta o}^{\delta s} (P_s - P_e)\, d\delta - \int_{\delta s}^{\delta m} (P_e - P_s)\, d\delta$$

$$= \int_{\delta o}^{\delta s} (P_s - P_e)\, d\delta + \int_{\delta s}^{\delta m} (P_s - P_e)\, d\delta$$

$$= \int_{\delta o}^{\delta m} (P_s - P_e)\, d\delta \qquad (10\cdot12)$$

If the system is stable after disturbance (transient disturbance), then

$$\frac{d\delta}{dt} = 0$$

i.e. from equation (10.11)

$$\int_{\delta o}^{\delta} (P_s - P_e)\, d\delta = 0$$

And hence from equation (10.12) we have for stability,

$$A_1 - A_2 = \int_{\delta o}^{\delta m} (P_s - P_e)\, d\delta = 0$$

i.e., $$A_1 = A_2 \qquad (10.13)$$

The above relation (equation 10.13) is very important as it helps us to determine conceptually the maximum angle of deflection (i.e. torque angle δ) of the rotor without loosing the stability limit. This is also referred to as equal area criterion for stability and can be applied to the simple cases only. However, if the load is changed suddenly or changes in load is too large, it may not be possible to make area $A_1 = A_2$ i e. $d\delta/dt \neq 0$ and in such case the system will loose stability as restoring force is never encountered in such cases. Fig. 10.1b represents a case where $A_1 \neq A_2$ and 10.1c a case where $A_1 = A_2$. Evidently case represented by Fig. 10.1c gives transient state stability limit.

Equal area criterion for stability can also be applied to other cases such as a single synchronous generater supplying power to the infinite bus through a transformer and double circuit three-phase transmission line as shown in Fig. 10.1d which shows reactance diagram of the system. For the sake of simplicity, resistances and shunt parameters have been neglected and hence, synchronous generator is represented by a constant voltage source E_m behind transient reactance, transformer by its leakage reactance, transmission line by reactances and infinite bus as an ideal voltage source. Initially, the synchronous generator is supplying power $P_e = P_{max} \sin \delta_0$ and the corresponding power input to the *sy*, generator is P_0. Now suppose a three phase bolted fault occurs in one of the transmission line close to the bus (near transformer end). Naturally, the real power output and thus power transfer will remain zero till the fault is not

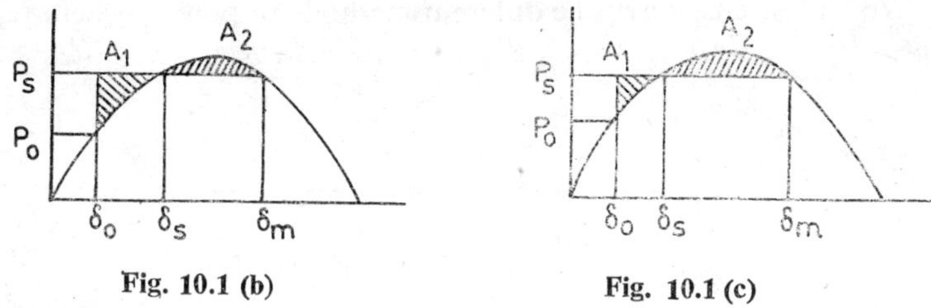

Fig. 10.1 (b) Fig. 10.1 (c)

cleared. Now suppose the fault is cleared by opening one of the transmission line on which the fault has occurred earlier, the power transfer i.e. power output of the synchronous generator to be delivered to the infinite bus is maintained through healthy transmission line. But since one line is out, reactance will increase and hence power output of the synchronous generator will decrease to XP_{max} sin δ as shown by lower curve in Fig. 10:1e. It is clear from Fig. 10.1e, that the fault must be cleared at

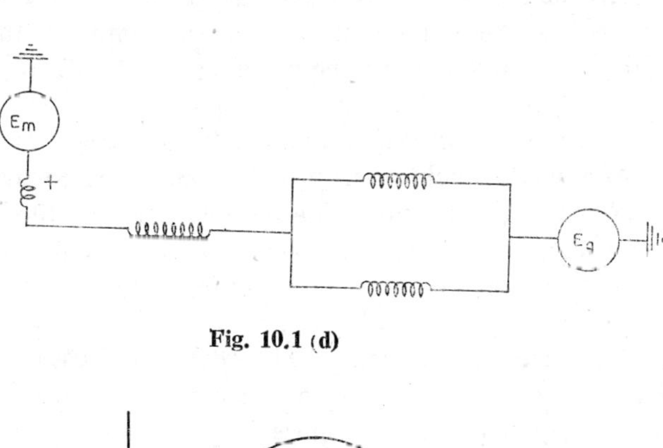

Fig. 10.1 (d)

Fig. 10.1 (e)

δ_c so as to make area $A_1 = A_2$. If the fault is cleared after δ_c, it is evident that $A_1 \neq A_2$ and hence, restoring torque is never encountered with the result system will loose stability. Angle δ_c is known as the critical clearing angle and the corresponding time is known as critical clearing time.

Before we go to the formulation of the power system problem for the purposes of transient stability studies for a large scale power system, we shall discuss the different types of representations of the power system.

10.4 Power System Representations

We will discuss now, the different methods of power system representations.

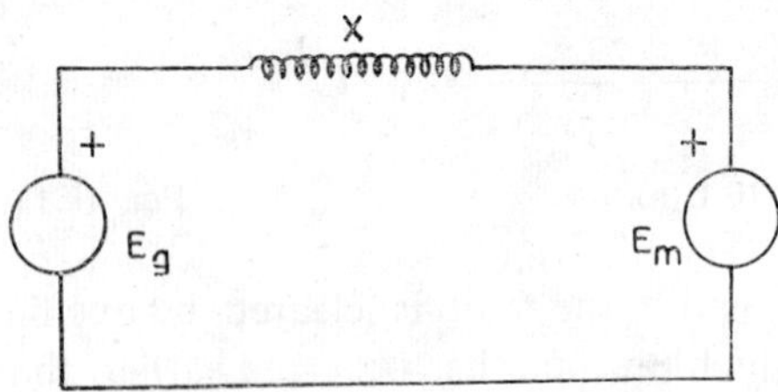

Fig. 10.2 Sample power system.

10.4.1 Most Simplified Representation of Power System

Most simplified representation of power system is a single synchronous machine which may be a synchronous generator or a synchronous motor connected to the infinite bus. Almost as simple as this, may be a a two machine system where in one machine operates as a synchronous generator and the other as a synchronous motor.

In most of the cases, even a multimachine system can be converted to a system of a single synchronous machine connected to the infinite bus or to a two machine system. This is because, all the synchronous machines connected to the same bus normally act together during disturbances and hence for the purpose of transient stability studies, they can be represented by a single equivalent synchronous generator. Even though, the synchronous machines are connected to different buses, but if they are not far away from each other, i.e. if the reactances between buses are not very high, these synchronous machines can be represented by a single (large) equivalent gererator.

From the above, we come to the conclusion, that, in most of the cases, multi-machine system can be transformed to a system of a single synchronous machine connected to the infinite bus or to a two machine system since all the synchronous machines connected at the same bus or at buses not far away from each other, act together during disturbances and hence they can be represented by an equivalent synchronous machine for the purpose of transient stability studies.

10.4.2 *Large Scale Power System Network*

For the purpose of transient stability studies, proper representations of a large scale power system involving multimachine systems located for far away from each other, are needed. Large scale power system network consists of several generating stations of different types such as hydel stations and thermal stations, complex transmission networks operating at different voltage levels and static loads, rotating loads such as synchronous motors or induction motors and composite loads. Therefore, representations of all these components such as synchronous generators, transmission network including transformers, and static, rotating and composite loads are necessary in order to carry out the transient stability studies of a large scale power system networks involving multimachine systems.

We have discussed in detail the representation i.e. mathematical model of synchronous generators and synchronous motors including the modelling of exciter and automatic voltage regulator, prime mover and governing systems and induction motors in the previous chapter. We will discuss here only the representation, i.e. mathematical modelling of loads.

10.4.3 *Load Representations*

Composite load representations in stability studies are quite complicated not so much because of difficulties in solution process but rather because not enough informations regarding data is available. Much efforts are required before satisfactory representation of load is possible. Actual system tests will be essential for realistic conclusion because load representations have a pronounced effect on stability studies.

Most stability programmes permit the user the form of load, a voltage dependent function, in any combinations of three options:

(1) costant current magnitude and constant phase angle. In this case, the real and reactive power are a linear function of voltage magnitude, i.e. $P - jQ = IV^* = \text{constant } XV^*$.

(2) constant real and reactive power. In this case, $P - jQ =$ constant or if represented as current injection,

$$I = \frac{P - jQ}{V^*} = \frac{\text{constant}}{V^*}$$

(3) constant shunt impedance. In this case, the real and reactive power are quadratic function of voltage magnitude.

The various load representations are shown in Fig. 10.3.

Note: In some power companies viz. B.P.A. (U.S.A.) the following assumptions are made for load representations:

(i) 80% of nominal power as constant P

(ii) 20% to vary linearly with V i.e. constant current, and

(iii) All reactive powers i.e. Q_s to vary as a quadratic function of V i.e. constant Z.

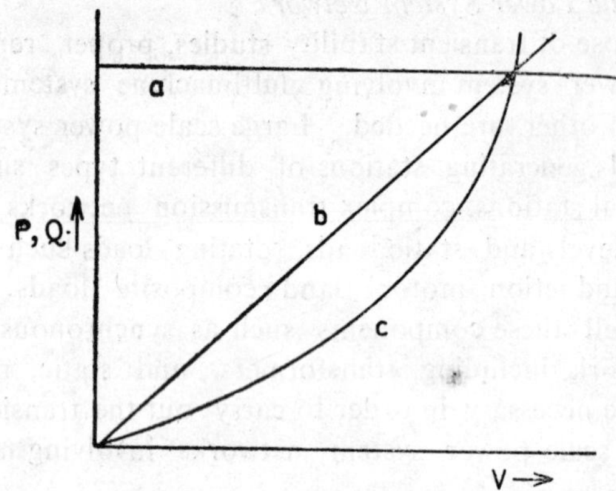

a—constant power; b—constant current; c—constant impedance

Fig. 10.3 Load representations.

Specific loads such as lighting, induction motors or synchronous motors can be represented fairly accurate (as discussed in the Chapter 9). However, the representation of composite load is difficult, because usually there is no or only little information is available about the composition. 'Venikov' suggests to take into account the load as frequency dependent and 'McAllister' suggests to take the action of protective relays also into account.

10.5 Transient State Stability Studies

Transient stability program is a natural extension of load flow program which has proved so successful. In order to conduct transient stability studies of a large scale power system network, we first of all conduct load flow studies of the power system in order to obtain system conditions prior to the disturbance. The power system representation for the purpose of load flow studies as discussed in Chapter 6, include a set of buses, transformers and transmission network. Hence after conducting the load flow studies, the representation of the power system must be modified to include remaining elements in addition to the above i.e. in addition to buses, transformers and transmission network. The remaining elements are rotating machines and loads. Hence the original bus impedance/admittance matrices must be modified to include the representation of rotating machines and static loads. Normally, rotating machines are represented by their equivalent circuit and static loads, by shunt impedances to ground. The operating characteristics of rotating machines such as synchronous or induction machines, are described by a set of algebraic and differential equations. Number of equations needed for the representations of the rotating machines depend upon the datas available and accuracy in the representation desired.

The next step in the transient stability studies is to create the disturbance and then solve a set of algebraic and differential equations of the power system to obtain system conditions during and immediately after the disturbance. Transient stability program, actually, combines the solution of the network (algebraic) equations with the machine dynamic (i.e. differential) equations.

10.5.1 Solution Technique for Conducting Transient Stability Studies

The following steps in sequence are followed for carrying out transient stability studies on a large scale power system network consisting of multimachines and composite loads.

1. The first step in transient stability study is to conduct the load flow analysis of the system. For the purpose of load flow studies, the power system representation includes all the buses, transformers and transmission lines. With the help of load flow studies, system conditions prior to disturbance is obtained. The data obtained from the load flow studies are the bus voltage magnitude, phase angle of the bus voltage and line flows.

2. Power system representation for the transient stability studies, contains, in addition to the buses, transformers and transmission lines, all the remaining elements such as synchronous generators, static and composite loads, synchronous motors and induction motors etc. Hence the original bus impedance/admittance matrices used for the load flow studies, must be modified to include these elements.

We have discussed in the previous chapter, the different types of representations i.e. modelling of synchronous generators. Its most simplified model is a constant voltage source behind proper reactance. Here the synchronous generator is represented by a constant voltage source behind transient reactance whose voltage magnitude remains the same, only the phase angle of the voltage (i.e. internal machine angle) is changed. In this model, of course, the saliency and changes in the flux linkages are neglected.

After modifying the network impedance/admittance matrices to include the representation of rotating machines and composite loads, the initial machine current for this simplied representation is calculated as follows:

$$I_i = \frac{P_i - jQ_i}{E_i^*} \quad \text{for} \quad i = 1, \ldots, m \tag{10.14}$$

where m is the number of machines at m buses. P_i and Q_i are the scheduled or calculated power at the bus i and E_i is the scheduled or calculated bus voltage. The datas for power and bus voltages are either scheduled i.e. specified or obtained from the load flow studies.

After calculating the initial machine current I_i for $i = 1, \ldots, m$, the voltage back machine impedance (i.e. transient reactance) is calculated from the following equation,

$$E'_{ig} = E_i + r_{ai}I_i + jx'_{di}I_i \tag{10.15}$$

where $E'_{ig} = e'_{ig} + jf'_{ig}$.

Here E'_{ig} is initial voltage behind transient reactance of ith machine and r_{ai} and x'_{di} are the armature effective resistance and direct axis transient reactance of the machine. The initial machine angle is calculated as follows:

$$\delta_i = \tan^{-1} \frac{f'_{ig}}{e'_{ig}} \tag{10.16}$$

In this formulation, it is assumed that there are m machines and each of them is represented by a constant voltage source behind proper i.e. transient reactance. Composite loads, as mentioned in the Section 10.4.3, are normally assumed to vary as a certain function of voltages.

When the load is represented as constant power (i.e. constant P and constant Q), then P and Q are either scheduled bus power or calculated from load flow studies. Similarly, when the load is represented as constant current magnitude and constant phase angle, the initial current I_p for this representation is obtained as follows:

$$I_p = \frac{P_p - jQ_p}{E_p{}^*} \tag{10.17}$$

where P_p and Q_p are the scheduled power or obtained from load flow studies and E_p is calculated bus voltage from the load studies. The current I_p which flows from bus p towards ground, will remain constant i.e. magnitude and phase angle of I_p obtained from the equation (10.17) will remain constant. When the load is represented as constant shunt impedance or admittance to ground, then the constant shunt admittance Y_p used to represent load at bus p, is calculated as follows:

$$Y_p = \frac{I_p}{E_p} \tag{10.18}$$

where I_p is obtained from the equation (10.17) and E_p is the calculated bus voltage.

In this way, the bus impedance/admittance matrix is modified to include new network elements representing rotating machines and also to account for changes in load representations. For this, the procedure described in Chapter 2 (Sections 2.7 and 2.8) for building algorithm for Z_{BUS} matrix is used. Here, each element representing a machine is a branch to a new bus and each element representing a load is a link to ground.

The initial speed ω of the rotor is assumed to be equal to the synchronous speed $\omega_0 = 2\pi f$ where f is the frequency in cycles/second. The initial mechanical power input for the ith synchronous generator, P_{mig}, is assumed to be equal to the electrical air gap power P_{ei}, where,

$$P_{ei} = P_i + I_i^2 R_{ai} \tag{10.19}$$

Here P_i is the scheduled or the calculated power.

3. The system parameters are changed to simulate disturbances which may be loss of generations, transmission facilities or loads or

switching operation or faults. The loss of generation, transmission facilities or loads are simulated by removing appropriate network elements. Faults are simulated according to their effects say a 3-phase fault at any bus p is simulated by setting the voltage of pth bus equal to zero. Then the modified network equations as shown below are solved to obtain system conditions immediately after the disturbance. Say for any bus p, which is a load bus,

$$P_p - jQ_p = E_p^* I_p = E_p^* \sum_{q=1}^{n} Y_{pq} E_q$$

$$= E_p^* [Y_{pp} E_p + \sum_{\substack{q=1 \\ q \neq p}}^{n} Y_{pq} E_q]$$

i.e.

$$E_p = \frac{1}{Y_{pp}} \left[\frac{P_p - jQ_p}{E_p^*} - \sum_{\substack{q=1 \\ q \neq p}}^{n} Y_{pq} E_q \right] \quad (10.20)$$

The term $\frac{P_p - jQ_p}{E_p^*}$ in equation (10.20) represents the load current at any bus p, and hence for the constant current representation, this term will be constant. It is natural that for the constant power representation, $P_p - jQ_p$ is constant. Similarly, when the load is represented by constant (static) impedance/admittance to ground, then the term $\frac{1}{Y_{pp}} \left[\frac{P_p - jQ_p}{E_p^*} \right]$ will be zero.

Hence if the load is represented as a constant static admittance to ground, which is normally the case, then the network equation (10.20) becomes

$$E_p = \frac{1}{Y_{pp}} \left[- \sum_{\substack{q=1 \\ q \neq p}}^{n} Y_{pq} E_q \right] \quad (10.21)$$

The network performance equation including n load buses (equation 10.21) and m machines (i.e. generation) buses will be as shown below,

$$E_p = \frac{1}{Y_{pp}} \left[- \sum_{\substack{q=1 \\ q \neq p}}^{n} Y_{pq} E_q - \sum_{i=1}^{m} Y_{pi} E_i' \right] \quad (10.22)$$

Here E_i' is the initial machine voltage for the ith machine obtained from equation (10.15).

From the network equation (10.22), the new estimate for bus voltages are obtained. In this model, where synchronous generators are represented by constant voltage source behind transient reactances, E_i' will be held constant for the entire iterations.

4. Finally, the machine differential equations are solved to obtain the changes in the internal machine voltage angle say δ_i and the speed ω_i for the ith machine. Even in the present case, where the synchronous generators are represented by constant voltage source behind transient reactances, at least two differential equations (swing equations (9.131) and 9.132)) per machine have to be solved and hence for m machine system,

there will be $2m$ such equations. These equations (swing equation 9.131 and 9.132) as developed in Chapter 9, are as shown below,

$$\frac{2H}{\omega_0}\frac{d\omega}{dt} + K_d(\omega - \omega_0) = P_{mech} - P_{elec} - P_{damp}$$

and $$\omega - \omega_0 = \frac{d\delta}{dt} \tag{10.23}$$

Equation (10.23) for the ith machine, will be,

$$\frac{d\delta_i}{dt} = \omega_i - \omega_o = \omega_i - 2\pi f$$

and

$$\frac{d\omega_i}{dt} = \frac{\omega_o}{2H}[P_{i(mech)} - P_{i(elect)} - P_{i(damp)} - K_d(\omega_i - \omega_o)]$$

$$= \frac{\pi f}{H}[P_{i(mech)} - P_{i(elect)} - P_{i(damp)} - K_d(\omega_i - 2\pi f)] \tag{10.24}$$

There are different solution techniques to solve these differential equations (equation 10.24) viz. modified Euler's method and classical Runge-Kutta method of order 4 etc. as discussed in Chapter 5. In the application of modified Euler's method, the initial estimate of internal machine angle δ and machine speed ω at time $t + \Delta t$ are obtained from the following equation,

$$\delta_{i(t+\Delta t)} = \delta_{i(t)} + \left.\frac{d\delta_i}{dt}\right|_t \Delta t$$

and $$\omega_i(t + \Delta t) = \omega_{i(t)} + \left.\frac{d\omega_i}{dt}\right|_t \Delta t \tag{10.25}$$

where $\delta_{i(t)}$ is obtained from the equation (10.16) and initial value of ω_i is equal to $2\pi f$. The derivatives are evaluated from equation 10.24. The second estimate is obtained from the derivatives at time $t + \Delta t$. For this, the initial estimates of the internal machine voltage behind machine impedance, has to be calculated. After this, the final estimate of internal machine angle δ and machine speed ω at $t + \Delta t$ is calculated. Then the network equation (equation (10.20)) is solved again to obtain system voltage at $t + \Delta t$. After this, the machine currents and power and line flows are calculated with the help of equation (10.20). The process is repeated till the time t equals specified time. The sequence of steps for solving transient stability problem is shown in the flow diagram (Fig. 10.4).

10.5.2 Effect of Exciter and Governer on Transient Stability

Till now, we have considered the most simplified model of synchronous machines for the purpose of transient stability studies. In this model, synchronous generator has been represented by a constant voltage source behind appropriate reactance. During the period of analysis, the voltage magnitude is held constant, only its phase angle (i.e. internal machine angle) is changed. In this representation, changes in flux linkages and saliency have been ignored. In addition, P_{mech} is also assumed to be

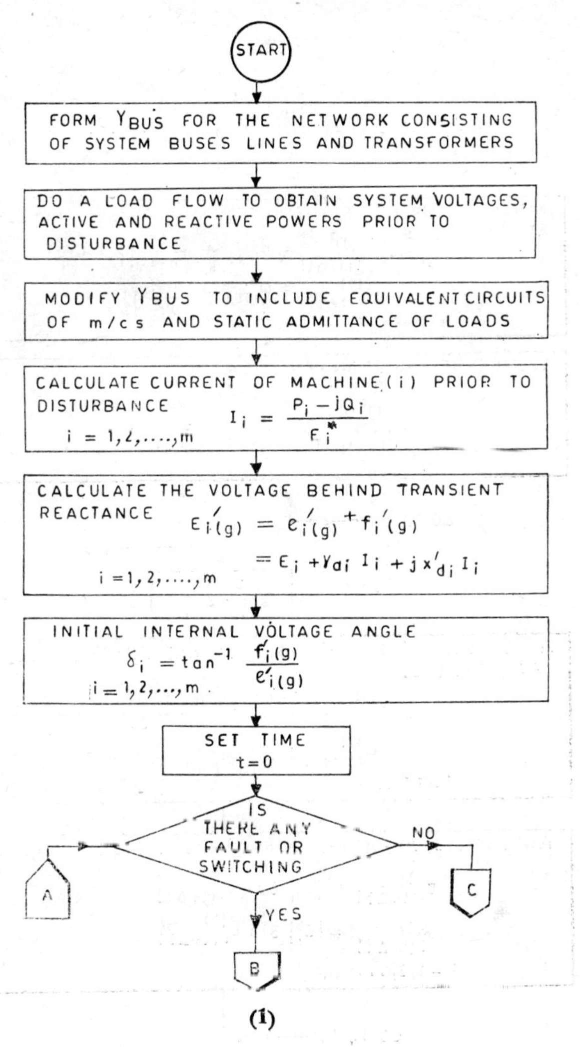
START
FORM YBUS FOR THE NETWORK CONSISTING OF SYSTEM BUSES LINES AND TRANSFORMERS
DO A LOAD FLOW TO OBTAIN SYSTEM VOLTAGES, ACTIVE AND REACTIVE POWERS PRIOR TO DISTURBANCE
MODIFY YBUS TO INCLUDE EQUIVALENT CIRCUITS OF m/cs AND STATIC ADMITTANCE OF LOADS
CALCULATE CURRENT OF MACHINE (i) PRIOR TO DISTURBANCE
$I_i = \frac{P_i - jQ_i}{E_i^*}$
$i = 1, 2, \ldots, m$
CALCULATE THE VOLTAGE BEHIND TRANSIENT REACTANCE
$E'_{i(g)} = e'_{i(g)} + f'_{i(g)}$
$= E_i + r_{ai} I_i + j x'_{di} I_i$
$i = 1, 2, \ldots, m$
INITIAL INTERNAL VOLTAGE ANGLE
$\delta_i = \tan^{-1} \frac{f'_{i(g)}}{e'_{i(g)}}$
$i = 1, 2, \ldots, m$
SET TIME
$t = 0$
IS THERE ANY FAULT OR SWITCHING
NO
YES
A
B
C

(1)

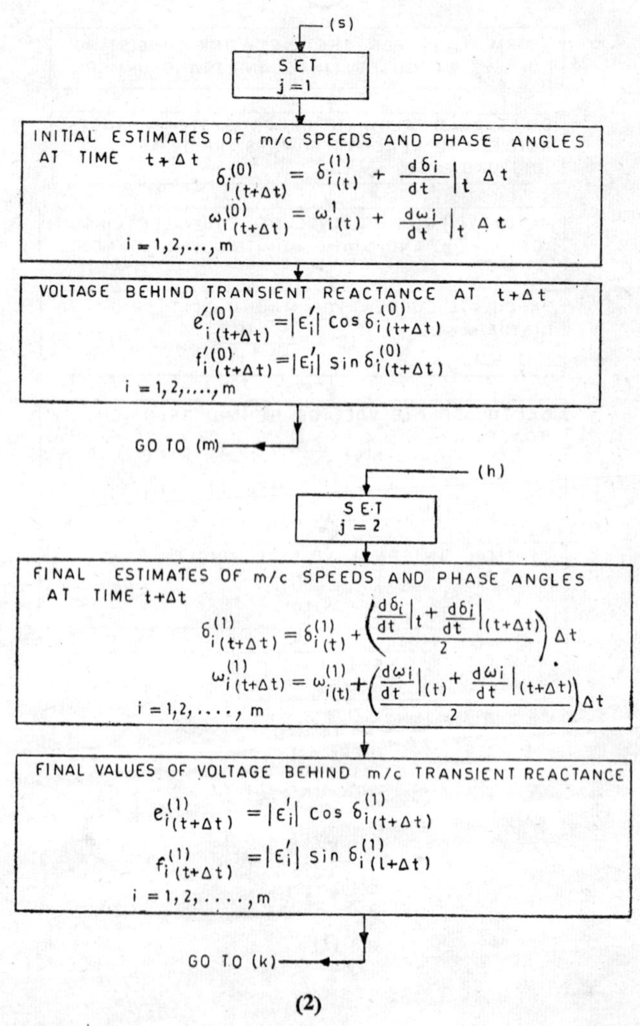
(s)
SET
j = 1
INITIAL ESTIMATES OF m/c SPEEDS AND PHASE ANGLES AT TIME t + Δt
$\delta_{i(t+\Delta t)}^{(0)} = \delta_{i(t)}^{(1)} + \left.\frac{d\delta_i}{dt}\right|_t \Delta t$
$\omega_{i(t+\Delta t)}^{(0)} = \omega_{i(t)}^{1} + \left.\frac{d\omega_i}{dt}\right|_t \Delta t$
i = 1, 2, ..., m
VOLTAGE BEHIND TRANSIENT REACTANCE AT t + Δt
$e_{i(t+\Delta t)}^{\prime(0)} = |E_i'| \cos \delta_{i(t+\Delta t)}^{(0)}$
$f_{i(t+\Delta t)}^{\prime(0)} = |E_i'| \sin \delta_{i(t+\Delta t)}^{(0)}$
i = 1, 2, ..., m
GO TO (m)
(h)
SET
j = 2
FINAL ESTIMATES OF m/c SPEEDS AND PHASE ANGLES AT TIME t + Δt
$\delta_{i(t+\Delta t)}^{(1)} = \delta_{i(t)}^{(1)} + \left(\frac{\left.\frac{d\delta_i}{dt}\right|_t + \left.\frac{d\delta_i}{dt}\right|_{(t+\Delta t)}}{2}\right)\Delta t$
$\omega_{i(t+\Delta t)}^{(1)} = \omega_{i(t)}^{(1)} + \left(\frac{\left.\frac{d\omega_i}{dt}\right|_{(t)} + \left.\frac{d\omega_i}{dt}\right|_{(t+\Delta t)}}{2}\right)\Delta t$
i = 1, 2, ... , m
FINAL VALUES OF VOLTAGE BEHIND m/c TRANSIENT REACTANCE
$e_{i(t+\Delta t)}^{(1)} = |E_i'| \cos \delta_{i(t+\Delta t)}^{(1)}$
$f_{i(t+\Delta t)}^{(1)} = |E_i'| \sin \delta_{i(l+\Delta t)}^{(1)}$
i = 1, 2, ... , m
GO TO (k)

(2)

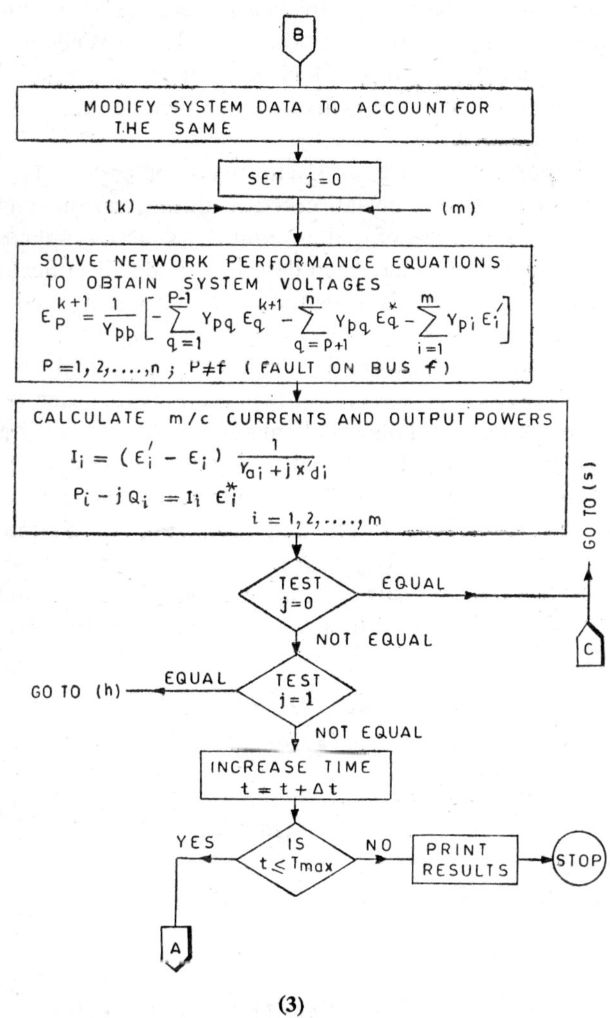

(3)

Fig. 10.4 Flow diagram for transient stability studies.

constant. However, if changes in flux linkages and saliency are also included, synchronous machine is represented by a number of algebraic and differential equations (equation 9.99) which must be solved simultaneously along with equations (10.24).

In the transient stability problems discussed so far, the effects of Exciter and Governor have not been considerd. When a more detailed analysis of system representation is desired, or when the period of analysis exceeds beyond one second, it is necessary to include the effect of Exciter and Governor control.

The important feature of Exciter control system is its ability to respond rapidly to voltage deviation both during normal and emergency conditions. Various types of exciter control system is employed in power system. One such system has been described in detail in Section 9.3. As shown, excitation system is essentially an automatic feedback control system whose components are as shown in Fig. 9.26. In order to represent an excitation system in digital computer dynamic stability analysis, it is necessary to derive transfer functions for each of the element in the form of first order differential equation. A block diagram representating continuously acting automatic voltage regulator and excitation system is shown in Fig. 9.26. The representation includes automatic voltage regulator, amplifier, exciter and stabilizing circuit loop. In order to include the effect of exciter, the following equations representating excitation system must be solved along with machine equations (equations 9.99 and 10.24). These equations are obtained directly from the transfer function of Fig. 9.26.

$$\frac{dV(t)}{dt} = \frac{V_T - V(t)}{T_R}$$

$$T_F \frac{dV_F(t)}{dt} - K_F \frac{dE_{FD}(t)}{dt} + V_F(t) = 0$$

$$V_R = \frac{K_A}{1 + ST_A} (V - V_{REF} - V_F)$$

$$V_{R\,MIN} \leqslant V_R \leqslant V_{R\,MAX}$$

$$\frac{dV_R(t)}{dt} = \frac{K_A[V(t) - V_{REF}(t) - V_F(t)] - V_R(t)}{T_A}$$

$$E_{FD}(t) = \frac{dS_E(t)}{dt}$$

$$T_E \frac{d^2E_{FD}(t)}{dt^2} + K_F \frac{dE_{FD}(t)}{dt} + E_{FD}(t) = \frac{dV_R(t)}{dt} \tag{10.26}$$

For less severe transients, the effect of modern fast acting excitation system on first swing is rather marginal. However, for more severe transients or when duration of fault is longer, the modern exciters have considerable pronounced effect on the transient stability of the system. Similarly during large disturbances, P_{mech} is also not constant, and therefore governing system including turbine, effect the transient stability of the power system. The effect of governor control on the transient stability is taken into account by its simplified representation as shown in Fig. 9.30 for steam turbine and in Fig. 9.31 for Hydroelectric system. In the case of reheat type steam governing system, the differential equations relating input and output derived from Fig 9.30 will be as shown

$$\frac{dE(t)}{dt} = \frac{1}{T_g}\left[\frac{W_{SHAFT} - W_{REF}}{r} - E(t)\right]$$

$$\frac{dP_s(t)}{dt} = \frac{P_0(t) - P_s(t)}{T_s}$$

$$T_{rh}\frac{dP_{mech}}{dt} - CT_{rh}\frac{dP_s}{dt} = P_s - P_{mech} \qquad (10.27)$$

The different symbols of equations (10.26 and 10.27) are described in Sections 9.3 and 9.4. These equations must be solved simultaneously with the machine equations to include the effect of governing system on transient stability.

Transient stability of a large scale power system can be improved by:

(i) increasing system voltages,

(ii) decreasing series reactance by duplicate lines or series capacitors,

(iii) using high speed protective relays and circuit breakers including high speed reclosure,

(iv) by having high speed excitation system including automatic voltage regulator along with supplementary derivative type control signals. The excitation system of synchronous machine have an extreme effect on the system stability and these are the most economical method of increasing stability limit, and

(v) by having fast acting governing system. The effects on the system performance of prime movers and governing system is usually not apparent until 1 second after a disturbance. However, when the system is marginally stable or when the stability is decided after the second swings, governor and prime movers play an important role.

10.6 Numerical Example

The load flow data is as given in Table 10.1 on 100 MVA base Tables 10.2 and 10.3 show a schedule generation and loads and synchronous machine data respectively on 100 MVA base. The sample system is as follows.

Table 10.1

Bus code $p-q$	Impedance Z_{pq}	Line charging $y'_{pq}/2$
1—2	$0.035+j0.12$	$0+j0.12$
2—3	$0.03+j0.12$	$0+j0.06$
1—3	$0.04+j0.12$	$0+j0.05$

Table 10.2

Bus code p	Assumed voltage	Generation Megwatts	Generation Megavars	Loads MW	Loads MVAR
1	$1+j0$	0	0	75	−20
2	$1+j0$	0	0	0	0
3	$1+j0$	150	−25	50	−20

Table 10.3

Bus code p	Inertia constant H	Direct axis transient react (x_d')
2	150	0.1
3	10	0.1667

The loads are to be represented as fixed impedances to ground. For time increment of 0.01 second calculate the changes in phase angle and speeds of the generators for a 3φ fault at the bus 2 for a duration of 0.1 second and a maximum time is 0.2 seconds. Solve the problem using modified Euler's method.

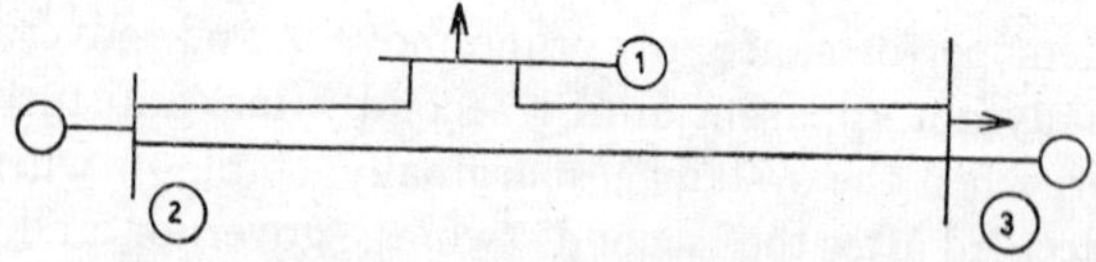

Solution: To get the pre-fault condition of the system load flow is done by using Gauss-Seidel method. The pre-fault bus voltages are

$$E_1 = 0.99769\ \underline{/-4.83^\circ}$$

$$E_2 = 1\ \underline{/0^\circ}$$

$$E_3 = 1.0225\ \underline{/0.852^\circ};\quad P_{G_2} = 0.569 + j1.237$$

For the purpose of fault calculation loads and generators are taken. Thus those are to be suitably represented. Generally, the loads are represented as a constant impedance or admittances to the ground.

$$y_{li} = P_{Li} - jQ_{li} / |E_i|^2 \quad E_i = 1, 2, 3 \ldots$$

$$y_{l_1} = 0.753 - j0.2$$

$$y_{l3} = 0.478 - j0.209$$

The generator bus voltage is modified into an equivalent voltage source in series with the direct axis transient reactance. We assume that the equivalent voltage source's magnitude do not change but the phase angle changes.

Thus the voltage behind the transient reactance is got by

$$|E_t| = E_i + r_{ai} I_t + jX_{ai} I_t$$

$$|I_t| = P_{Gt} - jQ_{Gi}/E_i$$

So we get,

$$|E_{t2}| = 0.8778 \qquad \delta_2 = 3.7° = 0.06457 \text{ rad.}$$

$$|E_{t3}| = 1.008 \qquad \delta_3 = 13.95° = 0.2434 \text{ rad.}$$

The system data is modified, thus

$$y_{ij \text{ New}} \neq y_{ij \text{ Old}} \quad \text{when } i \neq j$$

$$y_{ij \text{ New}} = y_{ii \text{ Old}} + 1/jX_{di} + y_{li}$$

$$1/jxd_2 = -j10; \quad 1/jXd_3 = -j6.$$

The modified system performance equations are given in matrix form.

$$\begin{bmatrix} 5.493 - j15.21 & -2.24 + j7.68 & -2.5 + j7.5 \\ -2.24 + j7.68 & 4.2 - j25.343 & -1.96 + j7.843 \\ -2.5 + j7.5 & -1.96 + j7\ 843 & 4.94 - j21.442 \end{bmatrix}$$

$$\begin{bmatrix} E_1 \\ E_2 \\ E_3 \end{bmatrix} = \begin{bmatrix} -j0E_{t2} \\ -j10E_{t2} \\ -j6E_{t3} \end{bmatrix}$$

The 3φ fault at Bus 2 is simulated by setting the voltage at this bus to zero. Then the performance equations are solved to get the fault voltage just at the fault instant.

$$E_1 = 0.1626 \angle 3.90\ 17$$

$$E_3 = 0.333 \angle 2.44 = 0.3326 + j0.01417$$

The machine currents and power of the machines can be found by

$$I_{ti} = (E_{ti} - E_i)\, y_{pi}$$

$$P_{ei} = R_e \{I_{ti}\, E^*_{ti}\}^2$$

The m/c currents and power at fault instant

$$I_{t2} = 0.568 - j8.76 \qquad P_{e2} = 0.9937$$

$$I_{t3} = 1.374 - j3.87 \qquad P_{e3} = 2.284$$

Initial conditions $\delta_1(o) = 0.06457$, $w_1(o) = 314.16$ rad/s, $\delta_2(o) = 0.2434$ rad, $w_2(o) = 314.16$ rad/s.

For 1st estimates

$$\frac{dw_i}{dt} = \frac{\pi f}{H_i}(P_{mi} - P_{ei(t)})$$

$$P_{m2} = 0.569 + \text{Real}(P_{G_2})$$
$$P_{m3} = 1.5$$

$$\left.\frac{dw_2}{dt}\right|_{(o)} = -0.444$$

$$\left.\frac{dw_3}{dt}\right|_{(o)} = -12.31$$

Initial estimation of speed

$$W_{i(t+\Delta t)}^{(0)} = W_{i(t)}^{1} + \left.\frac{dw_i}{dt}\right|_t \Delta t$$

$$\Delta t = 0.01$$

$$W_{2(0.01)}^{(0)} = 314.16 - 0.444 \times 0.01$$
$$= 314.155 \text{ rad/s}$$

$$W_{3(0.01)}^{(0)} = 314.03 \text{ rad/s.}$$

Internal voltage angles changes are calculated by

$$\frac{d\delta_i}{dt} = W_{i(t)} - 2\pi f$$

$$_{i(t+\Delta t)}^{(0)} = \delta_{i(t)}^{1} + \left.\frac{d\delta_i}{dt}\right|_t$$

Since $\frac{d\delta_i}{dt}$ is zero, internal angle remains the same.

$$\delta_{2(0.01)}^{(0)} = 0.06457; \; \delta_{3(0.01)}^{(0)} = 0.2434$$

Again the performance equations are solved with these new datas. Since the δ are same, the system voltages, machine currents and powers are the same as obtained at the instant of fault. So,

$$\frac{dw_i}{dt} \text{ at } t + \Delta t \text{ is equal to } \frac{dw_i}{dt} \text{ at } t.$$

Final estimates for the speed of the machine are given by

$$W_{i(t+\Delta t)}^{1} = W_{it}^{1} + \left(\left.\frac{dW_i}{dt}\right|_t + \left.\frac{dW_i}{dt}\right|_{t+\Delta t}\right)\Delta t/2$$

$$W^1_{2\,(0.01)} = 314.155 \text{ rad/s}$$

$$W^1_{3\,(0.01)} = 314.03 \text{ rad/s}$$

The $\frac{d\delta_i}{dt}$ at $t + \Delta t$ are calculated

$$\frac{d\delta_i}{dt} = W^1_{1(t+\Delta t)} + 2\pi f$$

$$\left.\frac{d\delta_2}{dt}\right|_{0.01} = -0.005$$

$$\left.\frac{d\delta_3}{dt}\right|_{0.01} = -0.13$$

The final estimates are done for the internal voltage angles by

$$\delta^1_{i(t+\Delta t)} = \delta^1_{i(t)} + \left(\left.\frac{d\delta_i}{dt}\right|_t + \left.\frac{d\delta_i}{dt}\right|_{t+\Delta t}\right)\Delta t/2$$

$$\delta^1_{2\,(0.01)} = 0.064545 \text{ rad}$$

$$\delta^1_{3\,(0.01)} = 0.24275 \text{ rad}$$

Again the performance equations are solved and this completes the iteration. After every iteration, power is found out. And after every iteration the time is advanced and seen that fault condition is changed or not. If it changes, then the network equations are solved without the fault constraint before proceeding in the normal course. This is continued until t reaches T_{max}.

Problems

10.1 The load flow data in p.u. on a 500 MVA base for a sample power system as shown in the figure below, is given in Tables 10.4 and 10.5. The slack bus is 1. The machine data in p.u. on 500 MVA base is given in Table 10.6. The loads are to be treated as constant impedances to the ground. For a time increment of 0.02 second and a maximum specified time of 1 second, calculate the changes in phase angles, and speeds of generators for a 3-phase fault at bus 2 for a duration of 0.01 second using modified Euler's method.

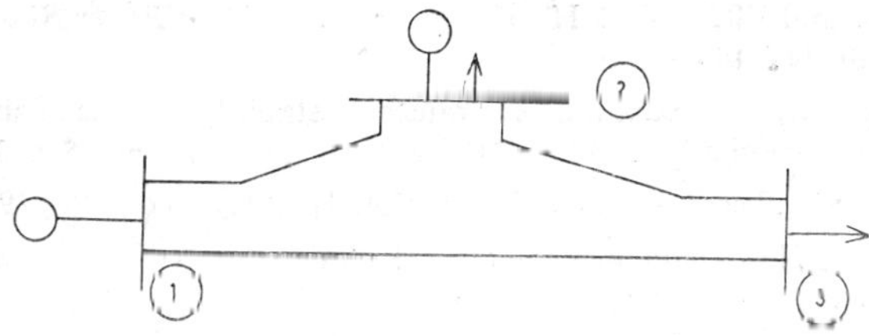

Figure for problem 10.1.

Table 10.4 Impedances of sample system

Bus code $p-q$	Impedance Z_{pq}	Line charging $y'_{pq}/2$
1—2	$0.05 + J0.22$	0.08
2—3	$0.025 + J0.25$	0.06
1—3	$0.1 + J0.18$	0.11

Table 10.5 Scheduled generations, loads and assumed bus voltages

Bus code p	Assumed bus voltage	Generations		Loads	
		MW	MVARS	MW	MVARS
1	$1.00 + J0.00$	0	0	0	0
2	$1.00 + J0.00$	100	75	50	25
3	$1.00 + J0.00$	0	0	80	40

Table 10.6 Synchronous machine data

Bus code p	Inertia constant	X_d'
1	100	0.15
2	0ı	0.2

10.2 What do you understand by steady-state, dynamic state and transient state stability of a power system. Discuss briefly. Discuss also the factors upon which they depend.

10.3 Solve problem 10.1 by classical Runga-Kutta method of order 4 and compare the result thus obtained with the result obtained in problem 10.1.

References

1. Brown, H.E., Happ, H.H., Person, C.E. and Young, C.C., "Transient stability solution by an impedance matrix method", *Trans. IEEE PAS*, Vol. 84. pp. 1204–1214, 1965.
2. Stagg, G.W. and El-Aliad, A.H., *Computer Methods in Power System Analysis*, McGraw-Hill, NY 1968.
3. Stagg, G.W. etc., "Calculation of transient stability problems using a high speed digital computer", *Trans. AIEE* Vol. 78, pt III A, pp, 566–574, 1959.
4. Kimbark, E.W., *Power system stability*, Vol. 1, Wiley, Inc., NY, 1948
5. Olive, D.W., "New techniques for the calculation of dynamic stability" *Trans. IEEE PAS*, Vol. 85, pp. 767–777, 1966.

Appendix 1

Dynamic Programming

Introduction

Economic load scheduling of the hydro-thermal system essentially belongs to the category of problems known as the multi-stage decision process requiring a sequence of decisions and which can be tackled more effectively by the techniques of dynamic programming. Dynamic programming was developed during 1950's by R. Bellman and his associates to study the optimization arising in the industry, economics, defence and in the social services where the modern optimization technique of linear and nonlinear programming and calculus of variation and its generalization are not applicable. Such category of problems has been referred to as multi-stage decision processes. Dynamic programming is a technique to tackle such problems.

Consider a system which can be described at discrete times by a finite number of variables, $X_1, X_2, \ldots, xX_n$, usually referred to as state variables. The values of these state variables at time (stage) i will be denoted by $X_1(i), X_2(i), \ldots, X_n(i)$. It is rather convenient to use state vectors.

$$X = [x_1, x_2, \ldots, x_n]$$

and its value at stage i is given by

$$X(i) = [x_1(i), x_2(i), \ldots, x_n(i)]$$

At each stage $i = 1, 2, \ldots, N$, we have to make a decision $d(i)$ from amongst a number of possible choices, may be infinite. In practice we may have to take a number of decisions at each choice. Then $d(i)$ can be thought of as a vector. To simplify the matter, however, we assume that we have to take a single decision at each stage. The effect of taking decisions $d(i)$ at stage i is to change the stage of the system so that at the stage $i + 1$, the state vector is $X(i + 1)$. Dynamic programming is applicable whenever,

$$X(i + 1) = G[X(i), d(i), i]$$

The dynamic programming can be applied when the state of the system at the stage $i + 1$ depends only upon the state of the system at the stage i, the decision $d(i)$ taken at the stage i, and the stage i itself.

At each stage i, of the system, a return function R is prescribed which is of the form,

$$R(i) = R[X(i), d(i), i]$$

24(45-10/1981)

The return function R, because of taking decision $d(i)$ also depends upon the state of the system at the stage i, i.e. $X(i)$, decision taken at stage i, i.e. $d(i)$ and the stage i itself.

Certain restrictions can also be imposed to be satisfied at each stage i by the state vector $X(i)$, the decision variable $d(i)$, which can be in general of the form,

$$\phi\,[X(i), d(i)\, i] \leq O, \geq O, = O$$

The restriction can also restrict the state and/or decision variables to take integral value only.

If the total number of stages is N, then corresponding to each choice of sequence of decisions $d(1), d(2), \ldots, d(N)$, we can associate a total return, I, for the whole process, by a relation like $I = I(R(1), R(2), \ldots R(N))$. In most of the application of the dynamic programming, we take

$$I = \sum_{i=1}^{N} R(i)$$

Thus our problem is to choose a sequence of decisions $d(1), d(2), \ldots, d(N)$ which results in the optimization of I.

The above system is called Discrete Time Multistage Decision Process. If any of the function G, R, ϕ defining such a system do not involve any chance elements, the system is referred to as deterministic. Otherwise it is called a probabilistic system. If the probability distributions of all the random variables occurring in a probabilistic decision process are known in advance, we refer to the process as a stochastic decision process. Otherwise it is called an adaptive decision process. Also, such a process is called stationary if each of the function G, R and ϕ are independent of i. The process is called infinite or finite stage according as 'N' is infinite or finite.

Principle of Optimality

We will start with the simplest type of multi-stage decision process, namely discrete time, deterministic, stationary finite multistage decision process. Let us assume that the process is given i.e. the function G, R, ϕ and $I = \sum_{i=1}^{N} R(i)$ are prescribed and our problem is to find out the maximum value of I and the corresponding optimal sequence of decisions.

The techniques of dynamic programming depends upon the following trivially true principle called principle of optimality by Bellman.

Whatever be the initial state or initial decision, the remaining decisions must constitute an optimal sequence of decisions with regards to the state resulting from the first decision.

Dynamic Programming Approach

Let the function, G, R, ϕ and $I = \sum_{i=1}^{N} R(i)$ define some discrete time, deterministic stationary, N-stage process. Our problem is to find out a sequence of N decisions, one decision at each stage, for which the value of I is maximum.

It is obvious that the maximum value of I depends upon N, the total number of stages to be considered and the initial state of the system say C. In view of this let us define the following sequence of optimal functions:

$f_N(C)$ = Maximum value of I when N stages remain and initial state is C. If the first decision d (1) is taken arbitrarily, then corresponding to this d (1), we have

$$X(2) = G[C, d(1)]$$

The maximum return over remaining $N-1$ stages, the initial state being X (2), is

$$f_{N-1}[G\{C, d(1)\}]$$

and the return from the first stage when decision d (1) is taken, is

$$R[C, d(1)]$$

Thus when the first decision, d (1), is arbitrary and the remaining $N-1$ decisions are optimal (with regard to the state resulting from the first, i.e. initial decision d (1), which is arbitrary), the total return is

$$R[C, d(1)] + f_{N-1}[G\{C, d(1)\}]$$

Thus using the principle of optimality we must have

$$f_N(C) = \underset{d(1)}{\underset{\text{permissible}}{\text{Max}}} \{R[C, d(1)] + f_{N-1}[G\{C, d(1)\}]\}$$

$$= R[C, d_N(C)] + f_{N-1}[G\{C, d_N(C)\}] \qquad (1)$$

so that $d_N(C)$ is the first optimal decision to be taken when the system starts with the state C and N stages remain.

Furthermore,

$$f_1(C) = \underset{d(1)}{\underset{\text{permissible}}{\text{Max}}} [R\{C, d(1)\}]$$

$$= R[C, d(1)] \qquad (2)$$

Equation (2) can be used to compute $f_1(C)$ for any possible initia state C. Equation (1) for $N = 2$, gives,

$$f_2(C) = \underset{d(1)}{\underset{\text{permissible}}{\text{Max}}} \{R[C, d(1)] + f_1[G\{C, d(1)\}]\}$$

$$= R[C, d_2(C)] + f_1[G\{C, d_2(C)\}] \qquad (3)$$

Equation (3) can be used to compute $f_2(C)$ for any possible initial state C. Continuing like this, we can compute $f_N(C)$ and $d_N(C)$ for any value of

N and any possible initial state C. Thus, $f_N(C)$ gives us the maximum value of I for the N stage process when the initial state is C. The optimal sequence of decisions will be:

$$d_N(C), d_{N-1}(P_N), \ldots$$

where

$$P_N = G[C, d_N(C)]$$

We may observe here that at every iteration of the procedure described above, we have to solve a minimization problem in one variable only. This result for one of the major reasons for the popularity of the dynamic programming technique. Thus the technique of dynamic programming tries to do with the multistage decision processes what diagonalization of matrices tries to do for a system of linear equations, what separation of variables tries to do for partial differential equations, and what the use of orthonormal function tries to do for the eigen value problems. The principle used in all these cases are similar, instead of trying to meet N difficulties simultaneously try to meet these one at a time.

Allocation Problem (Integral Variables): Consider the following optimization problem:

$$\text{maximize} \sum_{i=1}^{N} g_i(x_i) \tag{4}$$

$$\text{subject to} \sum_{i=1}^{N} a_i x_i \leq C; \; x_i \text{ are positive integers.} \tag{5}$$

Many practical problems, where in a single resource is to be allocated optimally in a number of activities, lend themselves to these type of problems. Since the maximum value of the objective function depends upon the number of activities to be considered, n, and the amount of available resource, C, we define a sequence of optimal functions, $f_N(C)$, as follows:

$$f_N(C) = \text{max value of} \sum_{i=1}^{N} g_i(x_i) \qquad 6)$$

when the variables, X_i, takes positive integral values only and satisfy the condition

$$\sum_{i=1}^{N} a_i x_i \leqslant C \tag{7}$$

Using the principle of optimality we obtain:

$$f_N(C) = \underset{X_n}{\text{Max}} = 0, 1, 2, \ldots . [C/a_n] \{g_n(x_n) + f_{N-1}(C - a_n x_n)\} \tag{8}$$

where $[C/a_n]$ denotes the greatest integer C. Also

$$f_1(C) = \underset{X_1}{\text{Max}} = 0, 1, \ldots, [C/a_1] \{g_1(x_1)\} \tag{9}$$

Equations (8) and (9) give us the required recurrence relations which can be used recursively to solve for $f_N(C)$ and the corresponding optimal value of variables.

Reference

1. Bellman, R., *Dynamic Programming*, Princeton University Press, Princeton, N.J., 1957.

Appendix 2

Mathematical Theory of Groups

Group: Suppose we are given a set G with a binary operation, i.e. an operation between any two of its elements, defined on it. Let '.' denotes this operation which is for historical reasons referred to as multiplication The set G is said to be a *group* if it satisfies the following postulates:

(i) *Closure*: Given any two elements a and b of the set G, $a.b$, the result of the binary operation on a and b is also in G.

(ii) *Associativity*: For elements a, b and c of the set G we have the following relation,

$$a \cdot (b \cdot c) = (a \cdot b) \cdot c$$

(iii) *Existence of Identity*: Among the elements of the set G, there exists an element e called identity element such that

$$a \cdot e = e \cdot a = a$$

(iv) *Existence of Inverse*: Corresponding to any element a of the set G, there exists an element a^{-1}, called the inverse of the element a of the set G such that

$$a \cdot a^{-1} = a^{-1} \cdot a = e$$

If in addition to the above four group axioms, the following condition is also satisfied, then the group is known as commutative or abelian group.

(v) *Commutativity*: For any element a and b in the group G we have

$$a \cdot b = b \cdot a$$

If the number of elements in the group is finite, the group is said to be finite and then the number of distinct elements in the finite group is called the order of the group.

Two groups G, G^1 are said to be isomorphic (i.e. $G \sim G'$) if a unique correspondence $a \sim a'$ $(a \in G, a' \in G')$ can be set up between them in such a way that, when $a \sim a'$ and $b' \sim b'$, we have $ab \sim a'b'$. The correspondence itself is then called an isomorphism between G and G'.

Group Multiplication Table: Group elements may have a physical interpretation. However, if the physical meaning assigned to the group element is ignored, then the elements of the group can be represented by abstract quantities, viz. a, b, c ... etc. If we have a complete list of such elements of a finite group and we also know their all possible products which are the result of multiplying each abstract element viz. $a, b, c, \ldots$ etc. with every other abstract element of the group, then the group is completely and uniquely defined, at least in the abstract sense. For a finite group, such multiplication operations are also finite. These multiplications regarding group elements and their product can most conveniently be presented in a table known as a group multiplication table. The group elements are arranged along the column and the row of the table. The entry in the i-jth position of the table is the group element $p_i p_j$ which result from the multiplication of p_i, an element in the ith row and p_j an element in the jth column.

It can easily be verified that each row and each column in the group multiplication table lists each of the elements once and only once. From this, it follows that no two rows will be identical nor any two columns will be identical. Thus each row and each column in a group multiplication table is a rearranged list of the group elements. As an illustration, let us examine some abstract groups with the help of their multiplication table.

Group of order one, consists of a single element which is identity E and hence there is no multiplication table in this case. However, the group of order 2, 3 and 4 can be assigned the following tables:

G_2	E	A
E	E	A
A	A	E

G_3	E	A	B
E	E	A	B
A	A	B	E
B	B	E	A

and

G_4	E	A	B	C
E	E	A	B	C
A	A	B	C	E
B	B	C	E	A
C	C	E	A	B

or

G_4	E	A	B	C
E	E	A	B	C
A	A	E	C	B
B	B	C	E	A
C	C	B	A	E

and so on.

Subgroup : If any group G is given, and if the set H, consisting of certain elements of G, is itself a group with the same law of multiplication which holds in G, then H is called a subgroup of G. Therefore, a subset H of a group G is a subgroup of G if the following group axioms are satisfied :

(i) The product of two elements a and b of H is again an element of H.

(ii) The identity element of the group G is an element of H.

(iii) The inverse of each element of H is again an element of H.

From the above we conclude that a subgroup H of G is a subset which is itself a group under the group multiplication defined in G. The entire group G and likewise its identity element E by itself from subgroups called improper subgroups of G. These are trivial subgroups possessed by all the groups. All other subgroups are called proper subgroups. It can easily be verified that the order of any subgroup, h, of a group of order g must be a diviser of g.

Cyclic Group : A cyclic group is one in which all its elements are generated by a single element known as a generating element or simply a generator. To illustrate a cyclic group, let us take the example of a cyclic group of order 4 in which the element A is a generator. Then its remaining elements B, C and E where E is an identity are generated as follows:

$$A \cdot A = A^2 = B$$

$$A \cdot A \cdot A = A^2\ A = BA = AB = C$$

$$A \cdot A \cdot A \cdot A = A^3 A = CA = AC = E$$

Thus its groups multiplication table will be as shown,

G_4	E	A	B	C
E	E	A	B	C
A	A	B	C	E
B	B	C	E	A
C	C	E	A	B

From the above, we conclude that for a cyclic group of order 4 with A as a generator, we have

$$A^4 = E$$

where E is an identity element. In general, a cyclic group of order h consists of an element x as generator and all of its power upto $x^h = E$. An important property of the cyclic group is, it is commutative, i.e. abelian as shown above.

References

1. Cotton, F., *Chemical Applications of Group Theory*, Wiley, New Delhi, 1976.
2. Miller, W., *Symmetry Groups and their Applications*, Academic Press, 1972.
3. Tinkham, M., *Group Theory and Quantum Mechanics*, McGraw-Hill, New York, 1964.

Index